Lipid Chemistry

Lipid Chemistry

by

Prof. Neelam Khetarpaul
Ph.D. (Foods and Nutrition), FICN
Dean
College of Home Science
CCS Haryana Agricultural University
Hisar – 125 004 (Haryana), India

&

Dr. Vipul Khetarpaul
(M.B.B.S.)

2018
Daya Publishing House®
A Division of
Astral International Pvt. Ltd.
New Delhi - 110 002

First Published, 2008
Reprinted, 2018

ISBN: 978-81-7035-537-3 (Hardbound)

Publisher's note:

Every possible effort has been made to ensure that the information contained in this book is accurate at the time of going to press, and the publisher and author cannot accept responsibility for any errors or omissions, however caused. No responsibility for loss or damage occasioned to any person acting, or refraining from action, as a result of the material in this publication can be accepted by the editor, the publisher or the author. The Publisher is not associated with any product or vendor mentioned in the book. The contents of this work are intended to further general scientific research, understanding and discussion only. Readers should consult with a specialist where appropriate.

Every effort has been made to trace the owners of copyright material used in this book, if any. The author and the publisher will be grateful for any omission brought to their notice for acknowledgement in the future editions of the book.

Published by : **Daya Publishing House®**
A Division of
Astral International Pvt. Ltd.
– ISO 9001:2008 Certified Company –
4736/23, Ansari Road, Darya Ganj
New Delhi-110 002
Ph. 011-43549197, 23278134
E-mail: info@astralint.com
Website: www.astralint.com

Preface

The ambition behind writing this book is to provide an up-to-date description of the various aspects related to lipids including structure and classification of lipids, as a basis for understanding their functionality in biological and technical systems.

With regard to the role of lipids in health and disease, this book focuses on their role in atherosclerosis, coronary vascular disease and immune system. The book is truly and strongly recommended to anyone not only in the field of lipids but also to colloid chemists in general.

This is a book which treats the entire spectrum of lipid properties from a short introduction about the chemical structure of lipids to fat replacers. The book is divided into 18 chapters, and there are numerous figures and diagrams supporting the text in each chapter. These chapters are related to classification of lipids, composition of fatty acids, structure, functions and metabolism of cholesterol, chemistry of PUFAs, other sterols and waxes, emulsions and emulsifiers, analysis of lipids, use of microorganisms for the production of lipids, chemical composition of vegetable oils, crystallization of fats, fat substitutes, nutraceuticals and trends in production of fats and oils etc. The most extensive chapter of the entire book, deals with the extraction, cleaning and storage of fats and oils. For nutritionists teaching lipids, some of the chapters on polyunsaturated fatty acids, cholesterol and fat substitutes should be a gold-mine. The book has been written clearly with many illustrations to help the reader to gain an understanding about fats. A detailed reference list, is appended to the book for further detailed reading.

Since no single chemistry discipline (organic chemistry, biochemistry) teaches the properties of lipids to any greater extent than, described in this book, hence it is strongly recommended for anyone working with lipids both in fundamental as well as in applied research. This is a concise book which can also be read by those interested in understanding the behaviour of lipids and gaining a new insight to their applications. The book will be extremely useful for all branches of food. This book should have wide appeal to formulators, chemists, graduate students and anyone who needs detailed structural information of lipids and their polymorphs, etc. It provides a most comprehensive treatment of the subject, making it essential reading for all those working with or studying lipids and therefore, can be used to introduce lipids to undergraduate as well as postgraduate students with great success.

We hope to get great response from our readers. We are also grateful to our friends and colleagues who have directly or indirectly helped us in writing this book. We wish to express our sincere thanks to the publishers, especially to Mr. Anil Mittal for his continued interest in bringing out this edition.

Neelam Khetarpaul

Vipul Khetarpaul

Contents

Chapter 1

Introduction

Lipids are among the major components of foods of plant and animal origin. There is no precise definition available for the term "lipid," however it usually includes a broad category of compounds that have some common properties and compositional similarities. Lipids are materials that are sparingly soluble or insoluble in water, but soluble in selected organic solvents such as hexane, benzene, chloroform, diethyl ether, and methanol. Lipids, together with carbohydrates and proteins, constitute the principal structural components of tissues. However, the common and unique features of lipids relate to their solubility rather than to their structural characteristics.

Foods contain any or all of lipid compounds, however, triacylglycerols (TAGs) and phospholipids (PLs) are the most abundant and important ones. Liquid TAGs at room temperature are referred to as *oils* and are generally of plant origin (*e.g.*, vegetable oils). Solid TAGs at room temperature are termed *fats,* which are generally of animal origin (*e.g.*, lard, tallow).

No exact definition of *lipids* actually exists. Christie defines lipids as "a wide variety of natural products including fatty acids and their derivatives, steroids, terpenes, carotenoids and bile acids, which have in common a ready solubility in organic solvents such as diethyl ether, hexane, benzene, chloroform or methanol."

Kates says that lipids are "those substances which are insoluble in water; soluble in organic solvents such as chloroform, ether or benzene; contain long-chain hydrocarbon groups in their molecules; and are present in or derived from living organisms."

Gurr and James point out that the standard definition includes "a chemically heterogeneous group of substances, having in common the property of insolubility in water, but solubility in non-polar solvents such as chloroform, hydrocarbons or alcohols."

Despite common usage, definitions based on solubility have obvious problems. Some compounds that are considered *lipids,* such as Cl–C4 very short chain fatty acids (VSCFA), are completely miscible with water and insoluble in non-polar solvents. Some researchers have accepted this solubility definition strictly and exclude Cl–C3 fatty acids in a definition of lipids, keeping C4 (butyric acid) only because of its presence in dairy fats. Additionally, some compounds that are considered lipids, such as some *trans* fatty acids (those not derived from bacterial hydrogenation), are not derived directly from living organisms. The development of synthetic acaloric and reduced calorie lipids complicates the issue because they may fit into solubility-based definitions but are not derived from living organisms, may be acaloric, and may contain esters of VSCFAs. .

The traditional definition of *total fat* of foods used by the U.S. Food and Drug Administration (FDA) has been the "sum of the components with lipid characteristics that are extracted by Association of Official Analytical Chemists (AOAC) methods or by reliable and appropriate procedures." The FDA has changed from a solubility-based definition to "total lipid fatty acids expressed as triglycerides," with the intent to measure caloric fatty acids. Solubility and size of fatty acids affect their caloric values.

Lipid oxidation in foods is a serious problem, difficult to overcome problem and leading to loss of shelf life, palatability, functionality, and nutritional quality. Loss of palatability is due to the generation of off-flavors that arise primarily from the breakdown of unsaturated fatty acids during autoxidation. The high reactivity of the carbon double bonds in unsaturated fatty acids makes these substances primary targets for free radical reactions. Autoxidation is the oxidative deterioration of unsaturated fatty acids via an autocatalytic process consisting of a free radical chain mechanism. The chain of reaction includes initiation, propagation, and termination. Propagation reactions are primarily responsible for the autocatalytic nature of autoxidation.

Autoxidation must be induced by preformed or primary hydroperoxides. The source of preformed hydroperoxides is either photosynthesized oxidation or lipoxygenase catalysis. Photosynthesized oxidation (photo oxidation) involves direct reaction of light-activated, singlet oxygen with unsaturated fatty acid and the subsequent formation of hydroperoxides. Lipoxygenase catalysis involves enzymatic oxidation of unsaturated fatty acids to their corresponding hydroperoxides. Formation of primary hydroperoxides is catalyzed by either light, metals, singlet oxygen and sensitizers, or preformed hydroperoxide decomposition products.

In foods containing lipids, antioxidants delay the onset of oxidation or slow the rate at which it proceeds. These substances can occur as natural constituents of foods, but they also can be intentionally added to products or formed during processing. Their role is not to enhance or improve the quality of foods, but they do maintain food quality and extend shelf life. Antioxidants for use in food processing must be inexpensive, nontoxic, effective at low concentrations, stable, and capable of surviving processing (carry-through effect); color, flavor, and odor must be minimal. The choice of which antioxidant to use depends on product compatibility and regulatory guidelines.

Most discussions of fats and health focus on the deleterious effects of the essential nutrients. In such cases, we are really discussing the possibly harmful effects of an excess of fats and oils. Fats (lipids) supply energy, support structural aspects of the body, and provide substances that regulate physiological processes.

Adipose tissue, which is the repository of most of our body fats, serves as an energy reservoir (fat supplies 9 Kcal/g compared to 4 Kcal/g for protein or carbohydrate), as a heat conserver, and as a shock absorber. Lipids contain essential fatty acids such as linoleic and linolenic acids. These are metabolized eventually to provide eicosanoids, substances that possess hormone like activity and thus may regulate many body functions. Fat is also the transport vehicle for vitamins A, D, E, and K.

Cholesterol, which has absorbed the brunt of the antifat attack, is a compound that is required for life. It is not essential in the sense of essential fatty acids, since the body can synthesize it, but it is a crucially important component of our biological economy. Cholesterol comprises about 0.2 per cent of normal body weight. Most of it (about 33 per cent) is in the brain and nervous system, where its function has not been probed beyond suggesting that its major activity is as an insulator. Almost another one-third of the body's cholesterol is in muscle, where it is a structural component. Every cell membrane contains cholesterol and phospholipid, another fatty substance. The esterified cholesterol found in muscle may represent a storage compartment. The percentage of cholesterol ester in muscle increases with age. Cholesterol is the parent substance for vitamin D_2, bile acids, adrenocortical hormones, and sex hormones. Thus, it is one of the more important biological substances. Fat also contributes to the palatability and flavor of food, hence to the enjoyment of eating.

The two major causes of death in the developed world are heart disease and cancer. Both have been described as lifestyle diseases, and effects of diet fit under that rubric. Since diet is one of the easiest lifestyle factors to investigate and possibly change, its role has been pursued with vigor. However, valid dietary data are not as simple to obtain as one might expect. For example, recall may be flawed by habitual underreporting of intake.

There is a complex relationship among dietary fats, eicosanoids, and the immune system. Because long chain polyunsaturated fatty acids the precursors for eicosanoids, these dietary components have the potential to modify levels of the products in the body. This is especially true if there is heavy reliance on a single fat in the diet. While this is the standard approach in nutrition studies in animals, it is also true that in a number of countries a single fat source provides as much as two-thirds of the human population's total fat intake. In controlled feeding trials in humans, both the type and the amount of fat have been varied, to permit an examination of the effect on eicosanoid production or changes in immune status.

Eicosanoids are only one of many possible mediators through which diet can influence the immune response. Eicosanoids are a large group of cyclized derivatives of the essential fatty acids that have potent biological activities and always contain 20 carbon atoms.

These compounds usually have very short half-lives (measured in seconds) and are derived from the precursor fatty acids via a series of enzymatic steps. In addition to the prostaglandins and leukotrienes, a variety of derivatives of 20-carbon fatty acids are produced. These include hydroxy and hydroperoxy fatty acids as well as hydroxylated or epoxydized derivatives. Cyclooxygenases add two oxygen molecules, lipoxygenase adds a single oxygen molecule, and cytochrome P450s add one atom of oxygen to the fatty acid.

The principal characteristic of the essential fatty acids (EFAs) is the presence of two or more cis double bonds in the families of ω3 or ω6 fatty acids, which must be derived from the diet. Fatty acids containing trans double bonds do not have EFA activity and may increase the requirement for EFA. Endogenously synthesized unsaturated fatty acids are of the ω9 series and cannot be converted to the EFA precursors. EFA-deficient animals have an increase in 20 : 3 ω9 (derived from oleic acid) in tissue, and the triene/tetraene ratio in plasma lipids is, often used clinically for diagnosis of this condition. This ratio essentially reflects the relative proportions of 20: 3 ω9 to 20 : 4 ω6. Individuals with fat malabsorption (especially cystic fibrosis patients) or those on long-term parenteral nutrition are most likely to display signs of EFA deficiency.

Many natural and processed foods exist either partly or wholly as emulsions, or have been in an emulsified state at some time during their existence. Milk is the most common example of a naturally occurring food emulsion. Mayonnaise, salad dressing, cream, ice cream, butter and margarine are all examples of manufactured food emulsions. Powdered coffee whiteners, sauces, and many desserts are examples of foods that were emulsions at one stage during their production, but subsequently were converted into another form. The bulk physicochemical properties of food emulsions, such as appearance, texture, and stability, depend ultimately on the type of molecules the food contains and their interactions with one another. Food emulsions contain a variety of different ingredients, including water, lipids, proteins, carbohydrates, minerals, sugars, and small molecule surfactants. By a combination of covalent and physical interactions, these ingredients form the individual phases and structural components that give the final product its characteristic physico-chemical properties. It is the role of food scientists to untangle the complex relationship between the molecular, structural, and bulk properties of foods, so that foods with improved properties can be created in a more systematic fashion.

Trans Fatty acids are present in a variety of food products; some are derived from natural sources like dairy products, but most come from products that contain commercially hydrogenated fats. The nutritional properties of *trans* fatty acids have been debated for many years, particularly with respect to the amounts of low density and high density lipoprotein (LDL, HDL) contained in serum. Some studies have shown that *trans* fatty acids elevate levels of serum LDL cholesterol and lower HDL cholesterol. Such results drew a great deal of media attention, which led to several requests for the U.S. Food and Drug Administration (FDA) to make labelling of *trans* fatty acids mandatory on food products. There have also been requests to either ban these substances or to impose strict limitations on their use. However, a comprehensive report of an expert panel convened by the International Life Sciences Institute concluded that food products containing partially hydrogenated fats are good substitutes for traditional fats rich in saturated fat; but are not good substitutes for unhydrogenated vegetable oils. The report also concluded that the overall effect of hydrogenated fat on serum cholesterol levels depends on the consumption of saturated, *trans*/*cis*-monounsaturated, and *cis*-polyunsaturated fatty acids. The authors recommended more research to better understand the effects of *trans* fatty acids on serum lipid concentrations and coronary heart disease. We recently discussed both the nutritional and food labelling issues associated with *trans* fatty acids.

Even after several years of research, there are still several laboratories in the world that have active research programs trying to improve the methodology to determine *trans* fatty acids in food products. The quantitation and identification of *trans* fatty acid isomers is difficult because of the wide range of positional monoene, diene, and triene fatty acid isomers present in hydrogenated oils. Moreover, cis positional isomers are also present, and commercial chromatographic standards are lacking for many fatty acid isomers.

Deep fat frying is a common method of food preparation that imparts desired sensory characteristics of fried food flavor, golden brown color, and crisp texture. During frying, at approximately 190°C, as oils thermally and oxidatively decompose, volatile and non-volatile products are formed which alter functional, sensory, and nutritional qualities of oils. During the past 30 years, scientists have extensively reported on the physical and chemical changes that occur during frying and on the wide variety of decomposition products formed in frying oils.

Fatty acids serve a wide variety of metabolic functions critical to all forms of life. They are a rich source of energy and carbon, and well designed as a convenient unit for energy storage. The importance of fatty acids in human nutrition and physiology, however, goes well beyond their role as a source of calories. Fatty acids provide the structure and hydrophobicity crucial to the maintenance of a semipermeable membrane barrier. Their structures can be modified by desaturation and elongation to produce a substantial variety of species with individual chemical and physical properties. Ester linkages to glycerides allow fatty acids to be easily exchanged for one another, and allow cells to manipulate the physical properties of their membranes. Fatty acids also serve as precursors to active signal molecules such as eicosanoids, which are capable of producing potent biological effects.

Evolution has produced a distinction between plants and animals in their capabilities for the metabolism of fatty acids. Higher animals are unable to synthesize all the fatty acids they require for certain tissue functions and are obligated to ingest fatty acids that have been synthesized by plants. Animals have evolved a separate and distinct series of metabolic modifications of fatty acids, but are still unable to alter the original modifications inserted by plants. As a result, the membrane, signal, and storage lipids of animals vary widely according to their dietary intakes. Additionally, an animal's ability to produce a specific fatty acid relies either on an inherent mechanism for desaturation of saturated fatty acids or on the ingestion of a convertible precursor. The ingestion and/or metabolism of particular unsaturated fatty acids is necessary for a great variety of physiological and cellular functions. Inadequate intakes or defective metabolism lead to various dysfunctions due to deficiencies of these fatty acids in particular cellular locations. Additionally, dietary fatty acids have been well correlated with metabolic and physiological alterations associated with heart disease and cancer. Unsaturated fatty acids in particular play an important role in these non-energy-producing metabolic functions.

Dietary fatty acids have the singular ability among macromolecules to become incorporated into tissue intact, thereby altering tissue acyl compositions. Proteins and nucleic acids, while providing energy and building blocks for metabolism, are incapable of remodeling the protein and nucleic acid compositions of tissues in their own image. Consequently, fatty acids occupy a unique and important role in human nutrition. With the rapid advances in basic knowledge of plant fatty acid biosynthesis and genetics, it will soon be possible to produce virtually any fatty acid in significant quantities. The availability of fatty acids for supplementation and the ability to engineer agricultural products provide opportunities to significantly modify the lipid content of the food supply. Unfortunately, knowledge concerning fatty acid function in physiology lags far behind abilities to modify dietary fatty acid compositions. Consequently, demand for reliable, applicable data on the effect of changes in dietary fatty acid consumption is increasing. Developing models to uncover data on unsaturated fatty acid metabolism mechanisms will prove key to nutrition and agriculture as "designer" foods are developed. More than likely, specific fatty acids will be beneficial to humans to varying degrees depending on age, metabolism, and disease states.

Although fat and oils are complex mixtures of fatty acids, each fat or oil has its characteristic fatty acid composition. Dairy fat, for example, is relatively rich in fatty acids with 14 or fewer carbon atoms, while olive oil has a high oleic content. Sunflower oil, on the other hand, is rich in linoleic acid, although certain varieties exist that contain large amounts of oleic acid. In normal, regular diets, palmitic and stearic acids are the most prevailing saturated fatty acids, while oleic and linoleic acids are, respectively, the most widespread dietary monounsaturated and polyunsaturated fatty acids. About 30 to 40 per cent of total energy intake is provided by fat. For a person consuming 10 MJ (2400 K.cal.) per day, this corresponds to a fat intake of 80 to 107g.

Many factors are associated with increased risk for coronary heart disease (CHD), a major cause of mòrbidity and mortality in prosperous Western countries. Some of these factors; like increasing age or a family history of premature CHD, are not amenable to preventive intervention, but other factors are: Three of these preventable factors–the distribution of plasma cholesterol over the low density and high density lipoproteins (LDL and HDL), the oxidizability of LDL, and hemostasis–can be modified by changing the sources and amount of fats and oils in the diet.

The development of methods to improve the nutritional and functional properties of fats and oils is of great interest to food processors. The molecular weight, unsaturation, and positional distribution of fatty acid residues on the glycerol backbone of triacylglycerols are the principal factors determining the physical properties of fats and oils. Chemical interesterification produces a complete positional randomization of acyl groups in triacylglycerols, It is used in the manufacture of shortenings, margarines, and spreads to improve their textural properties, modify melting behavior, and enhance stability. Interest in interesterification from the nutritional and functional standpoints is increasing because the process can be used to produce margarines with no trans-unsaturated fatty acids, to synthesize cocoa butter substitutes, and to improve the nutritional quality of some fats and oils. Recently, research efforts have been directed toward substituting some chemical interesterification applications with enzymatic interesterification because of the inherent advantages associated with the enzymatic process. Enzymatic reactions are more specific, require less severe reaction conditions, and produce less waste. Also, immobilized enzymes can be reused, which makes them economically attractive. Interesterification, whether chemical or enzymatic, is the exchange of acyl groups, between an ester and an acid (acidolysis), an ester and an alcohol (alcoholysis), or an ester and an ester (transesterification).

The major components of fats and oils are triacylglycerols, the composition of which is specific to the origin of each fat or oil. The physical properties of various fats and oils are different because of the structure and distribution of fatty acids in the triacylglycerols. In natural fats, acyl groups are distributed in a non-random fashion. During chemical or enzymatic interesterification, acyl groups are redistributed first intramolecularly, then intermolecularly, until a random distribution is achieved. With enzymatic interesterification, more control of final product composition is possible, and glyceride mixtures that cannot be obtained using chemical interesterification can be produced. At present, randomization of acyl group distribution by means of chemical interesterification is used to produce changes in crystal structure, solid fat content, and melting point of fats.

An effect of dietary fat on body weight has been well documented, especially in experimental animals. Warwick and Schiffman, in a thorough examination of the literature, reviewed the effects of diet type (fat vs. carbohydrate) and caloric density on caloric intake and weight gain reported in animal studies. They suggested that the caloric density of the diet plays a major role in promoting weight gain. If the density is 4.5 kcal/g or more, excess weight gain is more likely to occur. In several instances, an increase in body weight occurs without an increase in food intake.

Human studies, because they are more difficult to control, are less prevalent than studies with animals. However, human studies also demonstrate that dietary fat promotes weight gain. Lissner *et al.* demonstrated that subjects consuming a high-fat diet gained weight while those eating a low-fat diet lost weight. Others have reported that switching from a high-fat to a low-fat diet promotes relative weight loss. Finally, indirect evidence from epidemiological studies shows positive correlations between body weight and fat intake.

Fat is an important macromolecule component of plant and animal tissues. The various fats contribute to the physical and functional properties (solubility, viscosity, rheology, melting behavior, emulsification, body, creaminess, heat conduction, carrier of lipophilic vitamins and flavorants) of most food products, affecting as well sensory (appearance, taste, mouth-feel, lubricity, flavor) and nutritional (satiety, calories, essential fatty acid source, health benefits) aspects of food. These properties are difficult to duplicate in food formulations without adding fats. The amount and type of fat present in a food determine the characteristics of that food and consumer acceptance.

Fat is still the number one nutritional concern for most people in developed countries. The present estimate of average fat calories consumed by most Americans is 35 to 37 per cent. Recommendation by the U.S. Senate Select Committee on Nutrition and Human Needs and the Surgeon General is that fat consumption be reduced to 30 per cent of total calories of the diet. Excessive intake of fat in the diet has been linked to certain diseases, such as heart disease, cancer, obesity, and possibly gallbladder disease. Increased saturated fat intake is associated with high blood cholesterol and increased risk of coronary heart disease. It has been difficult for individuals to change their dietary habits to reduce or minimize fat intake while enjoying their favourite foods. This problem and the interest shown by consumers in alternative fats and foods low in calories or without calories led to the search by the food industry and scientific community for the "ideal" fat substitute.

A real fat substitute must be able to provide all the attributes of fats and replace the calories from fat on a 1 : 1 weight basis. Fats are the most concentrated source of energy: a given amount of fat contains more than twice the calories (9 kcal/g) of other macronutrients such as proteins and carbohydrates (4 kcal/g). An "ideal" fat substitute must look and function like fat and be able to substantially reduce caloric contributions to food.

Today people of all the ages are giving more importance towards their body weight. Now let us try to understand about the defect, or variation in the normal body weight.

Obesity is variably defined as having a body weight of greater than 20 per cent of ideal body weight or a body mass index [weight (kg)/height (m^2)] of 30 or more. Obesity is associated with many chronic diseases and alterations in physiologic function including cardiovascular disease, hypertension, diabetes mellitus, gallbladder disease, and certain types of cancer. Obesity is a major public health problem in the United States and Europe and is becoming increasingly important now in many other areas of the world. The prevalence of obesity in adults in the United States has increased by 30 per cent or more in the past decade, with increases in both genders and in all ethnic and racial populations and age groups. It is now estimated that over one-third of the adult population in the United States are obese.

The etiology of human obesity is quite complex, involving genetic, metabolic, behavioral, and environmental factors. Although obesity is believed to have a strong genetic component, the increased incidence of obesity in specific population groups undergoing Westernization indicates the importance of dietary and lifestyle changes in the manifestation of this disease. Among dietary factors, both total energy intake and fat intake are significantly correlated with body mass index in these population groups. However, increased intake of fat energy is associated with a greater per-unit increase in body mass than is increased intake of energy from non-fat sources. Therefore, much attention regarding dietary influences on obesity development or prevention has focused on high fat diets.

Chapter 2
Classification of Lipids

2.1 Lipid Nomenclature

Based on structure lipids can be classified as, derived, simple or complex lipids.

Classification of lipid structures is possible based on physical properties at room temperature (oils are liquid and fats are solid), their polarity (polar and neutral lipids), their essentiality for humans (essential and non-essential fatty acids) or their structure (simple or complex). Neutral lipids include fatty acids, alcohols, glycerides, and sterols, while polar lipids include glycerophospholipids and glyceroglycolipids. The separation into polarity classes is rather arbitrary, as some short chain fatty acid are very polar. A classification based on structure is, therefore preferable.

The fatty acids constitute the obvious starting point in lipid structures. However, a short review of standard nomenclature is appropriate. Over the years, a large number of different nomenclature systems have been proposed. The resulting confusion has led to a need for nomenclature standardization. The International Union of Pure and Applied Chemists (IUPAC) and International Union of Biochemistry (IUB) collaborative efforts have resulted in comprehensive nomenclature standards and the nomenclature for lipids has been reported. Only the main aspects of the standardized IUPAC nomenclature relating to lipid structures will be presented; greater detail is available elsewhere.

Standard rules for nomenclature must take into consideration the difficulty in maintaining strict adherence to structural-based nomenclature and elimination of common terminology. For example, the compound known as vitamin K_1 can be described as 2-methyl-3-phytyl- 1,4-naphthoquinone. *Vitamin K_1* and many other trivial names have been included into standardized nomenclature to avoid confusion arising from long chemical names. Standard nomenclature rules will be discussed in separate sections relating to various lipid compounds.

Fatty acid terminology is complicated by the existence of several different nomenclature systems. The IUPAC nomenclature, common (trivial) names and shorthand (omega) terminology will be discussed. As a lipid class, the fatty acids are often called free fatty acids (FFA) or non-esterified fatty acids (NEFA). IUPAC has recommended that fatty acids as a class be called fatty acids and the terms FFA and NEFA eliminated.

Before studying about nomenclature of fatty acids we must learn about systematic names of hydrocarbons.

In standard IUPAC terminology, the fatty acid is named after the parent hydrocarbon. Table 2.1 lists common hydrocarbon names.

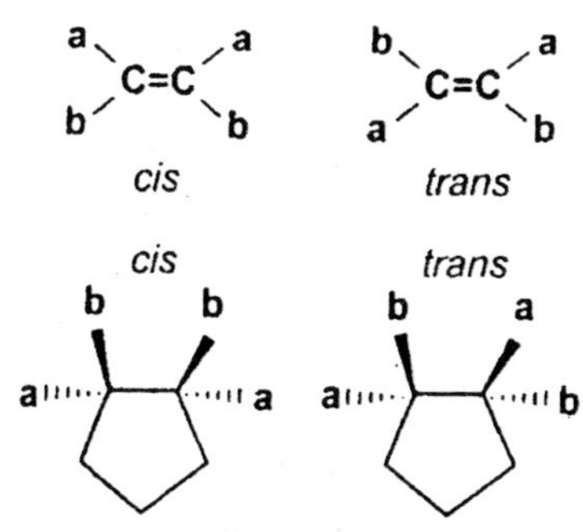

Figure 2.1: Examples of *cis/trans* Nomenclature

For example, an 18-carbon carboxylic acid is called octadecanoic acid, from octadecane, the 18-carbon aliphatic hydrocarbon. The name octadecanecarboxylic acid may also be used, but it is more cumbersome and less common.

Table 2.1: Common Hydrocarbon Names

Carbon Number	*Name*	*Carbon Number*	*Name*
1	Methane	19	Nonadecane
2	Ethane	20	Eicosane
3	Propane	21	Henicosane
4	Butane	22	Docosane
5	Pentane	23	Tricosane
6	Hexane	24	Tetracosane
7	Heptane	25	Pentacosane
8	Octane	26	Hexacosane
9	Nonane	27	Hepatacosane
10	Decane	28	Octacosane
11	Hendecane	29	Nonacosane
12	Dodecane	30	Triacontane
13	Tridecane	40	Tetracontane
14	Tetradecane	50	Pentacontane
15	Pentadecane	60	Hexacontane
16	Hexadecane	70	Heptacontane
17	Heptadecane	80	Octacontane
18	Octadecane		

2.2 Types of Fatty Acids

Saturated Fatty Acids

The saturated fatty acids begin with methanoic (formic) acid. Methanoic, ethanoic, and propanoic acids are uncommon in natural fats and are often omitted from definitions of lipids. However, they are found non-esterified in many food products. Omitting these fatty acids because they are water soluble would argue for also eliminating butyric acid, which would be difficult given its importance in dairy fats. The simplest solution is to accept the very short chain carboxylic acids as fatty acids, while acknowledging the rarity in natural fats of these water-soluble compounds.

Table 2.2 indicates the shorthand names of saturated fatty acids.

Table 2.2: Shorthand Names of Saturated Fatty Acids

Systematic Name	*Common Name*	*Shorthand*
Methanoic	Formic	1 : 0
Ethanoic	Acetic	2 : 0
Propanoic	Propionic	3 : 0
Butanoic	Butyric	4 : 0
Pentanoic	Valeric	5 : 0
Hexanoic	Caproic	6 : 0
Heptanoic	Enanthic	7 : 0
Octanoic	Caprylic	8 : 0
Nonanoic	Pelargonic	9 : 0
Decanoic	Capric	10 : 0
Undecanoic	–	11 : 0
Dodecanoic	Lauric	12 : 0
Tridecanoic	–	13 : 0
Tetradecanoic	Myristic	14 : 0
Pentadecanoic	–	15 : 0
Hexadecanoic	Palmitic	16 : 0
Heptadecanoic	Margaric	17 : 0
Octadecanoic	Stearic	18 : 0
Nonadecanoic	–	19 : 0
Eicosanoic	Arachidic	20 : 0
Docosanoic	Behenic	22 : 0
Tetracosanoic	Lignoceric	24 : 0
Hexacosanoic	Cerotic	26 : 0
Octacosanoic	Montanic	28 : 0
Tricontanoic	Melissic	30 : 0
Dotriacontanoic	Lacceroic	32 : 0

Unsaturated Fatty Acids

By far the most common monounsaturated fatty acid is oleic

acid (18: 1ω9), although more than 100 monounsaturated fatty acids have been identified in nature. The most common double-bond position of monoenes is Δ9. However, certain families of plants have been shown to accumulate what would be considered *unusual* fatty acid patterns. For example, *Eranthis* seed oil contains Δ5 monoenes and non-methylene-interrupted polyunsaturated fatty acids containing Δ5 bonds. Erucic acid (22: 1ω9) is found at high levels (40–50 per cent) in Cruciferae such as rapeseed and mustard seed. Canola is a rapeseed oil that is low in erucic acid (< 3 per cent 22 : 1ω9).

Polyunsaturated fatty acids (PUFA) are best described in terms of families because of the metabolism that allows interconversion within, but not among, families of PUFA.

The essentiality of ω6 fatty acids has been known since the late 1920s. Signs of ω6 fatty acid deficiency include decreased growth, increased epidermal water loss, impaired wound healing and impaired reproduction. Early studies did not provide clear evidence that ω3 fatty acids are essential. However, since the 1970s, evidence has accumulated illustrating the essentiality of the ω3 PUFA.

Table 2.3: Shorthand Names of Unsaturated Fatty Acids

Systematic Name	*Common Name*	*Shorthand*
c-9-Dodecenoic	Lauroleic	12 : 1ω3
c-5-Tetradecenoic	Physeteric	14 : 1ω9
c-9-Tetradecenoic	Myristoleic	14 : 1ω5
c-9-Hexadecenoic	Palmitoleic	16 : 1ω7
c-7,*c*-10,*c*-13-Hexadecatrienoic	–	16 : 3ω3
c-4,*c*-7,*c*-10,*c*-13-Hexadecatetraenoic	–	16 : 4ω3
c-9-Octadecenoic	Oleic	18 : 1ω9
c-11-Octadecenoic	*cis*-Vaccenic (Asclepic)	18 : 1ω7
t-11-Octadecenoic	Vaccenic	a
t-9-Octadecenoic	Elaidic	a
c-9,*c*-12-Octadecadenoic	Linoleic	18 : 2ω6
c-9-*t*-11-Octadecadienoic acid	Ruminic	a
c-9,*c*-12,*c*-15-Ocladecatrienoic	Linolenic	18 : 3ω3
c-6,c-9,c-12-Octadecatrienoic	γ-Linolenic	18 : 3w6
c-6,*c*-9,*c*-12,*c*-15-Octadecatetraenoic	Stearidonic	18 : 4w3

Contd...

Table 2.3–Contd...

Systematic Name	*Common Name*	*Shorthand*
c-11-Eicosenoic	Gondoic	20 : 1ω9
c-9-Eicosenoic	Gadoleic	20 : 1ω11
c-8,*c*-11,*c*-14-Eicosalrienoic	Dihomo-γ-linolenic	20 : 3ω6
c-5,*c*-8,*c*-11-Eicosatrienoic	Mead's	20 : 3ω9
c-5,*c*-8,*c*-11,*c*-14-Eicosatelraenoic	Arachidonic	20 : 4ω6
c-5,*c*-8,*c*-11,*c*-14,*c*-17-Eicosapentaenoic	Eicosapentaenoic (EPA)	20 : 5ω3
c-13-Docosenoic	Erucic	22 : 1ω9
c-11-Docosenoic	Cetoleic	22 : 1ω11
c-7,*c*-10,*c*-13,*c*-16,*c*-19-Docosapentaenoic	DPA	22 : 5ω3
c-4,*c*-7,*c*-10,*c*-13,*c*-16,*c*-19-Docosahexaenoic	DHA	22 : 6ω3
c-15-Tetracosenoic	Nervonic (Selacholeie)	24 : 1ω9

A number of different fatty acids have been identified having triple bonds. The nomenclature is similar to double bonds except the *-ane* ending of the parent alkane is replaced with *-ynoic acid, -diynoic acid,* and so on.

Shorthand nomenclature uses a lowercase *a* to represent the acetylenic bond; 9*c*,12*a*- 18 : 2 is an octadecynoic acid with a double bond in position 9 and the triple bond in position 12. Since the ligands attached to triple-bonded carbons are 180° from one another, (the structure thrcugh the bond is linear).

The acetylenic fatty acids found in nature are usually 18-carbon molecules with unsaturation starting at Δ9 consisting of conjugated double-triple bonds. Acetylenic fatty acids are rare.

trans Fatty Acids

trans Fatty acids include any unsaturated fatty acid that contains double bond geometry in the E *(trans)* configuration. Nomenclature differs only from normal *le/X3* fatty acids in the configuration of the double bonds.

The three main origins of *trans* fatty acids in our diet are bacteria, deodorized oils, and partially hydrogenated oils. The preponderance of *trans* fatty acids in our diets are derived from the hydrogenation process.

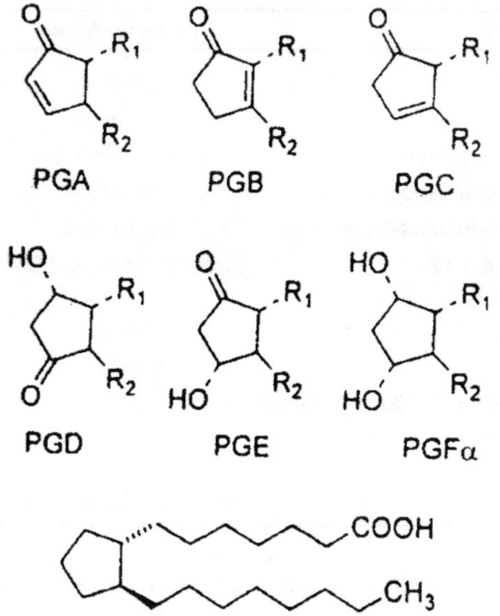

Prostanoic Acid

Figure 2.2: Prostanoic Acid and Prostaglandin Ring Nomenclature

trans Fatty acids are formed by some bacteria primarily under anaerobic conditions. It is believed that the formation of *trans* fatty acids in bacterial cell membranes is an adaptation response to decrease membrane fluidity, perhaps as a reaction to elevated temperature or stress from solvents or other lipophilic compounds that affect membrane fluidity (4- chlorophenol).

Branched Fatty Acids

A large number of branched fatty acids have been identified. The fatty acids can be named according to rules for branching in hydrocarbons. Beside standard nomenclature, several common terms have been retained, including *iso-*, with a methyl branch on the penultimate carbon, and *anteiso*, with a methyl branch on the antepenultimate carbon. The iso and anteiso fatty acids are thought to originate from a modification of the normal de novo biosynthesis, with acetate replaced by 2-methyl propanoate or 2-methyl-butanoate,

Tariaric (6-octadecyonic acid)

Stearolic (9-octadecyonic acid)

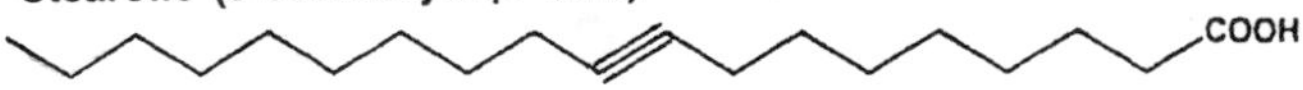

Ximenynic (*trans*-11-octadecene-9-ynoic acid)

Crepenynic (*cis*-9-octadecene-11-ynoic acid)

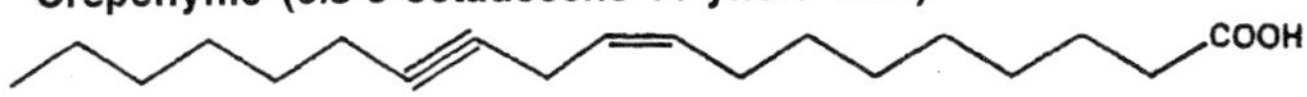

Crepenynic (*cis*-9-octadecene-11-ynoic acid)

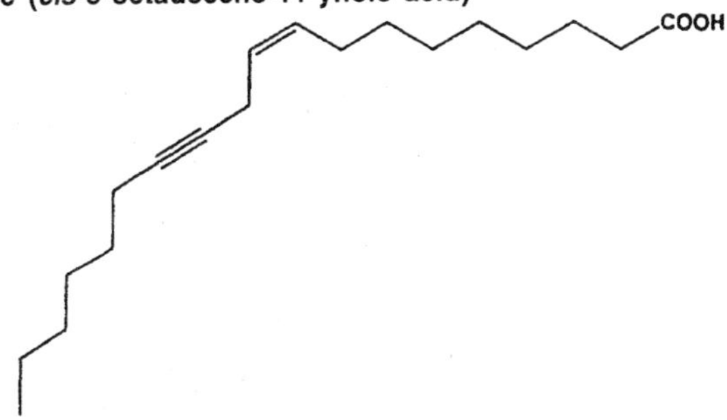

Figure 2.3: Some Acetylenic Fatty Acid Structures and Nomenclature

respectively. Other branched fatty acids are derived from isoprenoid biosynthesis including pristanic acid and phytanic acid.

Cyclic Fatty Acids

Many fatty acids that exist in nature contain cyclic carbon rings. Ring structures contain either three (cyclopropyl and cyclopropenyl), five (cyclopentenyl), or six (cyclohexenyl) carbon atoms and may be saturated or unsaturated. As well, cyclic fatty acid structures resulting from heating the vegetable oils have been identified.

In nomenclature of cyclic fatty acids, the parent fatty acid is the chain from the carboxyl group to the ring structure. The ring structure and additional ligands are considered a substituent of the parent

9-(2-butylcyclopentyl)nonanoic acid

A B C

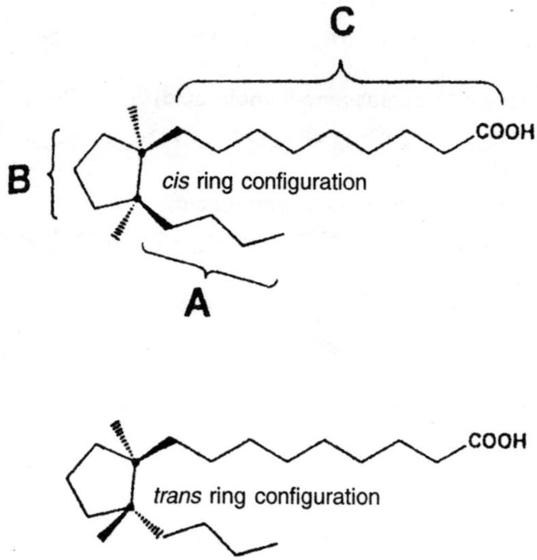

Figure 2.4: Nomenclature of Cyclic Fatty Acids

fatty acid. The parent is nonanoic acid (not pentadecanoic acid, which would result if the chain were extended through the ring structure). The substituted group is a cyclopentyl group with a 2-butyl ligand (2-butylcyclopentyl). Thus the correct standard nomenclature is 9-(2-butylcyclopentyl) nonanoic acid. The 2 is sometimes expressed as 2′ to indicate that the numbering is for the ring, and not the parent chain. The C1 and C2 carbons of the cyclopentyl ring are chiral, and two possible configurations are possible. Both the carboxyl and longest hydrocarbon substituents can be on the same side of the ring, or they can be on opposite sides. These are referred to as *cis* and *trans*, respectively.

The cyclopropene and cyclopropane fatty acids can be named by means of the standard nomenclature. They are also commonly named using the parent structure that carries through the ring

Cyclopropyl

10-(2-hexylcyclopropanyl)decanoic acid (lactobacillic or phytomonic acid)

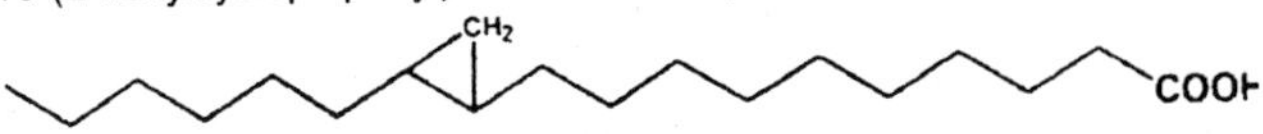

Cycloperopenyl

8-(2-octylcyclopropenyl)octanoic acid (sterculic acid)

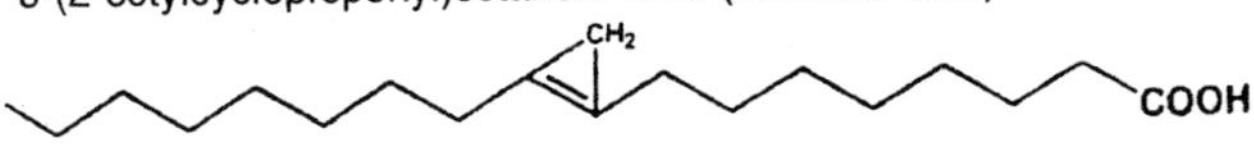

2-hydroxy-8-(2-octylcyclopropenyl)octanoic acid (2-hydroxysterculic acid)

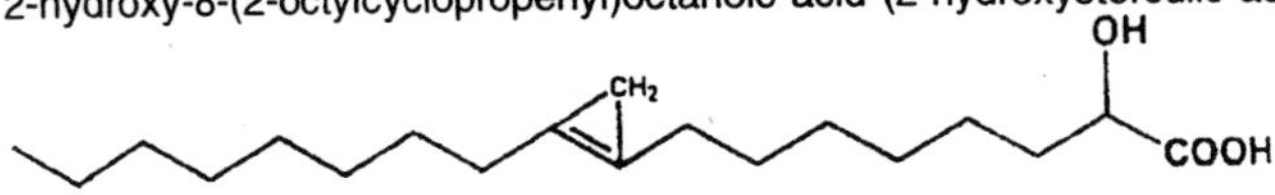

7-(2-octylcyclopropenyl)heptanoic acid (malvalic acid)

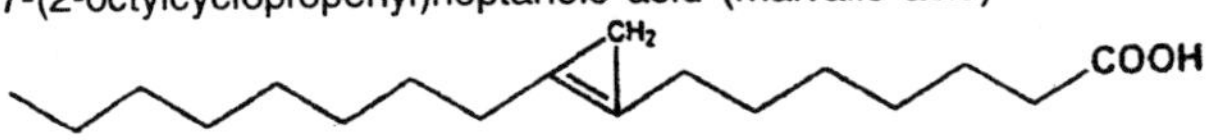

Cyclopentyl

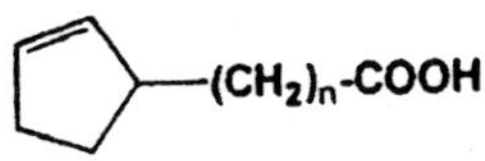

n =	
2	**3-(2-cyclopentenyl)propanoic acid (aleprolic acid)**
4	**5-(2-cyclopentenyl)pentanoic acid (aleprestic acid)**
6	**7-(2-cyclopentenyl)heptanoic acid (aleprylic acid)**
8	**9-(2-cyclopentenyl)nonanoic acid (alepric acid)**
10	**11-(2-cyclopentenyl)undecanoic acid (hyndocarpic acid)**
12	**13-(2-cyclopentenyl)tridecanoic acid (chaulmoogric acid)**

Figure 2.5: Cyclic Fatty Acid Structures and Nomenclature

structure. The fatty acid (commonly named lactobacillic acid or phycomonic acid) is named 10-(2-hexylcyclopropyl)-decanoic acid in standard nomenclature. An older naming system would refer to this fatty acid as cis-11,12-methylene octadecanoic acid, where *cis* designates the configuration of the ring structure. If the fatty acid is unsaturated, the term methylene is retained but the double bond position is noted in the parent fatty acid structure (*cis*-11,12-methylene-*cis*-octadec-9-enoic acid).

Hydroxy and Epoxy Fatty Acids

Saturated and unsaturated fatty acids containing hydroxy and epoxy funtional groups have been identified. Hydroxy fatty acids are named by means of the parent fatty acid and the hydroxy group(s) numbered with its Δ location. For example, the fatty acid with the trivial name *ricinoleic* is named R-12-hydroxy-*cis*-9-octadecenoic acid. Ricinoleic acid is found in the seeds of *Ricinus* species and accounts for about 90 per cent of the fatty acids in castor bean oil.

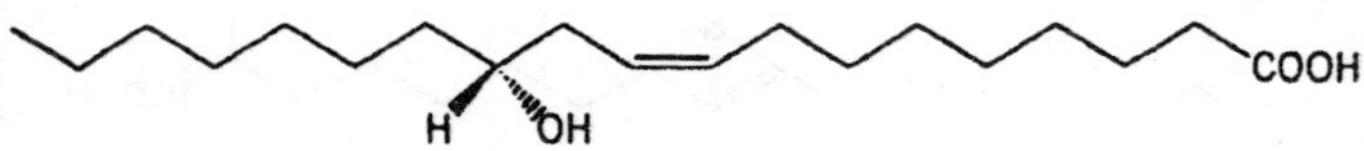

12(R)-hydroxy-*cis*-9-octadecenoic acid
Ricinoleic acid

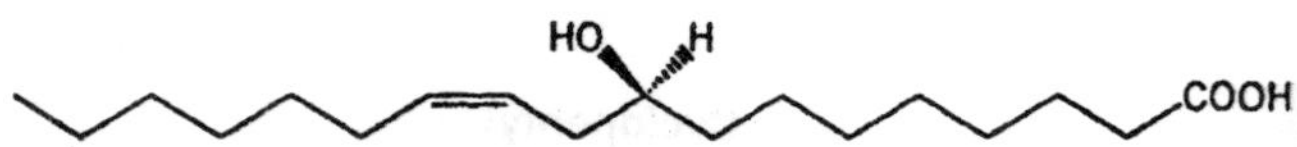

9(S)-hydroxy-*cis*-11-octadecenoic acid
Isoricinoleic acid

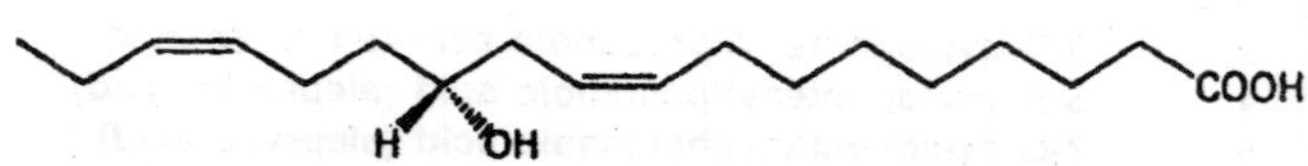

12(R)-hydroxy-*cis*-9,15-octadecenoic acid
Densipolic acid

Figure 2.6: Hydroxy Fatty Acid Structures and Nomenclature

Furanoid Fatty Acids

Some fatty acids contain an unsaturated oxolane heterocyclic group. They are more commonly called furanoid fatty acids because a furan structure (diunsaturated oxolane) is present in the molecule. Furanoid fatty acids have been identified in *Exocarpus* seed oils. They have also been identified in plants, algae, and bacteria and are a major component in triacylglycerols.

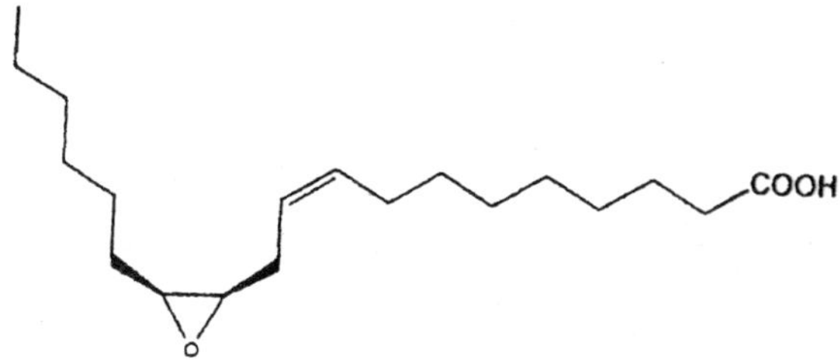

11-(3-pentyloxiranyl)-9-undecanoic acid
cis-12-13-epoxy-*cis*-9-octadecenoic acid
Vernolic acid
Both the (+) 12S, 13R and (–) 12R, 13S forms are found in nature

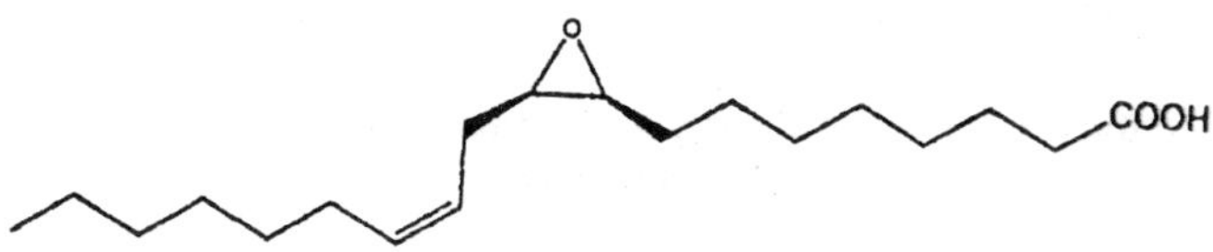

8-(3-*cis*-2'-nonenyloxiranyl)-octanoic acid
cis-9,10-epoxy-*cis*-12-octadecenoic acid
Coronaric acid

Figure 2.7: Epoxy Fatty Acid Structures and Nomenclature

2.3 The Steroid Class of Organic Compounds

The *steroid* class of organic compounds includes sterols of importance in lipid chemistry. Although the term "sterol" is widely used, it has never been formally defined. The following working definition was proposed some years ago: "any hydroxylated steroid that retains some or all of the carbon atoms of squalene in its side chain and partitions nearly completely into the ether layer when it is shaken with equal volumes of ether and water". Thus, for this definition, sterols are a subset of steroids and exclude the steroid hormones and bile acids. The importance of bile acids and their intimate origin from cholesterol makes this definition difficult. As

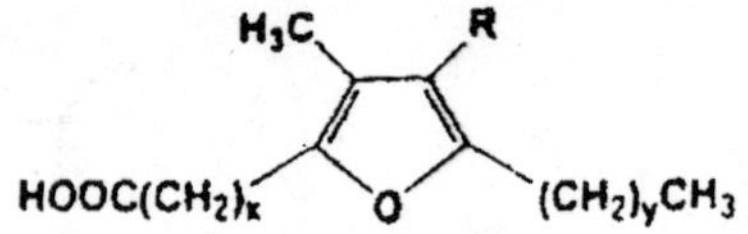

Name	*x*	*y*	*R*
F1	8	2	CH_3
F2	8	4	H
F3	8	4	CH_3
F4	10	2	CH_3
F5	10	4	H
F6	10	4	CH_3
F7	12	4	H
F8	12	4	CH_3

Figure 2.8: Furanoid Fatty Acid Structure and Shorthand Nomenclature

well, nonhydroxylated structures such as cholestane, which retain the steroid structure, are commonly considered sterols.

The sterols may be derived from plant (phytosterols) or animal (zoosterols) sources. They are widely distributed and are important in cell membranes. The predominant zoosterol is cholesterol. Although a few phytosterols predominate, the sterol composition of plants can be very complex. For example, as many as 65 different sterols have been identified in corn (*Zea mays*).

Now let us try to learn about nomenclature of the steroids. The nomenclature of the steroids is based on parent ring structures.

Sterol esters exist commonly and are named using standard rules for esters. For example, the ester of cholesterol with palmitic acid would be named cholesterol palmitate. The standard nomenclature would also allow this molecule to be named 3-*O*-palmitoyl-3β-cholest-5-en-3-ol.

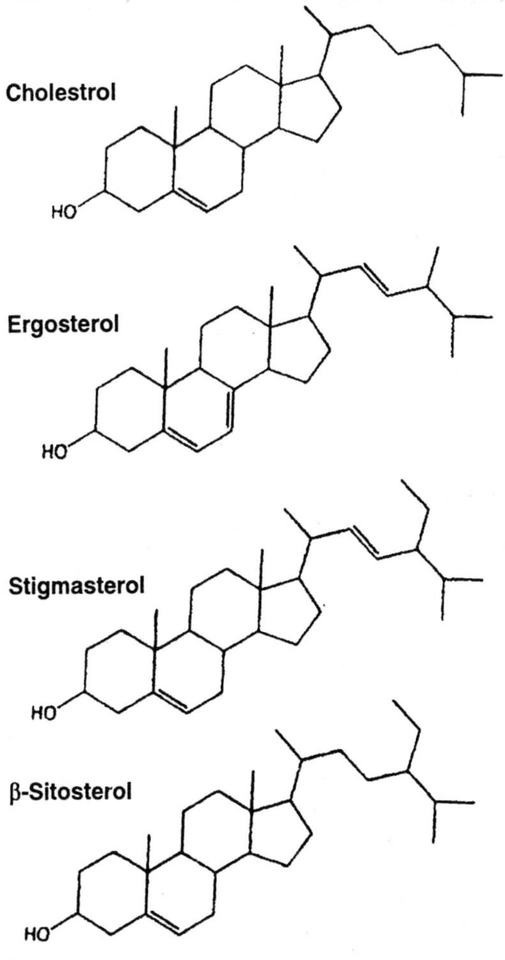

Figure 2.9: Common Sterol Structures

2.4 The Composition of Phospholipids

Phosphoglycerides (PL) are composed of glycerol, fatty acids, phosphate, and (usually) an organic base or polyhydroxy compound. The phosphate is almost always linked to the *sn*-3 position of glycerol molecule.

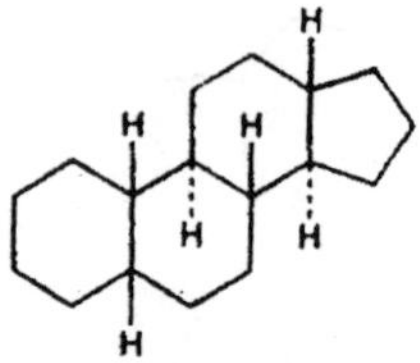

5β-Gonane

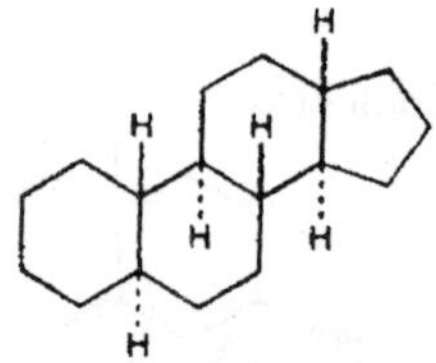

5α-Gonane

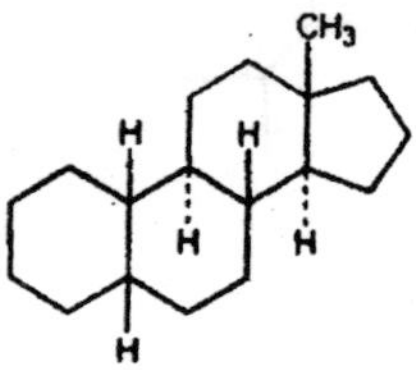

5β-Estrane

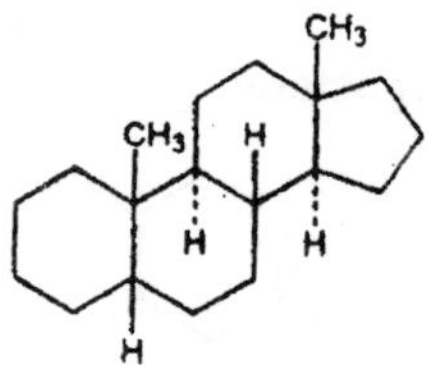

5β-Androstane

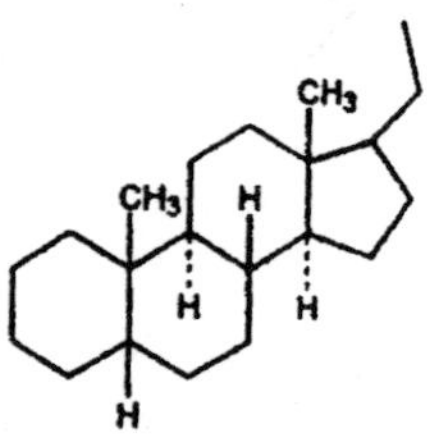

5β-Pregnane

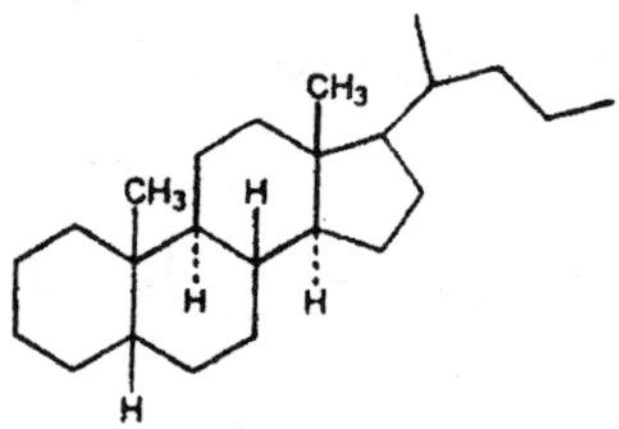

5β-Cholane

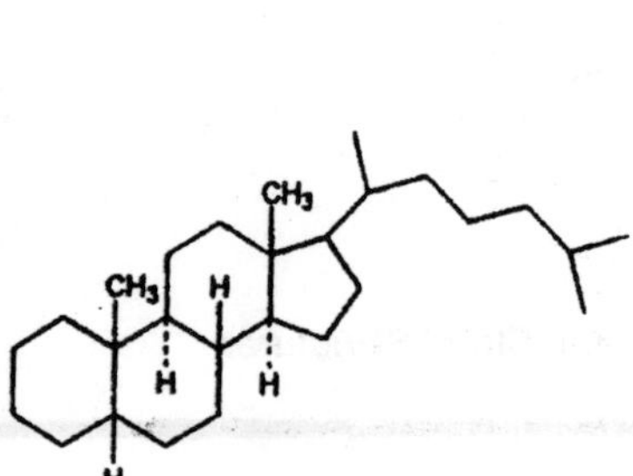

5β-Cholestane

Figure 2.10: Steroid Nomenclature

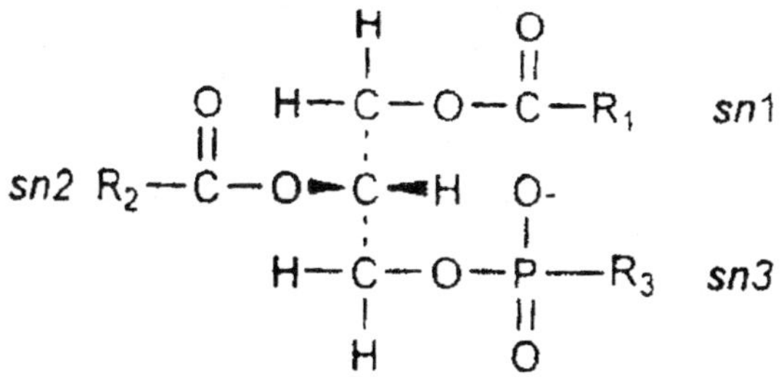

R_3 =	phospholipid class/nomenclature
-OH	phosphatidic acid (PA)
-OH-CH_2-CH(NH_2)-COOH	(3-*sn*-phosphatidyl)-L-serine (PS) 1,2-diacyl-*sn*-glycerol(3)phospha-L-senine
-OH-CH_2-CH_2-NH_2	(3-*sn*-phosphatidyl)ethanoiamine (PE) 1,2-diacyl-*sn*-glycerol(3)phosphocholine
-O-CH_2-CH_2-$N^+(CH_3)_3$	(3-*sn*-phosphatidyl)choline (PC) 1,2-diacyl-*sn*-glycerol(3)phosphacholine
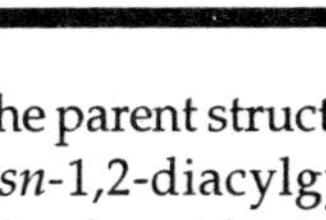	(3-*sn*-phosphatidyl)-L-*myo*-inositol (PI) 1,2-diacyl-*sn*-glycerol(3)phospho-L-*myo*-inositol

Figure 2.11: Nomenclature for Glycerophospholipids

The parent structure of the phosphoglycerides is phosphatidic acid (*sn*-1,2-diacylgycerol-3-phosphate). The terminology for phosphoglycerides is analogous to fat of acylglycerides with the exception of non-acyl group at *sn*-3. The prefix *lyso-*, when used for phosphoglycerides, indicates that the *sn*-2 position has been hydrolyzed and a fatty acid is esterified to the *sn*-1 position only.

Phospholipid classes are denoted using shorthand designation (PC = phosphatidylcholine, etc.). The standard nomenclature is based on the PL type. For example, a PC with an oleic acid on *sn*-1 and linolenic acid on *sn*-2 would be named 1-oleoyl-2-linolenoylsn-glycerol-3-phosphocholine. The name phosphorylcholine is sometimes used but is not recommended. The terms lecithin and cephalin, sometimes used for PC and PE, respectively, are not recommended.

Cardiolipin is a phosphoglyceride that is present in heart muscle mitochondria and bacterial membranes. Some cardiolipins contain the maximum possible number of 18 : 2ω6 molecules (4mol/mol).

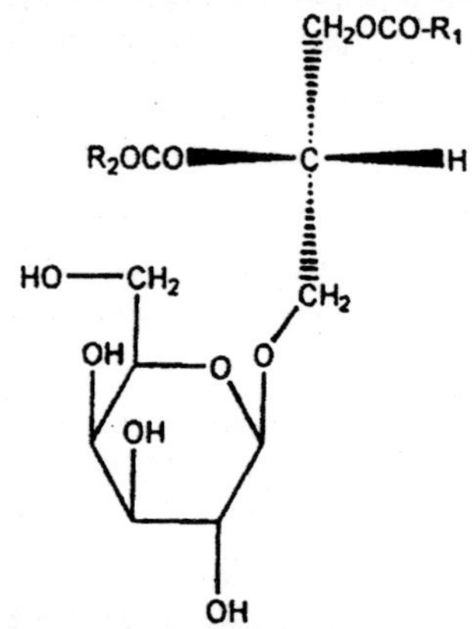

Monogalactosyldiacylglycerol (MGDG)
1,2-diacyl-3β-D-galactopyranosyl-L-glycerol

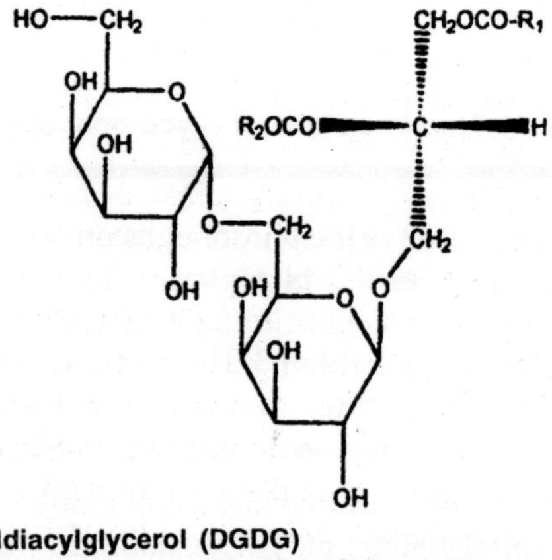

Digalactosyldiacylglycerol (DGDG)
1,2-diacyl-3(α-D-galactopyranosyl-1,6β-D-galactopyranosyl)L-glycerol

Figure 2.12: Glyceroglycolipid Structures and Nomenclature

The glyceroglycolipids or glycolipids are formed when a 1,2-diacyl-sn-3-glycerol is linked via the *sn-3* position to a carbohydrate molecule. The carbohydrate is usually a mono- or disaccharide, and less commonly a tri- or tetrasaccharide. Galactose is the most common carbohydrate molecule in plant glyceroglycolipids.

The names monogalactosyldiacylglycerol (MGDG) and digalactosyldiacylglycerol (DGDG) are used in common nomenclature. The standard nomenclature identified the ring structure and bonding of the carbohydrate groups.

The glycosphingolipids are also an important class of lipids which contain a long chain base fatty acids, and various other compounds, such as phosphates and monosaccharides.

2.5 Phospholipids as Components of Biological Membranes

Phospholipids are major components of biological membrane, and they are the integral part of prokaryotes and eukaryotes.

The predominant structures assumed by phospholipids in membranes are the bilayer and the H_{11} structure, which is dictated by the phase preference of the individual phospholipids. It is immediately apparent that a significant proportion of membrane lipids adopt or promote H_{11} phase structure under appropriate conditions. The most striking example is with PE, which may comprise up to 30 per cent of membrane phospholipids. Under such conditions, portions of the membrane that adopt an H_{11} phase would be expected to be incompatible with maintenance of a permeability barrier between external and internal compartments at those areas. Consequently, alternative roles for those structures must exist.

The ability of lipids to provide a bilayer permeability barrier between external and internal environments constitutes one of their most important functions in a biological membrane. Permeability coefficients of liquid-crystalline lipid bilayers for water are in the range of 10^{-2} to 10^{-2} cm/s, indicating a high permeability. The relative permeability of different membrane systems to water can be monitored by means of light-scattering techniques that measure swelling rates when osmotic gradients are applied. Results obtained from such studies indicate that increased unsaturation of the fatty acids of the membrane causes increases in water permeability. Since cholesterol reduces water permeability, the general conclusion has

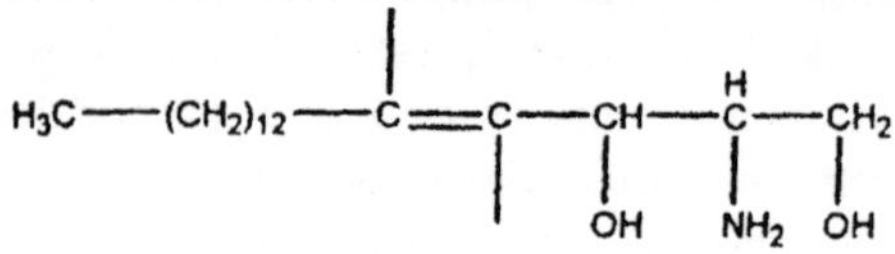

Sphingosine
D-*erythro*-1,3-dihydroxy-2-amino-*trans*-4-octadecene

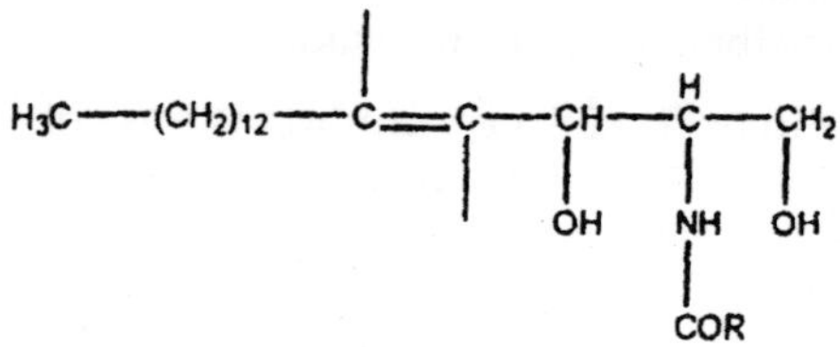

Ceramide
D-*erythro*-1,3-dihydroxy-2(N-acyl)-amino-*trans*-4-octadecene
(N-acyl sphingosine)

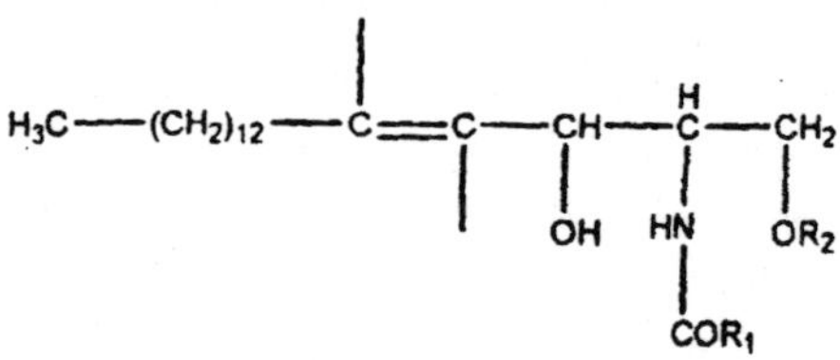

Cerebroside
1-O-β-Dgalactopyranosyl-N-acyl sphingosine

R_1 = fatty acid
R_2 = galactopyranose

Ganglioside

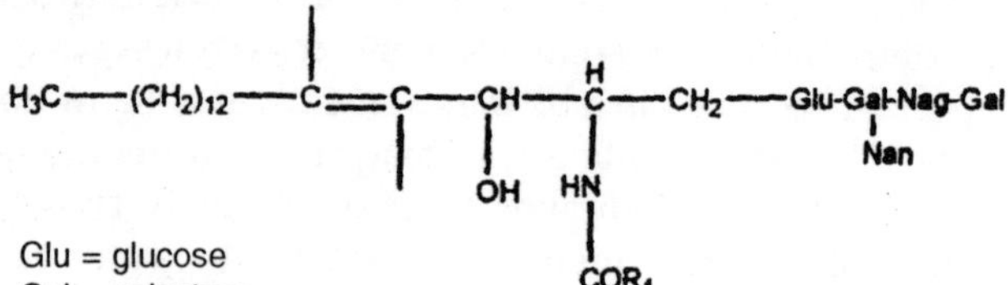

Glu = glucose
Gal = galactose
Nag = N-acetylgalactosamine
Nan = N-acetylneuraminic acid

Figure 2.13: Sphingolipid Structures and Nomenclature

been made that factors contributing to increased order in the hydrocarbon region decrease water permeability.

The diffusion properties of nonelectrolytes (uncharged polar solutes) also appear to depend on the properties of the lipid matrix in much the same manner as does the diffusion of water. That is, decreased unsaturation of phospholipids or increased cholesterol content results in lower permeability coefficients. In the case of nonelectrolytes, however, the permeability coefficients are atleast two orders of magnitude smaller than those of water. Furthermore, for a given homologous series of compounds, the permeability increases as the solubility in a hydrocarbon environment increases, indicating that the rate-limiting step in diffusion is the initial partitioning of the molecule into the lipid bilayer.

Table 2.4: Phase Preference of Membrane Phospholipids

Bilayer	*Hexagonal H_{11}*
Phosphatidylcholine	
Sphingomyelin	
	Phosphatidylethanolamine
Phosphatidylserine	Phosphatidylserine (pH < 3)
Phosphatldylglycerol	
Phosphatidylinositol	
Phosphatidic acid	Phosphatidic acid (+ Ca^{2+})
	Phosphatidic acid (pH < 3)

A remarkable impermeability of lipid bilayers exists for small ions, with permeability coefficients of less than 10^{-10} cm/s commonly observed. While permeability coefficients for Na^+ and K^+ may be as small as 10^{-14} cm/s, lipid bilayers appear to be much more permeable to H^+ or OH– ions, which have been reported to have permeability coefficients in the range of 10^{-4} cm/s. Measures of the permeability of membranes to small ions are complicated, since for free permeation to proceed, a counter flow of other ions of equivalent charge is required. In the absence of such a counterflow, a membrane potential is established that is equal and opposite to the chemical potential of the diffusing species. In terms of the relationship between lipid composition and membrane permeability, ion permeability appears to be related to the order in the hydrocarbon region, where increased order leads to a decrease in permeability. The charge on the phospholipid polar head group can also strongly influence permeability by virtue of the resulting surface potential. Depending on whether the surface potential is positive or negative, anions and cations could be attracted or repelled to the lipid-water interface.

The current concept of biological membrane is a dynamic molecular assembly characterized by the coexistence of structures with highly restricted mobility and components having great rotational freedom. These membrane lipids and proteins comprising domains of highly restricted mobility appear to exist on a micrometer scale in a number of cell types. Despite this heterogeneity, membrane fluidity is still considered as a bulk, uniform property of the lipid phase that is governed by a complex pattern of the components' mobilities. Individual lipid molecules can display motions of three different types: lateral, rotational, and transversal diffusion. Lateral diffusion of lipids in biological membranes refers to the two-dimensional translocation of the molecules in the plane of the membrane. Rotational diffusion of lipid molecules is restricted to the plane of biological membranes, while transverse diffusion is the out-of-plane rotation or redistribution of lipid molecules between the two leanets of the bilayer. Transverse diffusion or "flip-flop" motion is very low in lipid bilayers, and specific enzymes are required to mediate the process.

Two major components comprise membrane fluidity. The "order parameter" (S), also called the structural, static, or range component of membrane fluidity, is a measure of angular range of rotational motion, with more tightly packed chains resulting in a more ordered or less fluid bilayer. The second component of membrane fluidity is known as the "microviscosity," dynamic component of membrane fluidity, or rate. This component measures the rate of rotational motion and is a more accurate reflection of membrane microviscosity.

The many physical and chemical factors that regulate the fluidity properties of biological membranes include temperature, pressure, membrane potential, fatty acid composition, protein incorporation, and Ca^{2+} concentration. For example, calcium influences the structure of membranes containing acidic phospholipids by non-specifically cross-linking the negative charges. Consequently, increasing the calcium concentration may induce structural rearrangements and a decrease in membrane fluidity.

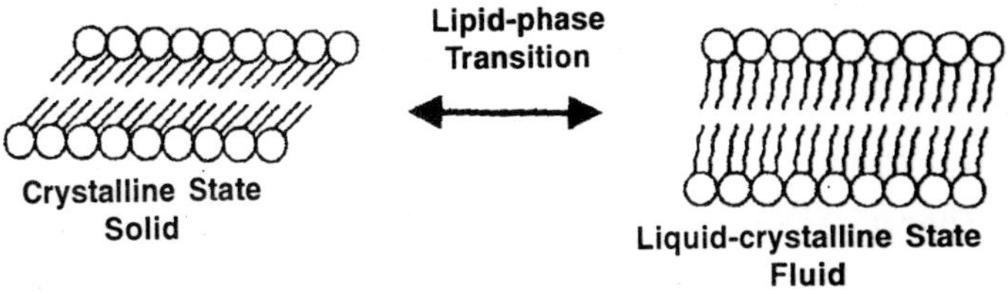

Figure 2.14: The Phospholipid Gel-liquid Crystalline Phase Transition

Fluidity is an important property of membranes because of its role in various cellular functions. Activities of integral membrane-bound enzymes, such as Na-K-ATPase, can be regulated to some extent by changes in the lipid portions of biological membranes. Changes in enzyme activities tightly connected to ion transport processes, in turn, could affect translocation of ions.

For complex mixture of lipids found in biological membranes, at temperatures above the phase transition, all component lipids are liquid crystalline, exhibiting characteristics consistent with complete mixing of the various lipids. At temperatures below the phase transition of the phospholipid with the highest melting temperature, separation of the component into crystalline domains (lateral phase separation) can occur. This ability of individual lipid components to adopt gel or liquid crystalline arrangements has led to the suggestion that particular lipids in a biological membrane may become segregate into a local gel state. This segregation could affect protein function by restricting protein mobility in the bilayer matrix, or it could provide packing defects, resulting in permeability changes. Exposure of plant food tissues to refrigerator temperatures could thus induce localized membrane phase transitions, upset metabolic activity, and create an environment that serves to reduce the quality product.

The complete functioning of biomembrane is generally controlled by both the protein and the lipid.

In a bilayer membrane that contains a heterogeneous distribution of both peripheral and integral proteins, there will be a certain proportion of the phospholipids interacting with the protein component to give the membrane its integrity at both the structural and the functional levels. Thus, the proportion of phospholipids in the bilayer interacting with protein at anyone time is dictated by protein density, protein type, protein size, and aggregation state of the proteins.

The major structural element of the transmembrane part of many integral proteins is the α-helix bundle, and the disposition and packing of such helices determine the degree of protein-lipid interactions. A single α-helix passing through a bilayer membrane has a diameter of about 0.8 to 1 nm, depending on side chain extension, which is similar to the long dimension of the cross section of a diacyl phospholipid (~ 0.9–1.0 nm). In the absence of any significant lateral restriction of such an individual peptide helix, the lateral and rotational motion of the peptide will be similar to that for the lipids. As the protein mass in or on the membrane increases, however, the motional restriction of the adjacent lipid also increases.

Quality losses in both plant and animal tissues may be attributed to membrane breakdown following slaughter or harvest. Postmortem changes in animal tissues, however, occur more rapidly than in plant tissues. In animals, cessation of circulation in the organism leads to lack of oxygen and accumulation of waste products, whereas in plants, respiratory gases can still diffuse across cell membranes and waste products are removed by accumulation in vacuoles.

Two different membrane breakdown pathways predominate in food tissues: free radical lipid oxidation, and loss of plasma and organelle membrane integrity. Some representative modifications that occur in membranes in response to lipid peroxidation include uncoupling of oxidative phosphorylation in mitochondria; alteration of endoplasmic reticulum function; increased permeability; altered activity; inactivation of membrane-bound enzymes, and polymerization, cross-linking, and covalent binding of proteins. Another consequence of lipid peroxidation is the formation of the volatile aldehydes that contribute to the aroma characteristics of many vegetables. With regard to loss of plasma and organelle membrane integrity, influx and efflux of solutes may occur, leading to' intimate contact among formerly separated catalytic molecules. Thus, in plants where small changes in calcium flux bring about a wide range of physiological responses, catastrophic changes may proceed in the event of loss of membrane integrity.

2.6 Structure of Fat-Soluble Vitamins

Vitamin A

Vitamin A exists in the diet in many forms. The most bioactive form is the all *trans* retinol, and cis forms are formed via light-induced iosmerization. The 13-cis isomer is the most biopotent of the mono- and di-cis isomers. The α- and β-carotenes have biopotencies of about 8.7 per cent and 16.7 per cent of the all-*trans* retinol activity, respectively. The recommended dietary allowance (RDA) for vitamin A is 600 retinol equivalents (RE), which represents 600 μg of all-*trans* retinol or 2400 of μg β-carotene. Vitamin A can be toxic when taken in levels exceeding the RDA. Some reports suggest that levels of 15,000 RE/day can be toxic.

Toxic symptoms of hypervitaminosis A include drowsiness, headache, vomiting, and muscle pain. Vitamin A can be teratogenic at high doses. Vitamin A deficiency results in night blindness and ultimately total blindness, abnormal bone growth, increased cerebrospinal pressure, reproductive defects, abnormal cornification, loss of mucus secretion cells in the intestine, and decreased growth. The importance of beef liver, an excellent source of vitamin A for cure of night blindness was known to the ancient Egyptians about 1500 BC.

Table 2.5: Approximate Biological Activity Relationships of Vitamin A Compounds

Compound	*Activity of All-*trans *Retinol (%)*
all-*trans* Retinol	100
9-*cis*-Retinol	21
11-*cis*-Retinol	24
13-*cis*-Retinol	75
9, 13-di-*cis*-Retinol	24
11, 13-di-*cis*-Retinol	15
α-Carotene	8.4
β-Carotene	16.7

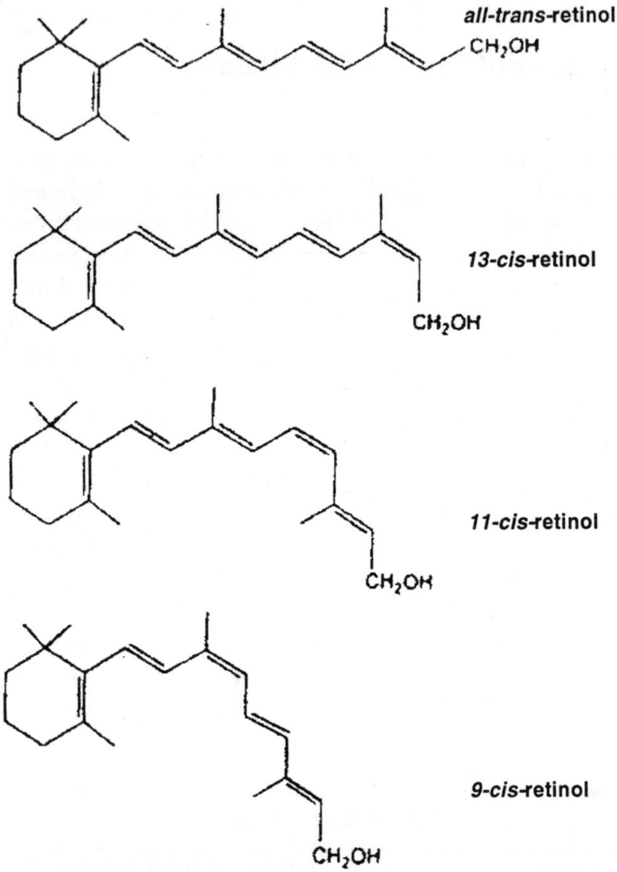

Figure 2.15: Structures of Some Vitamin A Compounds

Vitamin D

Although as many as five vitamin D compounds have been described, only two of these are biologically active, ergocalciferol (vitamin D_2) and cholecalciferol (vitamin D_3). Vitamin D_3 can be synthesized in man from 7-dehydrocholesterol, which occurs naturally in the skin via light irradiation.

The actual hormonal forms of the D vitamins are the hydroxylated derivatives. Vitamin D is converted to 25-OH-D in the

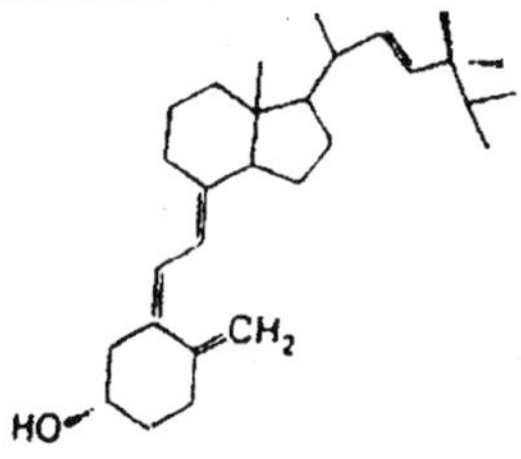

Ergocalciferol

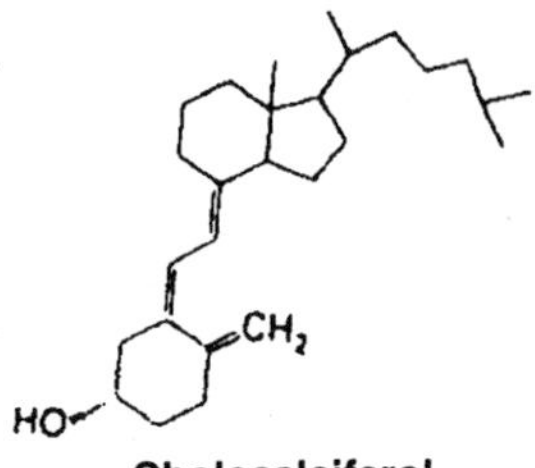

Cholecalciferol

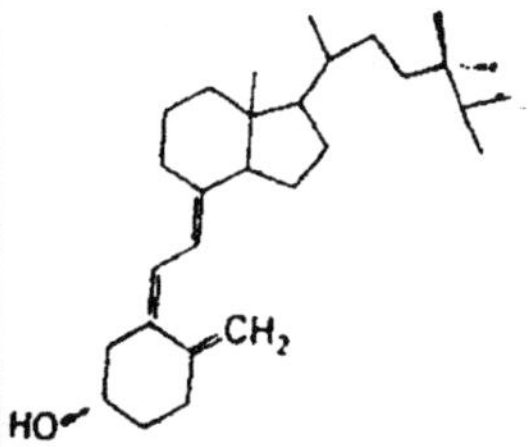

Vitamin D_4

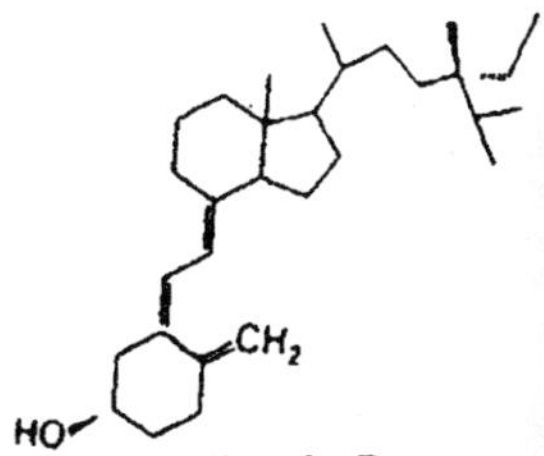

Vitamin D_5

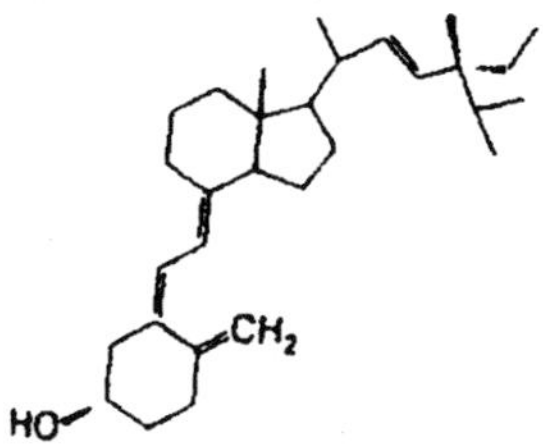

Vitamin D_6

Figure 2.16: Structures of Some Vitamin D Compounds

liver and further hydroxylated to 1,25-diOH-D in the kidney. The dihydroxy form is the most biologically active form in man.

Vitamin E

Vitamin E compounds include the tocopherols and tocotrienols. Tocotrienols have a conjugated triene double system in the phytyl side chain, while tocopherols do not. Methyl substitution affects the bioactivity of Vitamin E, as well as its *in vitro* antioxidant activity.

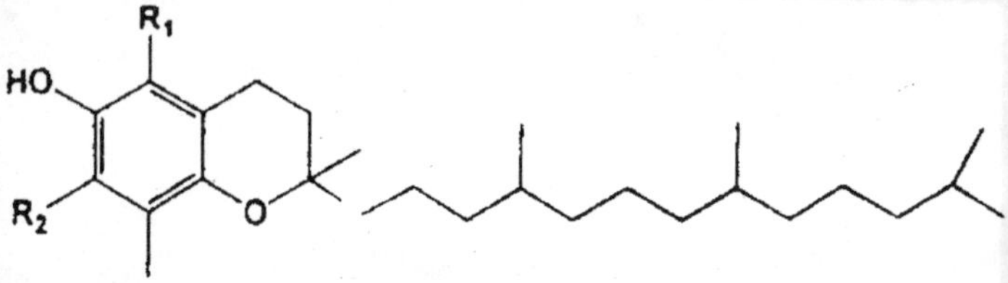

Tocopherol	R_1	R_2
alpha	CH_3	CH_3
beta	CH_3	H
gamma	H	CH_3
delta	H	H

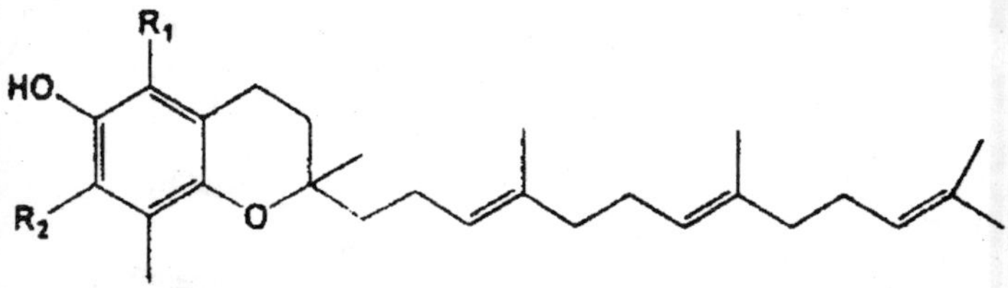

Tocotrienol	R_1	R_2
alpha	CH_3	CH_3
beta	CH_3	H
gamma	H	CH_3
delta	H	H

Figure 2.17: Structures of Some Vitamin E Compounds

Table 2.6: Approximate Biological Activity Relationships of Vitamin E Compounds

Compound	*Activity of D-α-tocopherol (%)*
D-α-Tocopherol	100
L-α-Tocopherol	26
DL-α-Tocopherol	74
DL-α-Tocopherol acetate	68
D-β-Tocopherol	8
D-γ-Tocopherol	3
D-δ-Tocopherol	–
D-α-Tocotrienol	22
D-β-Tocotrienol	3
D-γ-Tocotrienol	–
D-δ-Tocotrienol	–

2.7 Oxidation of Unsaturated Fatty Acids

Unsaturated fatty acids of phospholipids are susceptible to oxidation through both enzymatically controlled processes and random autoxidation processes. The mechanism of autoxidation is basically similar to the oxidative mechanism of fatty acids or esters in the bulk phase or in inert organic solvents. This mechanism is characterized by three main phases–initiation, propagation, and termination. Initiation occurs as hydrogen is abstracted from an unsaturated fatty acid of a phospholipid, resulting in a lipid free radical. The lipid free radical in turn reacts with molecular oxygen to form a lipid peroxyl radical. While irradiation can directly abstract hydrogen from phospholipids, in most foods initiation is frequently attributed to reaction of the fatty acids with active oxygen species such as the hydroxyl free radical and the protonated form of superoxide. These active oxygen species are produced when metal ion, particularly iron, interacts with triplet oxygen, hydrogen peroxide, and superoxide anion. Enzymatic abstraction of hydrogen from an unsaturated fatty acid, on the other hand occurs when Fe^{3+} at the active site of lipoxygenase is reduced to Fe^{2+}. While the majority of lipoxygenases require free fatty acids, there have been reports of lipoxygenase acting directly on fatty acids in phospholipids. Hence, enzymatic hydrolysis may not always be required prior to lipoxygenase action.

During propagation, lipid-lipid interactions foster propagation of free radicals produced during initiation by abstracting hydrogen from adjacent molecules; the result is a lipid hydroperoxide and a new lipid free radical. Magnification of initiation by a factor of 10 to 100 may occur through free radical chain propagation. Further magnification may occur through branching reactions (also known as secondary initiation) in which Fe^{2+} interacts with a hydroperoxide to form a lipid alkoxyl radical and hydroxyl radical, which will then abstract hydrogens from unsaturated fatty acids.

Hydroperoxides of fatty acids produced during propagation do not have a direct adverse effect on the flavor and aroma of foods. Rather, the decomposition of hydroperoxides to aldehydes and ketones is responsible for the characteristic flavors and aromas that collectively are often described by the term "rancid." Numerous studies, on the other hand, have shown that specific oxidation products may be desirable flavor components particularly when formed in more precise (*i.e.*, less random) reactions by the action of lipoxygenase enzymes and/or by the modifying influence of tocopherol on autoxidation reactions.

Through *in vitro* studies, membrane lipids have been shown to oxidize faster than emulsified triacylglycerols apparently because propagation is facilitated by the arrangement of phospholipids in the membrane. Proximity of the phospholipids to catalytic sites of oxidation (enzymic lipid peroxidation, heme-containing compounds), in addition to the high degree of polyunsaturation in phospholipids, has been advocated as contributing to the importance of membrane lipids in tissue oxidation. Evidence that phospholipids are the major contributors to the development of warmed-over flavor in meat from different animal species has been described in several sources. Similarly, during frozen storage of salmon fillets, hydrolysis followed by oxidation of the *n*-3 fatty acids in phospholipids was noted. Phospholipid importance, however, has not been restricted to animal and fish tissues. In an accelerated storage test of potato granules, both the amounts of phospholipids and their unsaturation decreased. Moreover, with pecans, a much stronger negative correlation was found between head space hexanal and its precursor fatty acid (18 : 2) from the phospholipid fraction ($R = -0.98$) than from the triacylglycerol fraction ($R = -0.66$) or free fatty acid fraction ($R = -0.79$). These results suggest that despite membrane lipid constituting only a small percentage of the total lipid (0.5 per cent), early stages of oxidation may actually be occurring primarily within the phospholipids.

The presence of phospholipids does not preclude acceleration of lipid oxidation. When present as a minor component of oil systems, solubilized phospholipids have limited the oxidation of the triacylglycerols. Order of effectiveness of individual phospholipids was as follows: SPH = LPC = PC = PE > PS > PI > PG. It has been postulated that Maillard reaction products formed from aldehydes reacting with the amino group of the nitrogen-containing phospholipid are responsible for the antioxidant activity. Alternatively, a nitroxide radical derived from the phospholipid has been observed in systems containing PE or PS.

Fatty acid composition is a major factor affecting the susceptibility of a phospholipid to assume the oxidized state, with more highly unsaturated fatty acids increasing the rate of oxidation.

The presence of an enol-ether bond at position 1 of the glycerol backbone in plasmalogen phospholipids, however, has led to inhibition of lipid oxidation, possibly through the iron-binding properties of these compounds. Variation within the phospholipid classes toward oxidation has also been attributed to the iron-trapping ability of the polar head group. For example, PS was shown to inhibit lipid peroxidation induced by a ferrous-ascorbate system in the presence of PC hydroperoxides. Stimulation of phospholipid oxidation by trivalent metal ions (Al^{3+}, Sc^{3+}, Ga^{3+}, In^{3+}, Be^{2+}, Y^{3+}, and La^{3+}), however, has been attributed to the capacity of the ions to increase lipid packing and promote the formation of rigid clusters, processes that bring phospholipid acyl chains closer together to favor propagation steps.

2.8 Hydrogenation of Fats, and Hydration of Phospholipids

Hydrogenation of fats involves the addition of hydrogen to double bonds in the chains of fatty acids. While hydrogenation is more typically applied to triacylglycerols to generate semisolid or plastic fats more suitable for specific applications, it may also be applied to phospholipid fractions. Hydrogenated lecithins are more stable and more easily bleached to a light color, hence more useful as emulsifiers than the natural, highly unsaturated lecithin from soybean oil. These advantages are exemplified by a report that hydrogenated lecithin functions well as an emulsifier and as an inhibitor of fat bloom in chocolate.

In practice, hydrogenation involves the mixing of the lipid with a suitable catalyst (usually nickel), heating, and then exposing the mixture to hydrogen at high pressures during agitation. Phospholipids are not as easily hydrogenated as triacylglycerols, requiring higher temperatures and higher hydrogenation pressures. Consequently, hydrogenation of lecithin is carried out at 75 to 80°C in at least 70 atm. pressure and in the presence of a flaked nickel catalyst. In chlorinated solvents or in mixtures of these solvents with alcohol, much lower temperatures and pressures can be used for hydrogenation, particularly when a palladium catalyst is used.

The amount of water absorbed by phospholipids has been measured by a number of different methods, including gravimetric, X-ray diffraction, neutron diffraction, nuclear magnetic resonance, and differential scanning calorimetry methods. For any measurement, however, Klose *et al.* caution that the morphology and method of sample preparation can induce the formation of defects in and between the bilayers, hence will influence the water content of lamellar phospholipids.

The amount of water absorbed by PC from the vapor phase increased monotonically from 0 water molecules per PC molecule at 0 per cent humidity to between 14 and 20 water molecules per lipid molecule at 100 per cent relative humidity. Observed differences may be due to the difficulty of exerting accurate control over relative humidities near 100 per cent when temperature gradients in the system are present. The results of X-ray diffraction studies indicated that when directly mixed with bulk water, PC imbibed up to 34 water molecules. Considerably less water was imbibed by PE, with a maximum of about 18 water molecules per lipid. From the saturated vapor phase, however, liquid crystalline egg yolk PE only absorbed about 10 water molecules per lipid molecule whereas for charged phospholipids, such as I, PI or PS, the phospholipid imbibed water without limit.

Neutron diffraction experiments on multilayers containing PC, PE, and PI have revealed that water distributions are centered between adjacent bilayers and overlap the head group peaks in the neutron scattering profile of the bilayer. These results imply that water penetrates into the bilayer head group region, but appreciable quantities of water do not reach the hydrocarbon core. By combining X-ray diffraction and dilatometry data, McIntosh and Simon were able to calculate the number of water molecules in the interbilayer space and in the head group region for dilauroyl-PE bilayers. They found that there are about 7 and 10 waters per molecule in the gel and liquid crystalline phases, respectively, with about half of these water molecules located between adjacent bilayers and the other half in the head group region.

The amount of water taken up by a given phospholipid depends on interactions between the lipid molecules, including interbilayer forces (those perpendicular to the plane of the bilayer) and intrabilayer forces (those in the plane of the bilayer). For interbilayer forces, at least four repulsive interactions have been shown to operate between belayer surfaces. These are the electrostatic, undulation, hydration (solvation), and steric pressures. Attractive pressures include the relatively long-range van der Waal's pressure and short-range bonds between the molecules in apposing bilayers, such as hydrogen bonds or bridges formed by divalent salts. Several of the same repulsive and attractive interactions act in the plane of the bilayer including electrostatic repulsion, hydration repulsion, steric repulsion, and van der Waal's attraction. In addition, interfacial tension plays an important role in determining the area per lipid molecule. Thus, as the area per molecule increases, more water can be incorporated into the head group region of the bilayer. Such a situation is found with bilayers having an interdigitated gel phase compared to the normal gel phase. At a given interbilayer fluid separation, there is more volume available for water in the interdigitated phase than in the normal gel phase.

The presence of monovalent and/or divalent cations in the fluid phase changes the hydration properties of the phospholipids. For example, the partial fluid thickness between dipalmitoyl PC bilayers increased from about 20 Å in water to over 90 Å in 1 mM $CaCl_2$. In contrast, monovalent cations, such as Na^+, K^+, or Cs^+, decrease the fluid spaces between adjacent charged PS or PG bilayers as a result of screening of the charge. Divalent cations, in addition, have a dehydrating effect on PS. The most extensively studied divalent cation, Ca^{2+}, binds to the phosphate group of PS, liberates water from between bilayers and from the lipid polar groups, crystallizes the lipid hydrocarbon chains and raises the gel to the liquid crystalline melting temperature of dipalmitoyl PS by over 100°C. In response to these changes, one would expect permeability of the membrane to be altered.

The interaction of phospholipids with water is critical to the formation, maintenance, and function of membranes and organelles. It is the low solubility of the acyl chains in water combined with the strong hydrogen bonding between the water molecules, that furnishes the "attractive" force that holds together polar lipids as supramolecular complexes (the "hydrophobic bond"). These ordered structures are generated when the phospholipid concentration exceeds its critical micelle concentration (cmc), which is dependent on the free energy gained when an isolated amphiphile in solution enters an aggregate. For diacyl phospholipids in water, the cmc in general is quite low, but it depends on both the chain length and the head group. For a given chain length, the solubility of charged phospholipids is higher, while the cmc of a single-chain phospholipid is higher than that of a diacyl phospholipid with the same head group and the same chain length.

Chapter 3

Composition of Fatty Acids and the Distribution of Cholestrol

3.1 Effects of Fatty Acids on Plasma Lipid

The study of effects of fatty acids on plasma lipid can be studied under three classes. They are as follows:

1. Medium chain fatty acids
2. Fatty acids with 12, 14, or 16 carbon atoms
3. Stearic acid

Saturated fatty acids with fewer than 12 carbon atoms are called short and medium chain saturated fatty acids (MCFAs) and are found in relatively large amounts in coconut fat, palm kernel oil, and butterfat, but also in certain structured lipids, parenteral nutrition, and sport drinks.

McGandy and coworkers have carefully compared the effects of MCFAs, mainly capric acid, on plasma total and LDL cholesterol levels. Eighteen physically healthy men were fed several diets, each for 4 weeks. Diets contained several low-fat food items to which the experimental fats were added. It was shown that modest dietary amounts of MCFAs had effects on the plasma total and LDL cholesterol and on triacylglycerols concentrations that were comparable to those of carbohydrates. Large amounts of MCFAs, however, increased triacylglycerols concentrations. Recent results of Cater *et al.* however, suggested that a mixture of MCFAs slightly decreases LDL cholesterol concentrations relative to palmitic acid, while increasing LDL cholesterol relative to oleic acid.

Palm kernel oil, coconut oil, and dairy fat, which are rich in lauric acid, contain also relatively high amounts of myristic acid. Therefore, it is hardly possible to study the specific effects of these two saturated fatty acids on plasma lipoproteins with natural fats. A diet enriched in palm kernel oil, for example, will contain high amounts of both lauric and myristic acids, and thus its effects on the plasma lipoprotein profile cannot be ascribed to either lauric or myristic acid.

To circumvent this problem, most studies have examined the cholesterolemic effects of these two saturated fatty acids by means of synthetic fats, which can be formulated to contain any desired fatty acid composition. For example, when one interesterifies trilaureate with a high oleic acid sunflower oil, the result will be a high lauric acid fat without any myristic acid.

Denke and Grundy compared the effects on plasma lipoproteins of lauric acid with those of palmitic acid and oleic acid. Each of three liquid formula diets, differing only in type of fat used, was fed for 3 weeks in random order to 14 men. It was concluded that lauric acid raised plasma total and LDL concentrations compared with oleic acid, but not as much as the palmitic acid diet had done. No effects were observed on plasma HDL cholesterol or triacylglycerols concentrations. Using mixed natural diets, however, Temme *et al.* found that lauric acid raises total cholesterol concentrations more than palmitic acid, partly as a result of an increase in HDL cholesterol.

The effects of myristic acid have been examined in several studies with mixed solid foods. Zock *et al.* concluded that myristic acid increased plasma total cholesterol concentrations relative to palmitic and oleic acids, citing an increase in both LDL and HDL cholesterol concentrations. Triacylglycerols concentrations, however, were not affected. Tholstrup *et al.* also found that myristic acid increased HDL cholesterol concentrations compared with palmitic acid. No differences, however, were observed in plasma LDL cholesterol concentrations.

Replacing palmitic acid in the diet for carbohydrates lowers LDL and HDL cholesterol concentrations. However, if the intake of palmitic acid is decreased at the expense of oleic acid, LDL cholesterol concentrations decrease, while HDL cholesterol is not affected. Recently, however, it has been postulated that in normolipidemic, nonobese subjects, palmitic acid may not always be a cholesterol-raising saturated fatty acid provided linoleic acid intake contributes at least 6 to 7 per cent of daily energy intake and daily dietary cholesterol intake is below 300 mg. These findings, of course, will be of great practical significance if it can be shown that under certain conditions palmitic acid can be exchanged for oleic acid without affecting LDL-cholestrol levels.

The most abundant monounsaturated fatty acid in the human diet, oleic acid, has 18 carbon atoms and one double bond. Although olive oil probably is the most well-known source of oleic acid, animal fats are in many countries a major contributor to the intake not only of total oleic acid, but also of palmitic and stearic acids.

Effects of monounsaturated fatty acids on plasma total cholesterol levels are often described as neutral, a term that is subject to misinterpretation. "Neutral" does not signify a plasma total cholesterol level that does not change when monounsaturates are added to the diet. Rather, the term indicates that monounsaturated fatty acids have the same effect on plasma total cholesterol as an isocaloric amount of carbohydrates. Although this definition may be correct with respect to plasma total cholesterol levels, oleic acid and carbohydrates do not have similar effects on the distribution of cholesterol of the various lipoproteins. This was also shown in a study with young healthy volunteers. Increasing the intake of oleic acid at the expense of carbohydrates increased plasma HDL cholesterol concentrations and decreased those of triacylglycerols, The increase in HDL cholesterol was compensated for by a decrease in VLDL cholesterol. Effects of carbohydrates and oleic acid on plasma total and LDL cholesterol concentrations were comparable.

Polyunsaturated fatty acids in the diet belong to either the *n-6* or the *n-3* family. About 90 per cent of all polyunsaturated fatty acids in the diet are linoleic acid, which is found in vegetable oils like sunflower oil, corn oil, and soybean oil. The mean daily intake of fatty acids from the *n-3* family is only 1 to 3 g. These polyunsaturated fatty acids are of either vegetable or animal origin. α-Linolenic acid is found in rapeseed and soybean oils, while the very long chain fatty acids timnodonic or eicosapentaenoic acid (EPA) and cervonic or docosahexaenoic acid (DHA) are present only in fish oils.

Most unsaturated fatty acids found in nature have the *cis* configuration. This means that the two carbon side chains attached to the double bond point into the opposite *(cis)* direction. However, in some fatty acids the carbon side chains point into the same (*trans*)

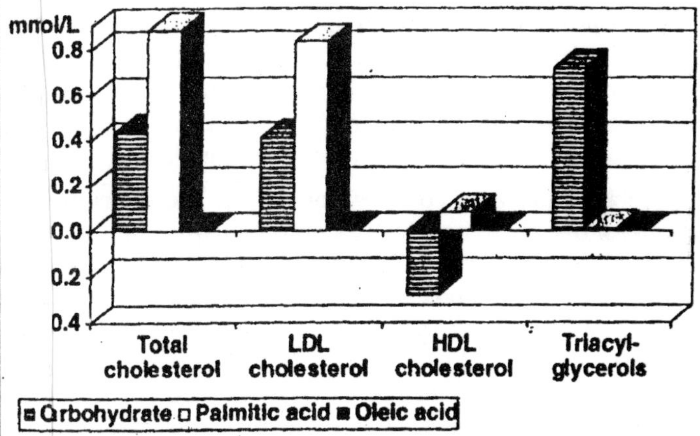

Figure 3.1: Effects of Carbohydrates and Palmitic Acid on Plasma Total, LDL Cholesterol, HDL Cholestrol, and Triacylglycerols Concentrations Relative to those of *cis*-monounsaturated Fatty Acids (Oleic Acid). Seven men received three liquid formula diets, each for 3 weeks, in random order. The diets were identical in composition except for 19 per cent of daily energy intake, which was provided as carbohydrates, palmitic acid (C16 : 0), or oleic acid (*cis*-C18 : 1).

direction. In this way, two compounds are formed that have exactly the same number and type of atoms, but different chemical, physical, and physiological characteristics.

Trans Fatty acids are formed when vegetable oils are hardened by hydrogenation. These hydrogenated fats are used for the production of certain types of margarine and frying fat, and foods prepared with these fats. Most *trans* fatty acids in the diet have 18 carbon atoms one double bond (*trans*-C18 : 1). *Trans* fatty acids, however, are found not only in hydrogenated oils, but also in milk fat and body fat from ruminants, being formed from dietary polyunsaturated fatty acids by the action of bacteria in the rumen of these animals. *Trans* polyunsaturated fatty acids in the diet mainly originate from hydrogenated fish oils.

LDL cholesterol concentrations increase when *cis*-monounsaturated fatty acids in the diet are replaced by *trans*-monounsaturated fatty acids. In most studies, a decrease in HDL cholesterol was also observed. Although the LDL cholesterol rasing effect of *trans*-monounsaturated fatty acids is less than the effect of a mixture of saturated fatty acids, *trans*-monounsaturated fatty acids also lowered HDL cholesterol relative to a mixture of saturated fatty acids. Therefore, it was concluded that both types of fatty acid have an unfavorable effect on the plasma lipoprotein profile.

Dietary fatty acid composition affects the distribution of cholesterol over LDL and HDL. Compared with an isoenergetic amount of carbohydrates, myristic acid (C14 : 0) is the most potent plasma total and LDL cholesterol raising saturated fatty acid. From the other saturated fatty acids in the diet, lauric acid (C12 : 0) and–to a lesser degree-palmitic acid (C16 : 0), also have a hyper-cholesterolemic effect, whereas stearic acid (C18 : 0) seems to be neutral. Linoleic (*cis,cis*-C18 : 2) and probably also oleic acid (*cis*-CI8 : 1) have a small LDL cholesterol lowering effect. *Trans* fatty acids have a strong plasma total and LDL cholesterol increasing effect. Effects on LDL cholesterol levels are positively related to those on HDL cholesterol concentrations. There is an exception, however: *trans* fatty acids that do not have an effect on HDL compared with carbohydrates. Thus, a reduction in the intake of the cholesterol-raising saturated fatty acids and *trans* fatty acids is more important for optimizing the plasma lipoprotein profile than a reduction in total fat intake per se.

3.2 Characteristics of Lipoproteins and Dietary Fatty Acids

Lipoproteins are globular high molecular weight particles, which are complex aggregates of lipid and protein molecular. Now let us try to understand more details about lipoproteins and dietary fatty acids.

A lipoprotein consists of a hydrophobic core that mainly contains triacylglycerols and cholesterylesters, and a polar, hydrophilic coat composed of phospholipids, unesterified cholesterol, and specific apolipoproteins. This construction protects the hydrophobic core from the watery surroundings and allows transport of large amounts of cholesterol and triacylglycerols through the blood vessels.

Lipoproteins are a heterogeneous group, which can be divided into five major classes: chylomicrons, very low density lipoproteins (VLDL), intermediate density lipoproteins (IDL), low density lipoproteins (LDL), and high density lipoproteins (HDL). Each class has a characteristic lipid and apolipoprotein composition, size, and density, while each apolipoprotein has its own specific metabolic functions.

Table 3.1: Major Fatty Acids in Some Edible Fats and Oils

Formula	*Fatty Acid*	*Source*
Saturated fatty acids		
	Medium chain fatty acids	Dairy fat, coconut oil, palm kernel oil
C12 : 0	Lauric acid	Dairy fat, coconut oil, palm kernel oil
C14 : 0	Myristic acid	Dairy fat, coconut oil, palm kernel oil
C16 : 0	Palmitic acid	Palm oil, meat
C18 : 0	Stearic acid	Meat, cocoa butter
Monounsaturated fatty acids		
C18 : 1,*n*-9	Oleic acid	Olive oil, rapeseed oil, high oleic acid sunflower oil
Polyunsaturated fatty acids		
C18 : 2,*n*-6	Linoleic acid	Sunflower oil, corn oil, soybean oil, corn oil
C18 : 3,*n*-3	α-Linolenic acid	Rapeseed oil, soybean oil
C20 : 5,*n*-3	Timnodonic acid	Fatty fish, fish oil capsules
C22 : 6,*n*-3	Cervonic acid	Fatty fish, fish oil capsules

Table 3.2: Some Physical Characteristics and Mean Composition of Lipoprotein Fractions from Normotriglyceridemic Subjects

	Chylomicrons	*VLDL*	*IDL*	*LDL*	*HDL*
Density (g/mL)					
Lower limit	–	096	1.006	1.019	1.063
Upper limit	0.96	1.006	1.019	1.063	1.21
Size (nm)	75–1200	30–80	25–35	19–25	5–12
Percent of total lipoprotein mass					
Core components					
Triacylglycerols	87	52	29	6	6
Cholesterylester	3	9	28	40	21
Surface components					
Phospholipids	6	23	22	22	24
Free cholesterol	2	6	7	9	3
Apolipoprotein	2	10	14	23	46

In the duodenum, dietary triacylglycerols are dissolved with the help of bile salts, as well as small quantities of fatty acids and monoglycerides. The enzyme pancreatic lipase then hydrolyzes the dietary triacylglycerols into mono- and diacylglycerol, free fatty acids, and glycerol. The so-formed emulsion of lipids, containing as well the dietary cholesterol, passes the mucous membrane of the intestinal cells. Within the cell, further hydrolysis of lipids takes place and new triacylglycerols are formed by reesterification of the free fatty acids with glycerol. The newly synthesized triacylglycerols and cholesterylesters, derived from the dietary cholesterol, are incorporated into chylomicrons, which enter the lymph and subsequently the blood circulation in the subclavian vein. In the blood, the triacylglycerols from the chylomicron core are hydrolyzed by lipoprotein lipase (LPL), an enzyme adhered to the endothelial cells of the blood vessels and activated by apolipoprotein C-11. The free fatty acids pass the endothelial cells and enter adipocytes or muscle cells. In these cells, the fatty acids are respectively stored as triacylglycerols or oxidized. The core of an emptied chylomicron mainly consists of cholesteryl esters and is called a chylomicron remnant. These remnant particles are removed from the circulation by the hepatic remnant receptor, which has a high affinity for apolipoprotein E (apoE) from the chylomicron surface.

The liver excretes cholesterol and triacylglycerols into the circulation by the formation of VLDL particles. These lipoproteins also interact with LPL, triacylglycerols become hydrolyzed, and VLDL remnants–also called IDLs–are formed. A part of the IDL is taken up by the liver, while the remaining part is converted in the circulation into LDL. LDLs are nearly devoid of triacylglycerols and carry about 60 to 70 per cent of the total amount of cholesterol in the plasma. Most of the LDL is now removed from the circulation through the hepatic LDL receptor pathway, which recognizes apolipoprotein 8–100 (apoB-100). LDL uptake from the blood by the LDL-receptor-mediated pathway is highly controlled, and this pathway will be down-regulated if the amount of cholesterol in the cell becomes too high. A smaller part, however, is removed via the scavenger pathway. Uptake via this pathway is not saturable and is positively related to the LDL cholesterol concentration.

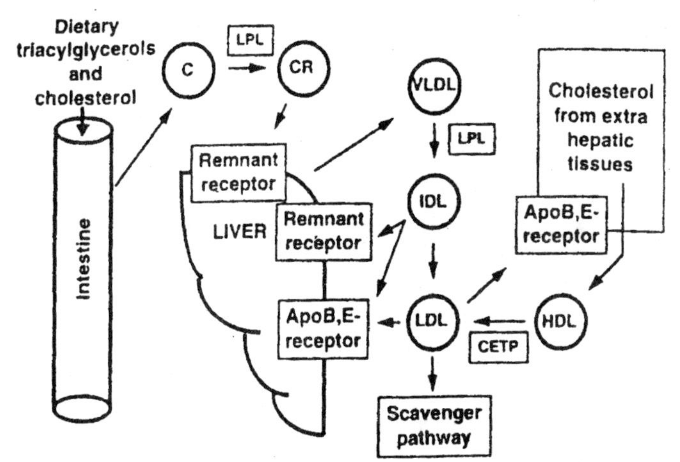

Figure 3.2: Overview of the Metabolism of Dietary Fatty Acids, Dietary Cholesterol, and Plasma Lipoproteins

3.3 LDL Oxidation and the Dietary Effects

LDL oxidation is a free radical driven process that may initiate a cascade of reactions.

Some experiments have suggested that cellular production of superoxide anions (O_2^-) or hydroxyl radicals (OH^+), which are intermediates in several metabolic processes from the mitochondrial respiratory chain or the cytochrome P450 system, initiate the lipid peroxidation reaction. Other experiments, however, have proposed that lipid peroxidation is initiated by lipoxygenase activity, since this enzyme forms radicals as intermediate products in the formation of eicosanoids. These hypotheses are not necessarily contradictory, because lipoxygenase activity might be particularly important in endothelial cells and in peroxide initiation by superoxide anions and hydroxyl radicals in smooth muscle cells.

In vitro oxidation of LDL results in alterations in both the lipid and the protein components of LDL. The unsaturated cholesterylester content decreases, especially the amount of cholesteryl-arachidonate and cholesteryl-linoleate. In addition, phosphatidylcholine, the main phospholipid in LDL, is converted into lysophosphatidylcholine after cleavage of a fatty acid from the *sn*-2 position by phospholipase A_2. It has been postulated that the released fatty acid is readily oxidized and may then become responsible for the propagation of the lipid peroxidation chain reaction.

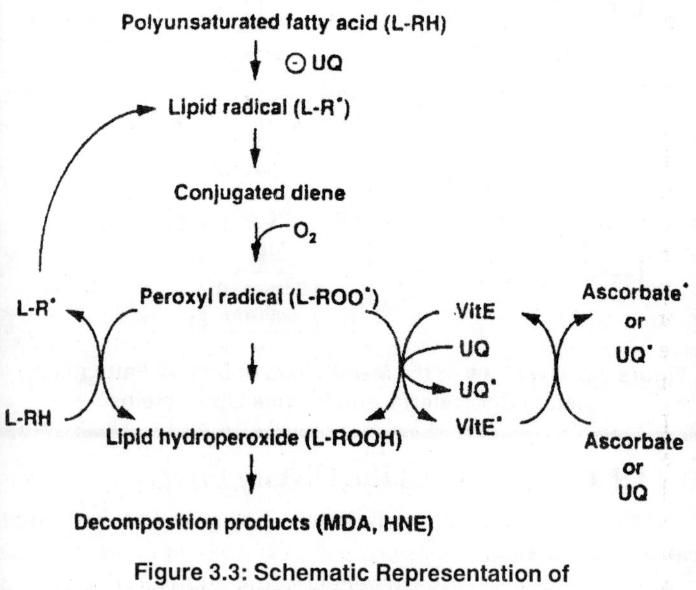

Figure 3.3: Schematic Representation of Lipid Peroxidation

After peroxidation, lipid peroxides decompose and breakdown products are formed: malondialdehyde (MDA), for example, and several aldehydes including 4-hydroxynonenal (4-HNE), These products can react with the e-amino groups of apoB-100, causing an irreversible modification of the apolipoprotein as the number of free cysteine and charged lysine residues of apoB-100 decreases. This series of reactions results in a reduced recognition and uptake of LDL by the LDL receptor, since the affinity of the LDL receptor is based on binding of positively charged apoB-100 to the negatively charged binding domain of the LDL receptor. The reduced uptake is compensated for by an increased affinity of these modified LDLs to the acetyl or scavenger receptors on the cell surface of macrophages, This uptake is not down-regulated and may lead to extensive lipid loading and the transformation of macrophages into foam cells.

LDL oxidation is thought to be initiated and to proceed primarily in the endothelial layer, while oxidized LDL particles are rapidly removed from the circulation. Thus, it is very difficult to quantify the LDL oxidation process *in vivo*, and several laboratory assays have been developed to measure LDL oxidation tendency *in vitro*.

Oxidative *in vivo* modification of LDL can be mimicked by exposure of LDL particles *in vitro* to redox-active metal ions (*e.g.*, Cu^{2+}) or to reactive oxygen species such as superoxide anion.

Esterbauer *et al.* have developed a widely used method to determine *in vitro* the susceptibility of LDL to oxidation by continuous monitoring of the formation of conjugated dienes, products formed after oxidation of the polyunsaturated fatty acids from the LDL particle. After LDL isolation, copper is added to the test tube to initiate the oxidation process, and the formation of conjugated dienes is quantified by measuring the change in absorbance at 234 nm. This curve can be divided into three consecutive phases: the lag phase, the propagation phase, and the decomposition phase.

During the lag phase, LDL-bound lipophilic antioxidants protect the polyunsaturated fatty acids from oxidation. Tocopherols and β-carotene, for example, scavenge lipid peroxide radicals, thereby breaking the chain reaction. The antioxidant has now become a relatively stable radical that does not induce lipid peroxidation but also is not regenerated, as may happen *in vivo*, Addition of water-soluble ascorbic acid (vitamin C) that is lost during

LDL isolation leads to an increase in the lag phase because this antioxidant can regenerate tocopherols. Thus, at a certain stage the LDL particle becomes depleted of antioxidants, and at this point the peroxidation reaction shifts to an autocatalytic mode.

The oxidation profile of LDL is described by the duration of the lag time, the maximal amount of dienes formed, and the maximal rate of oxidation. A short lag time, and/or a large amount of dienes are considered to reflect a high oxidative susceptibility of the LDL particle. Interpretation of the oxidation rate, however, is not clear.

The LDL oxidation causes coronary heart diseases in many cases.

Palinsky *et al.* have demonstrated the existence of autoantibodies directed against MDA-LDL in the human circulation, Further, Yiä Herttualla *et al.* have shown that atherosclerotic lesions contain compounds that react with antibodies directed against MDA-modified LDL and 4-HNE-lysine, while those antibodies did not react with native LDL. The same study reported that LDL, isolated from atherosclerotic lesions, showed a substantial correspondence with the characteristics of *in vitro* oxidized LDL (*e.g.* increased amount of lysophosphatidylcholine, chemotactic activity for monocytes). Finally, in patients with acute myocardial infarcts and in patients with carotid atherosclerosls, the amount of MDA-modified LDL in plasma was significantly higher compared with control subjects.

Several studies now strongly suggest that oxidation of LDL takes place *in vivo*, resulting in even more harmful LDL particles. Therefore, LDL oxidation may play an important role in the formation of atherosclerotic lesions.

Replacing saturated fatty acids in the diet with linoleic acid reduces plasma LDL cholesterol concentrations, However, this dietary intervention will also increase the proportion of linoleic acid in the LDL particle, and thereby its *in vitro* oxidizability, Since monounsaturated fatty acids are less readily oxidized and have comparable beneficial effects on the plasma lipoprotein profile, oleic acid might be preferred over linoleic acid. Increased *in vitro* oxidizability of the LDL particle, which might also be observed when *n*-3 polyunsaturated fatty acids from fish oils are added to the diet, can be overcome by simultaneously increasing the intake of vitamin

E. Although these studies clearly demonstrate that diet affects *in vitro* LDL oxidizability, the importance of the findings for the *in vivo* situation is less evident.

3.4 The Hemostatic System and Platelet Aggregation

Hemostasis, derived from the Greek words for blood and standing, is a complex, delicately balanced system of interactions to keep the blood circulating as a fluid through the blood vessels. In case of imbalance–for example, when a vessel is damaged–the blood stands and starts to clot, whereupon a stable thrombus forms. This, of course, is necessary to stop a wound from bleeding. However, if a stable thrombus is formed in a small coronary artery, the artery becomes occluded, the system's supply of blood and oxygen is hampered, and a heart attack will be the result.

The hemostatic system involves interacting, processes for the formation of a stable thrombus–platelet aggregation and blood clotting–but also a mechanism to dissolve the thrombus, fibrinolysis. *in vivo*, these processes are associated, and the interplay defines the prethrombotic state of the blood.

Hemostatic factors are difficult to measure. Venipuncture and subsequent blood sampling and plasma preparation may cause platelets to become activated, making it very difficult to obtain a true reflection of the *in vivo* situation. In addition, measurements are usually made in venous fasting blood, hence tell us little about

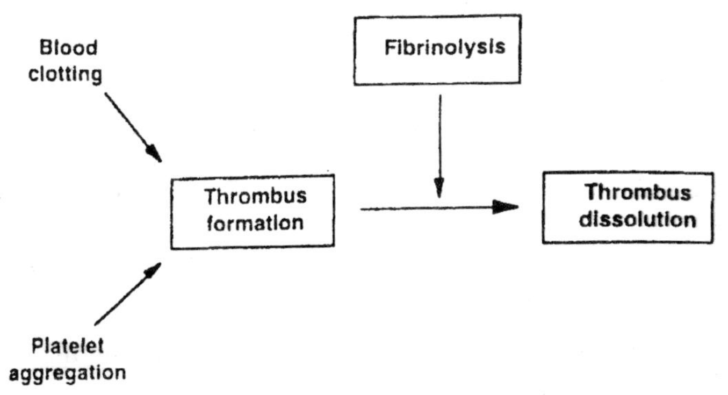

Figure 3.4: Processes Involved in Thrombus Formation

thrombotic tendency in the arteries. Also, many different methods are used, hindering comparisons between studies.

The activity of blood platelets is an important factor for thrombus formation. Aggregated platelets perform several functions that are important for blood coagulation: they adhere to the injured blood vessel to form a hemostatic plug; they excrete substances like thrombin and calcium; and they provide phospholipid surface.

Several mechanisms are being proposed to explain the effects of fatty acids on platelet aggregation. Differences in fatty acid composition can change the arachidonic acid content of platelet and endothelial phospholipids. Arachidonic acid acts as a substrate for thromboxane A_2 (TxA_2) in platelets and prostacycline (PGI_2) in endothelial cells, and the balance between these two eicosanoids affects platelet aggregation. Fatty acids have also been reported to affect directly TxA_2 receptors on platelet membranes, Finally, differences in fatty acid composition can affect the cholesterol content of membranes, with consequent effects on the fluidity of platelet membranes and platelet activation.

A broad scale of methods is available to measure platelet aggregation *in vitro*. First of all, the blood sample needs to be anticoagulated to avoid clotting of the blood in the test tube or in the aggregometer. Different anticoagulants are used. Citrate, for example, depletes the sample of calcium, whereas either heparin or hirudin will inhibit the conversion of prothrombin to thrombin. *In vitro*

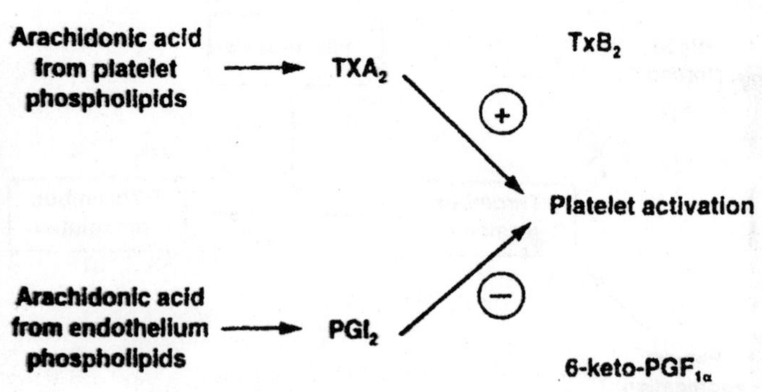

Figure 3.5: Schematic Representation of Platelet Activation

platelet aggregation can then be measured in whole blood, in platelet-rich plasma, or–to remove the influence of possible interfering constituents from the plasma–in a washed platelet sample. Finally, the aggregation reaction in the test tube can be triggered by introducing many different compounds, including collagen, adenosine diphosphate (ADP), arachidonic acid, and thrombin.

In vitro platelet aggregation has been reported to be an important maker for the prediction of reoccurrence of coronary events. In addition, platelet aggregation measured in whole blood was strongly associated with the prevalence of ischemic heart disease.

3.5 Coagulation Cascade

The blood coagulation is an important process, which was previously known as the extrinsic pathway of blood coagulation. Let us try to understand more things about coagulation cascade.

The coagulation cascade is further regulated by the action of coagulation inhibitors. Two important inhibitors of coagulation are the tissue factor pathway inhibitor (TFPI), which inhibits the activity

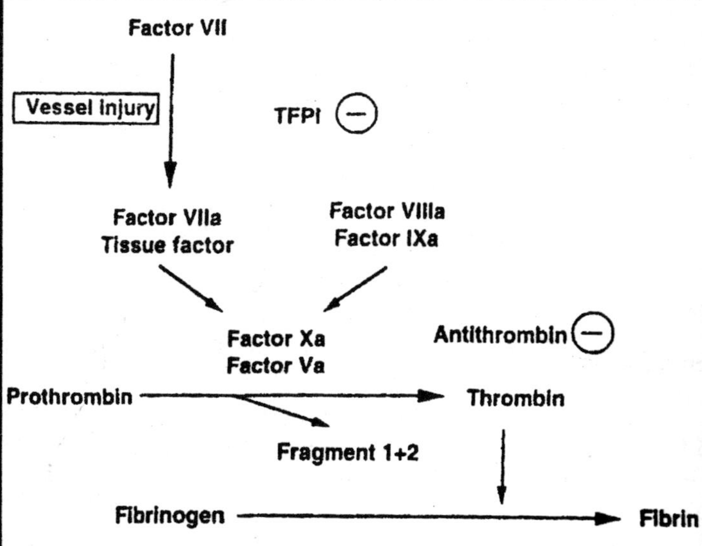

Figure 3.6: Schematic Representation of the Coagulation Cascade

of the tissue factor/factor VIIa complex; TFPI, therefore, prevents further activation of the coagulation cascade. The tissue factor-factor VIIa complex is also inhibited by antithrombin III, which also suppresses the action of thrombin and other activated coagulation factor.

Most assays measure the total amount (*e.g.*, factor VII antigen or fibrinogen concentrations) or the activity of circulating coagulation factors. However, these factors are normally present in large excess in the blood, and only a small percentage is converted into active enzymes under *in vivo* situations. Other assays now available reflect actual *in vivo* coagulation. The plasma fragment 1 + 2 concentration reflects the amount of prothrombin actually converted into thrombin, while concentrations of fibrinopeptides A and B reflect the conversion of fibrinogen into fibrin. However, in healthy subjects, concentrations of most of these markers are very low, just above detection limits, which makes it sometimes difficult to implement them in dietary studies.

Many methods are available to measure factor VII. In many studies, factor VII is measured with a coagulant assay (factor VII coagulant activity). Factor VII activity, however is measured with a two-step chromogenic assay that depends on the rate of generation of factor Xa from factor X by factor VIIa. The methods may give different results and do not differentiate between factor VIIa and factor VII antigen concentrations.

Fibrinogen concentrations are usually measured with the method of Clauss. A fixed surplus of thrombin is added to diluted platelet-poor plasma samples, the clotting time recorded. and the fibrinogen concentration read from a calibrator curve.

Long-term prospective epidemiological studies have consistently reported that in healthy males factor VII coagulant activity and fibrinogen concentrations were higher in subjects who developed cardiovascular diseases at a later stage of the study. Factor VII coagulant activity was in particular associated with an increased risk of dying from cardiovascular disease. In addition, the Northwick Park Heart Study (NPHS) reported that low and, unexpectedly, also high concentrations of antithrombin III were associated with increased deaths from CHD. It must be noted that these prospective studies used male subjects only; whether associations for females are similar awaits confirmation.

The dietary fats influence more on the blood coagulation process.

Marckmann and colleagues have investigated both short-term and longer-term effects of low-fat/high-fiber diets on human blood coagulation. In an 8-months study, it was found that a low-fat/high-fiber diet significantly decreased plasma factor VII coagulant activity with 5 to 10 per cent but only in the first 2 months and in the last months of dietary intervention. Absence of effects in the middle portion of the study was explained by the failure of the subjects to follow strictly the dietary guidelines during holidays that coincided with this period. Plasma fibrinogen concentrations were not changed. The results of this study were confirmed in a shorter duration.

In another study, Marckmann *et al.* investigated whether the low-fat or the high-fiber component of diets was responsible for decreased factor VII coagulant activity found in earlier studies. The experimental diets of this trial differed only in their fat content and

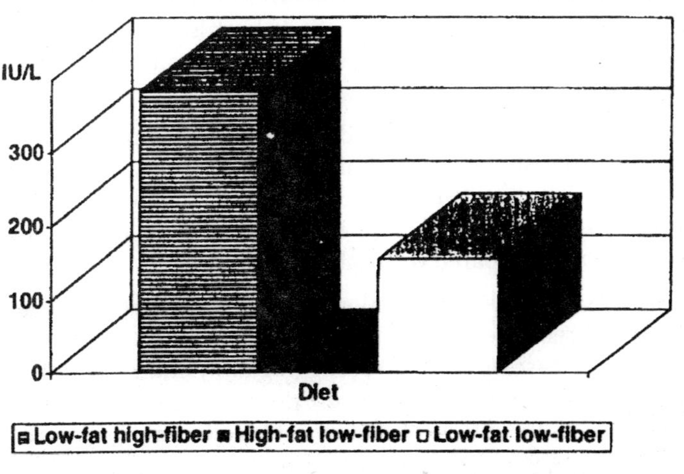

Figure 3.7: Effects of Fat and Fiber on Plasma Euglobolin Fibrinolytic Activity relative to those of a high-fat/low-fiber Diet

carbohydrate content. Factor VII coagulant activity was similar on the low-fat and high-fat diets. In addition no changes in fibrinogen concentrations were found. This limited number of studies indicates reducing effects on factor coagulant activity of low-fat/high-fiber diets. However, more controlled studies are needed to definitely address whether the reduced fat content of low-fat diets, the increased fiber content, or a combination of these two dietary factors, is responsible for the decreased factor VII coagulant activity reported in the studies just described.

Effects of different fatty acid compositions on fibrinolytic capacity have been investigated extensively for fatty acids from fish, although recently more data have become available for the effects of other fatty acids.

High intakes of fish may affect fibrinolytic capacity. Brown and Roberts studied the effects of daily consumption of 200 g of lean fish, with or without fish oil supplement.

They observed an apparent enhancement of plasma euglobulin fibrinolytic capacity, which tended to be accentuated with fish oil supplementation. However, compared with meat diets, diets enriched with fatty fish have also been associated with an unfavorable increase in PAI-1 activity. Increased PAI-1 activity has also been reported on diets enriched in partially hydrogenated soybean oil compared with a butterfat diet.

Tholstrup and co-workers did not report unfavorable effects of lauric plus myristic acid or palmitic acids diets relative to stearic acid on euglobulin fibrinolytic capacity, t-PA activity, or PAI-1 activity. Also myristic acid and palmitic acid diets did not change these variables compared with habitual diets.

Results of platelet aggregation studies are difficult to interpret because of large differences in the methods used. Lowering dietary fat intake decreases *in vitro* platelet aggregation tendencies induced by some agonists but increases *in vitro* platelet aggregation by other agonists. Also, the effects of dietary fatty acid composition on *in vitro* platelet aggregation are contradictory.

To affect coagulation and fibrinolytic factors, it appears that lowering fat intake must be accompanied by changes in dietary fatty acid composition and/or increased fiber intakes. Favorable effects on blood coagulation can be obtained by lowering the content of the major dietary saturated fatty acids and replacing them with unsaturated fatty acids.

Chapter 4

Polyunsaturated Fatty Acids

4.1 PUFAs in Human Nutrition

Plant fatty acids provide a seminal source of polyunsaturated fatty acids (PUFAs) in human nutrition. Fatty acid desaturation of nutritional importance to human is largely similar among plants algae and animals.

The desaturation of fatty acids produced de novo by plants involves an initial Δ9 desaturation, followed by the Δ12 and Δ15 desaturations. The insertion of double bonds on the methylene side of the previously unsaturated Δ9 bond is in opposition to animal metabolism, where double bonds are inserted only on the carboxyl side of the Δ9 double bond, This difference allows dietary fatty acids to be converted into forms not possible as a result of plant or animal metabolism alone. Plants express two pathways for the desaturation of fatty acids. The first, known as the *prokaryotic* pathway, is an array of desaturase activities that are present in the plastid. The second, *eukaryotic,* pathway involves the translocation of the fatty acids to the endoplasmic reticulum, where desaturases act on the acyl groups. The products of the eukaryotic pathway are either utilized directly by the cell or translocated back into the plastid for further processing. Harwood provides an extensive review on plant fatty acid desaturation. An interesting phenomenon of plant fatty acid biogenesis is the use of complex lipids as substrate during desaturation. Both plastid- and endoplasmic reticulum-based desaturations introduce a double bond into fatty acids esterified in a phospholipid. In contrast, animals desaturate acyl-CoA complexes.

Fatty acid desaturation in animals utilizes fatty acyl-CoA thioesters as substrate. Despite multiple desaturase activities cytochrome b_5 reductase acts as an electron donor common to all terminal desaturases. The desaturases are present on the cytoplasmic face of the endoplasmic reticulum. Animals possess Δ9, Δ6, Δ5 and Δ4 desaturase activity; however, the Δ4 desaturase complex has not been identified or localized. Alternative to the presence of a Δ4 desaturase, there is evidence to suggest that fatty acid products requiring the introduction of a Δ4 double bond may be the result of a three-step process not involving a Δ4 desaturase. The conversion of docosapentaenoic acid to docosahexaenoic acid, for example, has been shown to occur via Δ4-independent pathway utilizing a second Δ6 desaturase activity. Life desaturses in plants, each of the desaturases has preferred substrates which can display organism- and tissue-specific differences.

The first double bond inserted into the saturated açyl-CoA products of fatty acid synthetase is at the Δ9 position and is catalyzed by stearoyl-CoA desaturase. In contrast to desaturation in plants, further desaturation occurs only on the carboxyl side of the Δ9 position. Interestingly, most PUFAs are of plant origin, since Δ9 fatty acids produced de novo are not further desaturated by animals except in times of essential fatty acid deficiency. However, fatty acids that have been desaturated by plants are readily desaturated to form familiar fatty acids such as arachidonic acid and docosahexaenoic acid. These highly unsaturated fatty acids are products of both plant and animal desaturation, and they have double bonds inserted on both sides of the original Δ9 double bond.

Knowledge of tissue or even cellular fatty acid composition is not sufficient to predict the effects of PUFAs in cell physiology. The realization that fatty acid location plays an important part in the use and function of specific fatty acids has been a major advancement in lipid metabolism. Unsaturated fatty acids, as well as their saturated counterparts, are esterified in phospholipids of cell membranes. There are a variety of phospholipid types, and each type has its own unique compositional fatty acid identity. In addition to phospholipid identities, individual cell membranes have unique phospholipid and fatty acid compositions. Although there are myriad data on the fatty acid composition of plant and animal tissues, these fatty acid profiles are subject to important changes brought about by diet, disease state, or a variety of other factors. As a result of the variety of phospholipid "pools" in which fatty acids can be esterified, very slight changes in total cell fatty acid content can have significant effects on cell function. It is quite possible that lipids have been investigated and rejected as causal agents of a number of pathologies on the basis of largely unchanged fatty acid compositions. It is now clear that fatty acids exert their effects from specific locations and that their physiological effects are mediated in part by the movement of PUFAs into important phospholipid pools. The movement of fatty acids into phospholipid pools is catalyzed by fatty acid carrier proteins, and the acyl compositions of membranes and specific phospholipid pools are mediated by the specificities of these enzymes.

4.2 Bacteria and Algae as the Synthesizers of PUFA

Most bacteria are fully capable of synthesizing all the fatty acids required for their normal growth and reproduction.

The fatty acid synthetase enzyme complex responsible for this activity adds acetate units to a final chain length of 16 to 18 carbons. These can be further desaturated (normally once to a monounsaturated fatty acid), and then the saturated and unsaturated fatty acids are esterified to yield membrane phospholipids. Bacteria in general do not store energy as fats; hence triacylglycerols are not abundant forms of lipid in bacteria. They are thus not a quantitatively important fat source in foods. The unique metabolism in particular bacteria, especially the production of branched chain and odd-numbered fatty acids, can occasionally generate measurable quantities of unusual fatty acids in, for example, bovine milk, as a result of incorporation of the products of rumen fermentation. In the production of commercially viable oil sources, wax esters and eicosapentaenoic acid have been produced at a commercially relevant scale in bacteria.

Microbes as a broad class of single-celled (or relatively undifferentiated collections of cells) organisms have the obligate biosynthetic capability of generating fatty acids necessary for membrane synthesis, and similar activities. Historically, these organisms have not been considered an important source of fat in the diet, although it has been recognized that they add small quantities to certain foods, whose presence may have been an important contribution (*e.g.* for flavor). Nevertheless, even for microbes that produce large quantities of storage triacylglycerides, the economics of growing and obtaining the oils has been noncompetitive with the traditional sources of edible oils.

Very recently, however, this area has seen a considerable resurgence in interest both academically and in commercial applications. This change is largely due to the improved efficiency and capabilities of large-scale microbial fermentation, to the identification of therapeutically useful edible oils, and to the capability of microbes to produce unusual or unusual concentrations of particular fatty acids and glycerides. Fatty acids that have raised the ante, as it were, for edible oils include dihomo-γ-linoleic acid, eicosapentaenoic acid, and docosahexaenoic acid, for their ability to alter arachidonic acid metabolism, hence exerting an effect on thrombosis, innammation, cancer, and autoimmune diseases; docosahexaenoic acid for inclusion in infant formulas; nervonic acid, for its potential in treating neuropathies; long chain monounsaturated fatty acids for adrenoleukodystrophy; and sterculic acid a possible treatment for bowel cancer.

Several factors favor microbial production for high value lipids. The conspicuously greater potential for aggressive recombinant approaches to manipulating microbial lipid metabolism to obtain novel fatty acids is likely to increase the growth of this cottage industry for fatty acid production. Higher plants and animals are also somewhat limited in the glyceride forms they will produce. For example, most plants will not place a saturated fatty acid in the *sn*-2 position of a triglyceride. Similarly, fish tend to place virtually all the long chain *n*-3. PUFAs in the *sn*-2 position. Thus we find both limits on the total range of glycerides available when these plants and animals serve as sources of lipids and structural effects imposed on digestion and absorption of the fatty acids in nutritional applications. These limitations are both less well defined and more mutable in microbial fermentation applications. This area, though coming under intense regulatory scrutiny, may reach a significant segment of the food industry at least in the short-term.

Single-celled eukaryotes have, as a class, a remarkably wide variety of lipid metabolic capabilities. Some yeasts for example, produce only a single desaturase, the stearoyl or Δ9 enzyme, and neither produce nor require any PUFAs for growth. At the other end of the spectrum, some fungi and algae can produce very high levels of arachidonic, dihomo-γ-linolenic, eicosapentaenoic, and docosahexaenoic acids.

An additional and synthetically useful attribute of these organisms is their ability to take up fatty acids from the medium and either incorporate them into triacylglycerides and phospholipids (even with unusual stereospecificity) or further metabolize them prior to esterification. These various properties were known before but were not thought to warrant commercialization. This is increasingly changing. Already, the use of microbial lipid sources is proving to be a cost-effective feedstock for shrimp and fish aquaculture. The ability of the microbial feedstock to also elaborate valuable pigments and antioxidants is used to advantage, so this entire technology and its biotechnological elaborations likely will increase in impact in the future.

In addition to the differences in total quantity of fat and its tissue distribution, the composition of storage triacylglycerols in animal species differs as well. Red meats tend to be relatively higher in saturated FAs and lower in PUFAs than poultry or fish. Poultry and fish differ significantly in the chain length of monounsaturated fatty acids and in the high levels of *n*-3 PUFAs found fish.

A consistent observation of the lipid content in different animal tissues is the variability *within* species. Even within ruminant animals in which the dietary PUFAs are largely hydrogenated by rumen flora, there is a significant range of composition. Among monogastric animals, the variability in fat composition within species due to muscle type, diet, environment, and age is typically greater than the differences noted between them.

4.3 PUFA Biochemistry

The field of PUFA biochemistry has only begun to develop convincing molecular models for the effects of fatty acids on physiology. Having lagged somewhat behind, lipid biochemistry is now poised to develop in the same way that protein and nucleic acid biochemistry have over the last 20 years. There is a great deal known about fatty acid synthesis, as well as a massive collection of data on the fatty acid composition of foods. It has recently even become feasible to modify the fatty acyl content of foods through genetic manipulation. Yet, there are very few data on how and why particular fatty acids modulate physiology. An understanding of the molecular basis of fatty acid nutrition will be important for the design of diets appropriate to the individual. It is thought that PUFAs exert their physiological effects through a variety of mechanisms, ranging from acting as precursors for signal molecule formation to modulation of membrane structure. The remainder of this chapter reviews what is known about the non-energy-producing functions of PUFAs and how individual fatty acids modulate these functions.

Unsaturated fatty acids and particularly PUFAs can form a vast array of chemical structures, each of which has unique physico-chemical properties. Cells can use fatty acids to modulate their membrane properties and the activities of membrane- associated enzymes, as well as for the production of potent signal molecules. Knowledge of the effect of fatty acids on membrane properties has been impeded by the difficulty of greatly modulating membrane fatty acid composition. While the difficulty associated with modifying cell membrane composition does suggest that membrane homeostasis is critical to cells, it hinders investigation on the effects of individual fatty acids. The regulation of the ether-linked, arachidonate-containing phospholipid pools is the best-described example of how small changes in membrane compositions can have significant effects on cell physiology. Future research in this area will likely have to focus on developing controllable models in which the contributions of specific fatty acids are determined.

Fatty acids occupy a unique position in nutrition in that they have the ability to survive digestion intact, enabling them to replace the fatty acid content of the consumer. It is reasonable then to expect the lipid composition and, correspondingly, the physiology of individuals who consume particular fats and oils, to reflect the fatty acid composition of their respective diets. Interestingly, it has been postulated that the dietary PUFA composition of an average human diet has changed markedly with modern advances in agriculture.

Wild foods are typically much higher in *n*-3 PUFAs than products that have been successfully developed by agriculture. There is a variety of evidence to suggest that the changing ratio of *n*-3 to *n*-6 fatty acids has affected human physiology adversely and that humans may have developed major classes of pathologies as a result of this change. The lower rates of coronary heart disease and cancer in populations consuming a higher *n*-3 to *n*-6 PUFA ratio have been well documented. Despite mounting evidence that human populations would benefit from an increased consumption of *n*-3 fatty acids, it is unlikely that this change will occur in the near future, primarily because agricultural crops rich in *n*-6 fatty acids have been extraordinarily successful. *n*-6 Fatty acids are most

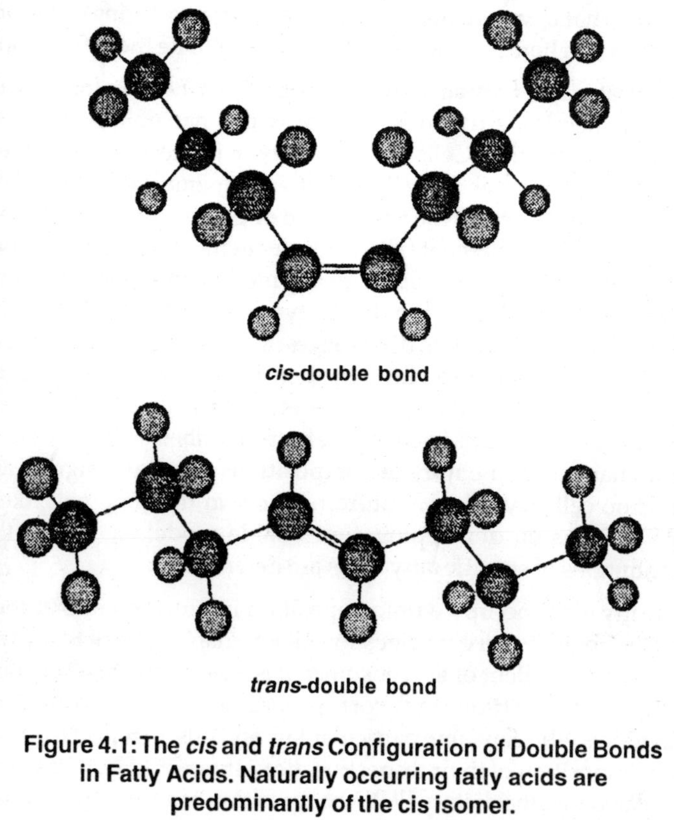

Figure 4.1: The *cis* and *trans* Configuration of Double Bonds in Fatty Acids. Naturally occurring fatly acids are predominantly of the cis isomer.

typically found in seed crops, which are not only consumed directly but also are used to produce food oils and for animal feed. Additionally, *n*-6-rich crops are generally more stable than *n*-3-rich crops, leading to their preferential cultivation. The increases in coronary artery disease and cancer may be a direct consequence of the advancement of modern agriculture.

Membranes are largely held together by London-van der Wall's forces between adjacent fatty acyl chains. Since these interaction forces are significantly diminished with even a slight increase in the distance between acyl chains, fatty acid packing plays a key role in membrane structure. The placement of unsaturated fatty acids into membranes is confined with few exceptions to the *sn*-2 position of phospholipids, and as a result, the acyl content of membranes rarely if ever exceeds 50 mol per cent unsaturated acyl chains. Considering their minor presence, PUFAs have a considerable effect on membrane properties. These effects can be attributed to two primary structural properties of the fatty acid: the length of the acyl chain and the degree of unsaturation. These two structural elements can significantly affect both the relative volume the fatty acid occupies and the distance between a fatty acid and its neighboring acyl chain. The degree of unsaturation is particularly important for membrane properties, since van der Waal's forces are acutely sensitive to the distance between the interacting acyl chains.

Estimates of the spatial widths of fatty acids show significant differences between cis and trans double bonds. While stearic acid is estimated to be 0.25 nm in diameter, Δ9-*cis*-octadecanoic acid (oleic) and Δ9-*trans*-octadecanoic acids have spatial widths of 0.72 and 0.31 nm, respectively. Additionally, because the acyl chain continues on the same side of the double bond, the cis configuration imparts a 30° bend in the fatty acid that is not relievable through any rotation of the single-bonded carbon atoms. In contrast, the structure of a *trans* fatty acid is relatively unaffected by the double bond. The cis-unsaturated bond thus interrupts a succession of London-van der Waal's forces between membrane fatty acids by increasing the distance between adjacent fatty acids and also lowers the membrane crystallization temperature. The more unsaturated the fatty acid, the less it is able to rotate around its carbon-carbon bonds, and consequently, the more it influences membrane acyl packing. Some calculations suggest that an increase in the number of double bonds provides diminishing returns in terms of the fatty acid's influence on van der Waal's forces. It is clear that acyl unsaturation can affect van der Waal's interactions either through changes in the length of association or by increasing the distance of acyl separation. Additionally, the position of a double bond in a fatty acid plays a role in the modification of membrane structure. Double bonds at the methyl end of acyl chains are not particularly effective at modulating membrane bulk properties because they induce a lesser degree of acyl separation. Interestingly, double bonds are typically first inserted at the Δ9 position, and fatty acids with double bonds only near the methyl end of the chain are conspicuously absent from nature.

Cells take advantage of the large variety of fatty acid structures to maintain consistent bulk membrane properties in the face of changing temperature and pressure. Clearly then, PUFAs are critical to cell function if viewed only from the perspective of membrane structure. Thus it is interesting that animals are largely incapable of producing PUFAs de novo.

4.4 Membrane-Associated Enzymes

A large number of membrane associated enzymes have been reported to be responsive to their fatty acid environment.

There is considerable evidence that membrane-unsaturated fatty acid content can modify the structure and therefore the functionality of membrane enzymes, and these effects are often cited when dietary fatty acids modulate physiological functions. This paradigm of fatty acid metabolism should be viewed with caution, however, since there are significant problems with its interpretation. In particular, effects on membrane enzyme activities are not easily ascribable to the direct action of membrane unsaturation. Fatty acids have both structural and energetic roles in cell physiology, and uncoupling these processes is essentially impossible. Many of the data indicating a link between membrane unsaturation and enzyme activity could be interpreted in multiple ways, including the direct effects of dietary oils and fatty acids on cell metabolism. Changes in membrane-associated enzyme activities as a result of membrane properties must be assigned on a best explanation basis and should be regarded as putative at best. However, an interesting line of experimentation that involves measuring membrane-associated activities as a function of temperature has provided compelling evidence that the degree of membrane unsaturation is critical to cell function. Membrane-associated enzyme activity, molecular transport and the insertion of proteins into membranes have all been shown to have maximum activity at temperatures just above the transition temperature of the membrane. While this clearly does not indicate that cells use the changes in membrane unsaturation to modulate enzyme activities, it does provide strong evidence that diet-induced changes in membrane composition could play an important role in the modification of cell physiology.

For all the difficulties in assigning a causal mechanism to changes in membrane associated enzyme activities, it is worth noting that the effects themselves are largely incontrovertible. Dietary oils and fats have significant effects on many metabolic parameters, including the activity of membrane-associated enzymes. An increasing body of literature has been a massed concerning the modulation of membrane-bound enzyme activities in both cell and reconstituted systems.

Some research foucuses on the activity of membrane-associated enzymes in reconstituted systems. While fraught with difficulties of their own, such investigations indicate that reconstituted systems largely eliminate the confounding factors associated with cell metabolism. Acyl coenzyme A-cholesterol acyltransferase (ACAT) has been found to have a significantly higher activity in reconstituted liposomes constructed of monounsaturated fatty acids than saturated fatty acids. Mathur *et al.* further showed that when ACAT-containing rat liver microsomes were altered by coincubation with liposomes and a phospholipid exchange protein, significant differences in ACAT activity were associated with the fatty acid content, and not with phospholipid type of the liposomes. Exchange with liposomes containing linoleic acid completely abolished ACAT activity, while oleic acid enhanced activity to a greater degree than palmitic acid. These changes in activity associated with the fatty acid composition of reconstituted ACAT-containing membrane systems clearly indicate that some enzymes may be acutely sensitive to their fatty acid environment.

In a similar set of experiments, Carruthers and Melchior showed that the activity of a hexose transporter-containing reconstituted vesicle system was dependent on the fatty acid composition of the vesicles. Some acyl-dependent activity changes have also been demonstrated for acyl-CoA synthetase in isolated, modified rat liver plasma membranes. It is useful to note that although the extent of change described in these experiments may not be feasible *in vivo*, these data may provide insight into the true effects of membrane unsaturation on enzyme activity, since they are performed in a cell-independent system.

A particularly interesting example of membrane-dependent enzyme regulation is that of the fatty acid desaturases. Some work by Martin *et al.* and Kasai *et al.* suggests that the activities of certain bacterial desaturase systems are correlated with the degree of unsaturation of the ER. While corroborating experiments have been performed neither *in vitro* nor in animal systems, this possible correlation would support proposals of an interesting self-regulatory mechanism for the production of unsaturated fatty acids.

Many of the enzymes for which the modulation of activity as a result of membrane properties has been described are directly involved in either lipid metabolism or cell energetics. While this creates confounding factors in terms of assigning causal mechanisms, it is exactly these enzymes that are likely to need lipid feedback. Nature, in fact, may have taken great care in preventing enzymes that are not involved in lipid metabolism or energetics from being affected by membrane properties. Significant progress in analytical capabilities must precede the determination of true causality and the direct effects of membrane unsaturation on enzyme activity.

Much as with changes in membrane-associated enzyme activities, receptors have been reported to be sensitive to the physical properties of cell membranes. Receptors provide cells with the ability to communicate with their environment, and alterations in this signal transduction pathway can cause cells to respond differently to the same stimuli. Thus, dietary fatty acids have a potentially powerful role in the modulation of cell- and tissue-level physiology. The most studied example of this phenomenon is that of the low density lipoprotein receptor and its modulation by dietary fatty acids. Other receptors reported to be affected by dietary and membrane fatty acid compositions include the insulin receptor, the interleukin receptor, and β-adrenergic receptors.

Table 4.1: The Effect of Oils and Fatty Acids on the Activity of Various Membrane-Associated Enzymes

Activity	*System*	*Modification*	*Effect*
Acyl CoA-cholesterol acyltransferase	Chick liver membranes	Increased PUFA	Increased activity
Adenylate cyclase	Rat heart	Essential fatty acid deficiency	Decreased activity
	Rat retina	Decreased *n*-3/*n*-6 ratio but constant PUFA content	No effect
Alkaline phosphatase	Rat small intestine	Long-term dietary fish oil supplementation	Increased activity
Cytochrome *c*	Chick liver membranes	Increased PUFA	No change in activity
	Reconstituted vessicles of cardiolipin	Linoleic-acid-rich cardiolipin	Increased activity
	Rat heart mitochondrial membranes	*n*-3 enrichment/*n*-6 depletion	Decreased activity
Δ6 and Δ9 desaturases	Rat lymphocyte EA and plasma membrane	Increases in the total saturated fatty acid content	Increased specific activity
F_1F_0-ATPase	Rat heart mitochondrial membranes	*n*-3 enrichment/*n*-6 depletion	Increased activity
Glycine uptake	Y79 retinoblastoma cells	Increased PUFA	Increased K'_m and V'_{max}
Na^+/K^+-ATPase	Myeline phospholipids of rat cerebella	Decreases in the long chain PUFA arachidonic acid and DHA	Increased activity
	Rat erythrocyte membranes	Increasing *n*-6 and total unsaturation	Decreased activity
5'-Nucleotidase	Rat liver plasma membrane	Increasing *n*-3 and decreasing *n*-6 fatty acids	Increased V'_{max} No effect on K'_m
	Piglet brush border membrane	Essential fatty acid deficiency	Increased activity
Nucleoside triphosphatase	Mouse nuclear envelope	Increased PUFA	Increased activity

4.5 The Primary Unsaturated Fatty Acids in Human Nutrition

The primary unsaturated fatty acids can be studied as follows.

Palmitoleic Acid

Palmitoleic acid is a minor component of both animal and vegetable lipids. Fish oil is particularly enriched in palmitoleic acid, while some seed oils also represent a significant source of the fatty acid. Palmitoleic acid is produced de novo by both plants and animals by the Δ9 desaturation of palmitic acid.

cis-Vaccenic

cis-Vaccenic acid is a major product of bacterial fatty acid synthesis, but is also present in some seed oils. *cis*-Vaccenic acid biosynthesis likely involves the elongation of palmitoleic acid. The true content of *cis*-vaccenic acid in the diet has been underestimated because it is so difficult to separate it from its *n*-9 isomer. It is possible that a significant fraction of the octadecanoaic monoene reported as oleic acid is really vaccenic acid. In animals, *cis*-vaccenic acid appears to be concentrated in the mitochondrial lipid cardiolipin.

Oleic Acid

Oleic acid is a Δ9 desaturase product of stearic acid and is produced de novo in plants, animals, bacteria, and algae. Oleic acid is the most common unsaturated fatty acid and is the precursor for the production of most other polyunsaturates. Plants produce both *n*-3 and *n*-6 PUPAs from oleic acid, and animals can elongate and desaturate oleic acid into a variety of *n*-9 fatty acids. Olive oil is a particularly rich dietary source, but most foods, especially nuts and butter, are rich in oleic acid.

Erucic Acid

Erucic acid is a long chain monounsaturated fatty acid found in plants, particularly in rapeseeds. This elongation product of oleic acid is an uncharacteristically long chain unsaturated fatty acid for plants. Mildly toxic, erucic acid has been bred out of rapeseeds used for food oil production. In animals, dietary erucic acid can be retroconverted to form oleic acid via β-oxidation.

Mead Acid

Mead acid is a hallmark of essential fatty acid deficiency and has the distinction of being the only major PUFA produced de novo by animals. In the absence of dietary n-6 and n-3 fatty acids, the Δ6 desaturase converts oleic acid to 18 : 2n-9, which is further elongated and desaturated (Δ5) to form mead acid. It has been speculated that mead acid compensates for the loss of n-3 and n-6 PUFAs by increasing the unsaturation of animal cell membranes.

Linoleic Acid

Linoleic acid along with α-linolenic acid, is a primary product of plant PUFA synthesis. Linoleic acid is produced de novo by plants, and in particular is enriched in seed oils. While nature produces linoleic acid at a level fairly equitable with α-linolenic acid, modern agriculture has greatly enriched linoleic acid in the food supply. While animals are incapable of producing linoleic acid, livestock are fed diets particularly rich in this fatty acid, and thus humans acquire a large portion of their linoleic acid from meats. Linoleic acid serves as a precursor for the production of the essential fatty acid arachidonic acid, as well as other n-6 acyl species.

γ-Linolenic Acid

γ-Linolenic acid is produced in animals and de novo by plants by the Δ12 desaturation of linoleic acid. Natural sources include evening primrose oil, borage oil, and black currant oil, while minute amounts can be found in animal tissue. In animals, dietary linoleic acid desaturated by the Δ12 desaturase to produce γ-linolenic acid is an intermediate in the production of arachidonic acid. Interestingly, dietary γ-linolenic acid is not substantially converted to arachidonic acid. There has been a great deal of recent interest in dietary γ-linolenic acid for its antagonistic action on arachidonic acid metabolism. In part this is due to the kinetics of the conversion of γ-linolenic acid to arachidonic acid; but a more significant role probably involves eicosanoids produced directly from the immediate conversion product of γ-linolenic acid, dihomo-γ-linolenic acid.

Dihomo-γ-linolenic Acid

The elongation product of linoleic acid, dihomo-γ-linolenic acid (DGLA) is a minor component of animal phospholipids. DGLA serves as a precursor to the formation of the essential fatty acid arachidonic acid as well as for the prostaglandin G_1 series, Dietary DGLA does not appear to be rapidly converted to arachidonic acid, and because prostaglandins of the G_1 series have anti-inflammatory properties, DGLA has received attention as a potential therapeutic agent.

Arachidonic Acid

Exceedingly rare as a plant lipid, arachidonic acid is primarily the product of desaturation and elongation of linoleic acid in animals. Arachidonic acid is also produced in quantity in marine algae. Dietary linoleic acid is converted to arachidonic acid in animals by the concerted activity of the Δ6 desaturase, a microsomal elongase and the Δ5 desaturase. Arachidonic acid is referred to as an essential fatty acid for its action as the precursor for the production of eicosanoids. It appears to be present in all tissues and is particularly enriched in ether-linked phospholipid membrane pools.

Docosatetraenoic Acid

Docosatetraenoic acid is the direct elongation product of arachidonic acid and is present in minimal amounts in animal tissues. It has been reported that docosatetraenoic acid is a substrate for peroxisomal retroconversion resulting in the formation of arachidonic acid.

4.6 Double Bonds in PUFAs

The vast majority of polyunsaturated fatty acids contain multiple double bonds in a 1,4-pentadiene structure in which a single methylene carbon is positioned between the two double bonds. More double bonds are added as a direct result of the positional selectivity of the subsequent desaturase enzymes. Hence, from fatty acids containing two double bonds (linoleic acid 9, 12 18: 2) to those containing six double bonds (docosahexaenoic acid 22: 6 4, 7, 10, 13, 16, 19), all exhibit methylene interruption over the entire length of the molecule. However, there are certain naturally occurring fatty acids in which single double bonds exist at a distance. These are the non-methylene-interrupted fatty acids (NMIFAs). Various NMIFAs have been described including allenic, conjugated, allylic, enoic, acetylenic, cyclic, branched, hydroxylated, iso, and anteiso fatty acids. The most common NMIFAs are fatly acids in which one of the double bonds is ostensibly missing from the middle of a double-bond system. Conifers have been shown to contain up to 20 per cent by weight in their seeds of the NMIFA 5, 11, 14-eicosatrienoic acid and 5, 11, 14, 17-eicosatetraenoic acids. In most cases studied, the basis of the synthesis of these fatty acids is through an active elongase enzyme that elongates 18 to 20 carbon fattyacids, bypassing the Δ6 desaturase. Indeed, it has been argued that the methylene-interrupted structure is not due to explicit enzyme specificity but rather to the predisposition of substrate fatty acids. Desaturation at the Δ5 position of the elongated fatty acid produces a 20-carbon fatty acid without the Δ8 double bond. This aggressive elongation has been argued to be the basis for the occurrence of small quantities of 5, 11, 14-eicosatrienoic acid found in animal tissues.

While most unusual fatty acids are not readily esterified into membrane phospholipids, for certain structures this is not true. 5, 11, 14-Eicosatrienoic acid, which is the structural analog to arachidonic acid but without the Δ8 double bond, is esterified into several membranes and even shows preference for incorporation into specific phospholipids including phosphatidylinositol. Since the absence of the Δ8 double bond makes it impossible to synthesize either prostaglandins or leukotrienes from NMIFA replacement of arachidonic acid with NMIFA in phospholipids would be predicted to have substantial effects on eicosanoid cellular signaling proportional to the extent of displacement of arachidonic acid. This was shown to be true for isolated tissues. Further, in an animal model of genetic autoimmunity, animals fed diets containing these fatty acids were found to exhibit a net reduced severity of disease consistent with a selective decrease in eicosanoid signaling.

Other examples of unusual non-methylene-interrupted fatty acid structures with potential biological actions include conjugated linoleic acid (CLA), which has been shown to exhibit potent anticancer properties in animal models of carcinogenesis. The true therapeutic value, as well as the potential toxicities of these fatty acids, will await the development of commercial sources in food-grade quantities for larger studies. Nevertheless, the potential for modifying the biosynthetic capabilities of crop plants is already being exploited for many nutritional and functional targets, and the possibility of producing fatty acids with unusual structures is likely to be limited only by the documented value of producing them.

The fatty acid nomenclature system based on *n* designation was developed to describe fatty acids in terms of their nutritional functions. While this approach is of some use, it may lead to confusion over the true basis by which fatty acids modulate physiology. There is mounting evidence that each fatty acid has its own role in nutrition that is not dictated by its omega designation. Chemically similar fatty acids often have widely ranging functionalities. This phenomenon is exemplified by the antagonistic relationship between arachidonic acid and EPA metabolism, even though the fatty acids differ only by the additional double bond in EPA. The fundamental relationship between members of the same ω family is one of interconvertibility. Families are grouped only by their potential to be converted to longer chain members of the same omega designation, or in a more practical sense, by their ability to act as precursors for the production of a particular fatty acid of interest. The enzymatic activities required for the conversion of fatty acids to longer, more unsaturated chain members of the same omega family are redundant. Consequently, if cells are not capable of recognizing fatty acids based on their omega designation and responding by modulating desaturase and elongase activities appropriately, animals are truly at the mercy of their diets.

The regulation of acyl content appears to be critical to the function of a cell, in as much as mammalian cells expend a substantial amount of energy in maintaining distinct and heterogeneous membrane fatty acid compositions. Cells are capable of preserving these compositional identities even under conditions of phospholipid diffusion and vesicular transport, suggesting that acyl composition is tightly regulated. In addition to overall membrane acyl content, phospholipid acyl composition is rigorously maintained in a positionally specific manner. The complex framework of enzymatic activities that upholds these compositions makes large-scale changes in membrane composition rare, and as a result, the effects of specific fatty acids on membrane structure and physiology are not well understood. However, the extraordinary selectivity of certain tissues for individual PUFAs strongly suggests a specific role for those fatty acids in cell function. For instance, brain and neural tissues are enriched in DHA, while adrenal glands have a high content of docosatetraenoic acid. Neither of these PUFAs can be produced de novo in humans, and so their accumulation in cell membranes must be mediated either by a preferential absorption or specific desaturation reactions within the given tissue.

Polyunsaturated fatty acid specificity is also prevalent in subcellular organelles–for example, the inordinate enrichment of vaccenic acid in mitochondria and the prominent occupancy of arachidonic acid of the *sn*-2 position of nuclear membrane ether-linked phospholipids. Selective incorporation of PUFAs implies that fatty acids have some degree of functionality in the cell. Additionally, these function must be monitored by the cell if regulatory adaptation is to occur. How a cell senses its acyl composition is not understood. To date, the best data on adaptive acyl regulation in mammalian cells involve the essential fatty acid deficiency response. Cells deficient in arachidonic acid or its precursor *n*-6 fatty acids convert oleic acid produced de novo to *n*-9 eicosatrienoic (mead) acid by the action of the Δ5 and Δ6 desaturases. The conversion mechanism is identical to that of the production of C20 *n*-6 and *n*-3 fatty acids. The unusual desaturation of oleic acid is commonly believed to be the result of a cell's attempt to replace arachidonic acid in the membrane with a similarly unsaturated species. This is an example of a cross-family compensation and provides further evidence for the lack of an association between omega designation and function. The regulation of mead acid production is intriguing in that it is not found in animals fed sufficient dietary *n*-6 fatty acids. This may be the consequence of a differential activity of the Δ6 desaturase on linoleic and oleic acids.

Given that each fatty acid is unique in terms of its chemistry and potentially its function, one interesting question concerning PUFA metabolism is, can animals recognize, omega designation? If not, how big a role do dietary PUFAs play in physiology? How can animals regulate the production of important acyl compounds in the face of changing dietary *n*-3/*n*-6 ratios, given that animals have a redundant system of desaturation for both ω families? It is more than likely that if the acyl-processing enzymes are not adaptable to changing ω ratios, such a diet will in fact have serious consequences for health. Changing *n*-3/*n*-6 PUFA ratios appears to have significant effects on many pathologies in humans, including cardiovascular disease, immune function, and cancer. Since dietary acyl compositions vary tremendously throughout the world and over time, key questions must be answered concerning the regulatory factors and signals associated with the production of fatty acid and their metabolic products.

Chapter 5

Sterols and Waxes

5.1 Introduction

Sterols constitute a large group of compounds with a broad range of biological activities and physical properties. They are common in eukaryotic cells but more in prokaryotes.

The natural occurring sterols usually possess the 1,2-cyclopentanophenthrene skeleton with a stereochemistry similar to the trans-syn-trans-anti-trans-anti configuration at their ring junctions, and have 27 to 30 carbon atoms with an hydroxy group at C3 and a side chain of at least seven carbons at C17. Sterols can exhibit both nuclear variations (differences within the ring system) and side chain variations. Sterols have been defined as hydroxylated steroids that retain some or all of the carbon atoms of squalene in the side chain and partition almost completely into an ether layer when shaken with equal volumes of water and ether.

Sterols are common in eukaryotic cells but rare in prokaryotes. Without exception, vertebrates confine their sterol biosynthetic activity to producing cholesterol. Most invertebrates do not have the enzymatic machinery for sterol biosynthesis and must rely on an outside supply. Sterols of invertebrates have been found to comprise most complex mixtures arising through food chains. In plants, cholesterol exists only as a minor component. Sitosterol and stigmasterol are the most abundant and widely distributed plant sterols, while ergosterol is the major occurring sterol in fungus and yeast. The plant sterols are characterized by an additional alkyl group at C-24 on the cholesterol nucleus with either α- or β-chirality. Sterols with methylene and ethylidene substitutes are also found in plants (*e.g.* 24-methylene cholesterol, fucosterol). The other major characteristics of plant sterols are the presence of additional double bonds in the side chain, as in porifeasterol, cyclosadol, and closterol.

Despite the diversity of plant sterols and sterols of invertebrates, cholesterol is considered the most important sterol. Cholesterol is an important structural component of cell membrances and is also the precursor of bile acids and steroid hormones. Cholesterol and its metabolism are of importance in human disease. Abnormalities in the biosynthesis or metabolism of cholesterol and bile acid are associated with cardiovascular disease and gallstone formation. Our discussion will mainly focus on cholesterol and its metabolites with a brief comparison of the biosynthesis of cholesterol and plant sterols.

The chemistry of sterols encompasses a large amount of knowledge relating to the chemical properties, chemical synthesis, and analysis of sterols. A detailed discussion on all these topics is impossible in one chapter. We consider the analysis of sterols to be

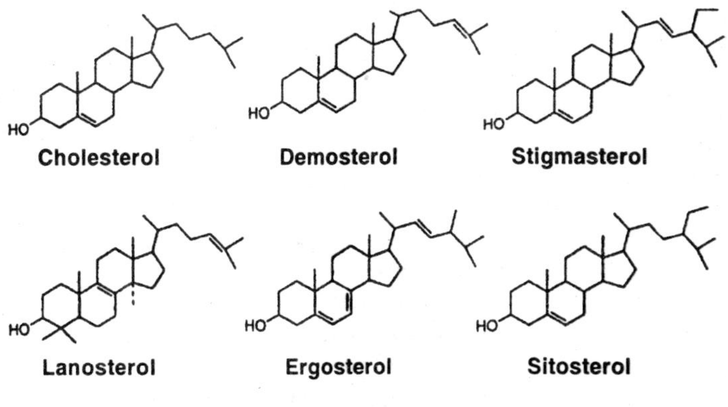

Figure 5.1: Naturally Occurring Sterols

of primary interest, and therefore our treatment of the chemistry of sterols is confined to the isolation, purification, and characterization of sterols from various sources.

The principle mammalian sterol is cholesterol, and it is the precursor of steroid hormones.

The elucidation of the cholesrerol biosynthesis pathway has challenged the ingenuity of chemists for many years. The early work of Konrad Bloch in the 1940s showed that cholesterol is synthesized from acetyl coenzyme A. Acetate isotopically labeled in its carbon atoms was prepared and fed to rats. The cholesterol that was synthesized by these rats contained the isotopic label, which showed that acetate is a precursor of cholesterol. In fact, all 27 carbon atoms of cholesterol are derived from acetyl coenzyme A (acetyl CoA). Since then many chemists have put forward enormous efforts to elucidate this biosynthetic pathway, and this work has yielded our present derailed knowledge of Sterol biosynthesis. This outstanding scientific endeavor was recognized by the awarding of several Nobel prizes to investigators in research areas related to sterol.

The cholesterol biosynthetic pathway can be generally divided into four stages; the formation of mevalonic acid from three molecules of acetyl CoA, the biosynthesis of squalene from six molecules and mevalonic acid through a series of phosphorylated intermediates, the biosynthesis of lanosterol from squalene via cyclizalion of 2,3-epoxysqualene, and the modification of lanosterol to produce cholesterol.

The first stage in the synthesis of cholesterol is the formation of mevalonic acid and isopenryl pyrophosphate from acetyl CoA. Three molecules of acetyl CoA are combined to produce mevalonic acid as shown in Scheme 1. The first step of this synthesis is catalyzed by a thiolase enzyme and results in the production of acetoacetyl CoA, which is then combined with third molecule of acetyl CoA by the action of 3-hydroxy-3-methylglutaryl-CoA (HMG-CoA) is its cleavage to acetyl CoA and acetoacetate. Acetoacetate is further reduced to D-3. hydroxybulyrate in the mitochondrial matrix. Since it is a β-keto acid, acetoacetate also undergoes a slow, spontaneous decarboxylation to acetone. Acetoacetare, D-3-hydroxybutyrate, and acetone, sometime referred to as ketone bodies, occur in fasting or diabetic individuals. Alternatively, HMG-CoA can be reduced to mevalonate and is present both in the cytosol and in the mitochondria of liver cells. The mitochondrial pool of this intermediate is mainly a precursor of ketone bodies, whereas the cytoplasmic pool gives rise to mevalonare for the biosynthesis of cholesterol.

The reduction of HMG-CoA to give the mevalonic acid is catalyzed by a microsomal enzyme, HMG-CoA reductase, which is of prime importance in the control of cholesterol biosynthesis. The biomedical reduction of HMG-CoA is an essential step in cholesterol biosynthesis. The reduction of HMG-CoA is irreversible and proceeds in two steps, each requiring NADPH as the reducing reagent. A hemithioacetal derivative of mevalonic acid is considered to be an intermediate. The concentration of HMG-CoA reductase is determined by rates of its synthesis and degradation, which are in turn regulated by the amount of cholesterol in the cell. Cholesterol content is influenced by the rate of biosynthesis, dietary uptake, and a lipoprotein system that traffics in the intercellular movement of cholesterol. During growth, cholesterol is mainly incorporated into the cell membrane. However, in homeostasis cholesterol is mainly converted to bile acids and is transported to other tissues via low density lipoprotein (LDL). High density lipoprotein (HDL) also serves as a cholesterol carrier, which carries cholesterol from peripheral tissues to the liver. The major metabolic route of cholesterol is its conversion to bile acids and neutral sterols, which are excreted from the liver via the bile. Kandutsch and Chen and others have shown that oxysterols regulate the biosynthesis of HMG-CoA reductase as well as its digression, which controls cholesterol biosynthesis; the regulation of HMG-CoA reductase by oxysterols is discussed in more detail in a later section. A number of substrate analogs have been tested for their inhibition of HMG-CoA reductase. Some of them (*e.g.*, compactin and melinolin) were found to be very effective in treating hypocholesterol diseases.

The coupling of six molecules of mevalonic acid to produce squalene proceeds through a series of phosphorylated compounds. Mevalonate is first phosphorylated by mevalonic kinase to form a 5-phosphomevalonate, which serves as the substrate for the second phosphorylation to form 5-pyrophosphomevalonate. There is then a concerted decarboxylation and loss of a tertiary hydroxy group from 5-pyrophosphomevalonate to form 3-isopentyl pyrophosphate, and each step needs to consume one molecule of ATP. 3-lsopentyl pyrophosphate is regarded as the basic biological "isoprene unit" from which all isoprenoid compounds are elaborated. Squalene is synthesized from isopentyl pyrosphate by sequence coupling reactions. This stage in the cholesterol biosynthesis starts with the isomerization of isopentyl pyrophosphate to dimethylallyl pyrophosphate. The coupling reaction shown in Scheme 2 is catalyzed by a soluble sulfydryl enzyme, isopentyl pyrophosphate-dimethylallyl pyrophos phate isomerase. The coupling of these two isomerics units yields geranyl pyrophosphate, which is catalyzed by farnesyl pyrophosphate synthetase. This reaction proceeds by the head-to-tail joining of isopentyl pyrosphosphate to dimethylallyl pyrophosphate. A new carbon-carbon bond is formed between the C1 of dimethylallyl pyrosphosphate and C4 of isopentyl pyrophosphate. Consequently, geranyl pyrophosphate can couple in a similar manner with a second molecule of isopentyl pyrophosphate to produce farnesyl pyrosphate (C_{15} structure). The last step in the synthesis of squalene is a reductive condensation of two molecules of farnesyl pyrophosphate.

5.2 Sterols in Plants and Animals

In higher plants and algae the first cyclic product is cycloartenol. The cyclization intermediate, tetracyclic protosterol cation, undergoes a different series of concerted 1,2-trans migrations of hydrogen and methyl groups. Instead of the 8,9-double bond, a stabilized C9 cation intermediate is formed. Following a trans elmination of enzyme-X^- and H^+ from C19, with the concomitant formation of the 9,19-cyclopropane ring, cycloartenol is formed. A nearby α-face nucleophile from the enzyme is necessary to stabilize the C-9 cation and allow the final step to be a trans elimination according to the isoprene rule. The biosynthesis pathway from acetyl CoA to 2,3-oxisqualene in plants is the same as that in animals.

The conversion of cycloartenol to other plant sterols can be generally divided into three steps, which are the alkylation of the side chain at C24, demethylation of the C4 and C14 methyl groups, and double-bond manipulation. Alkylation in the formation of plant sterols involves methylation at C24 with S-adenosyl methionine to produce C_{28} sterols. The further methylation of a C24 methylene substrate yields C24 ethyl sterols. The details of the mechanism of demethylation and double-bond manipulation in plants are not

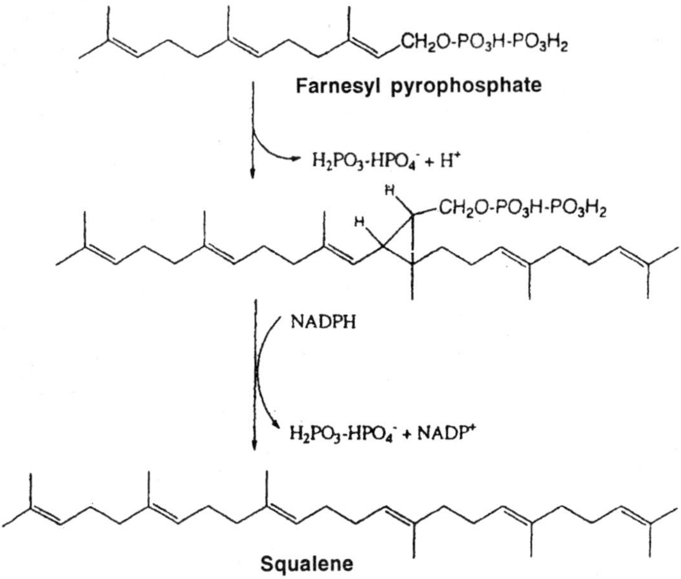

Figure 5.2: The Synthesis of Squalene from the Coupling of Two Molecules of Farnesyl Pyrophosphate

clear, but it is highly likely to be very similar to that in animals. In plants, C4 methyl groups are removed before the methyl group at C14, whereas in animals it is other way round. Sterols found in plants are very diversified.

Sterol biosynthesis in mammalian systems has been intensely studied for several decades.

More recently, the results of numerous clinical studies have indicated that lowering serum cholesterol levels may reduce the risk of coronary heart disease and promote the regression of atherosclerotic lesions. The total exchangeable cholesterol of the human body is estimated at 60 g for a 60 kg man. The cholesterol turnover rate is in the order of 0.8 to 1.4 g/day. Both the dietary uptake and the biosynthesis contribute significantly to body cholesterol. In western countries the daily ingestion of cholesterol ranges from 0.5 to 3.0 g, although only a portion of this sterol is

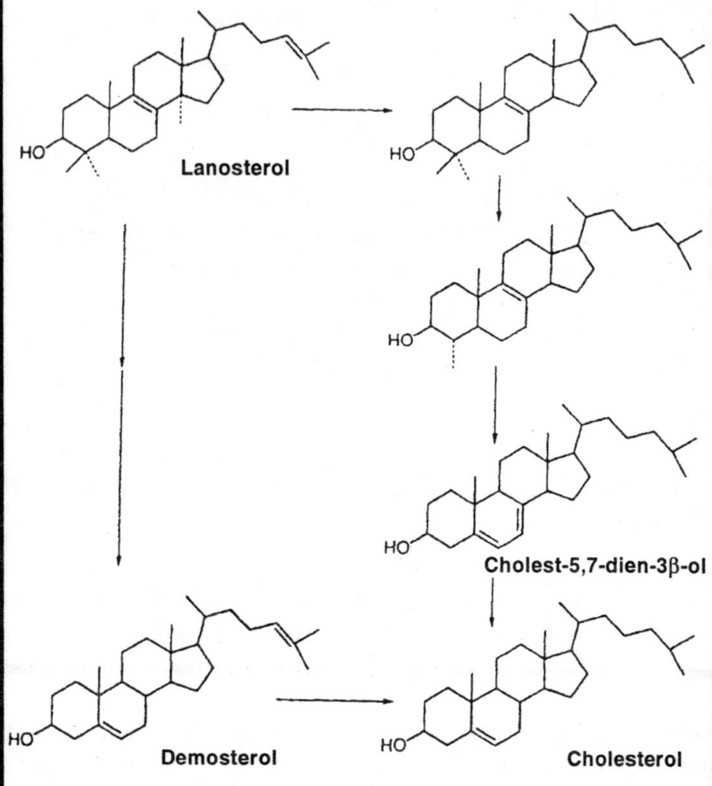

Figure 5.3: The Biosynthesis of Cholesterol from Lanosterol

absorbed from the intestine. The absorption of dietary cholesterol ranges from 25 per cent (high dietary sterol intake) to 50 per cent (low dietary sterol intake). The total body cholesterol level is determined by the interaction of dietary cholesterol, the excretion of cholesterol and bile acid, and the biosynthesis of cholesterol in tissue.

The major site of cholesterol synthesis in mammals is the liver. Appreciable amounts of cholesterol are also formed by the intestine. The rate of cholesterol formation by these organs is highly responsive to the amount of cholesterol absorbed from dietary sources. This feedback regulation is mediated by changes in the activity of HMG-

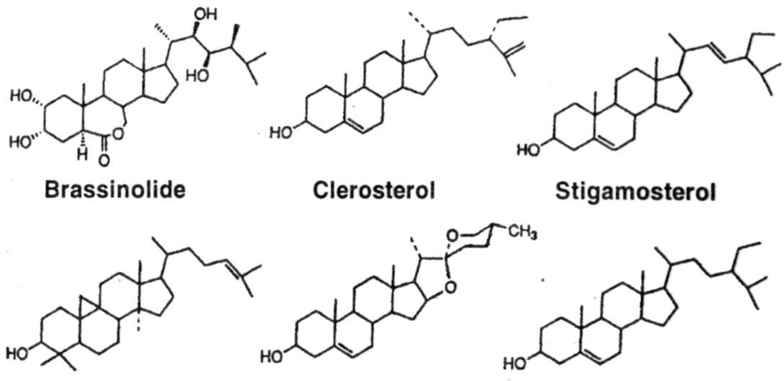

Brassinolide　Clerosterol　Stigamosterol

Cycloartenol　Diosgenin　Sitosterol

Figure 5.4: Plant Sterols

CoA reductase. As discussed in connection with pathway for the biosynthesis of cholesterol, this enzyme catalyzes the formation of mevalonate, which is the committed step in cholesterol biosynthesis. Dietary cholesterol suppresses cholesterol biosynthesis in these organs through the regulation of HMC-CoA reductase activity.

Cholesterol, triacylglycerols, and other lipids are transported in body fluids to specific targets by lipoproteins. A lipoprotein is a particle consisting of a core of hydrophobic lipids surrounded by a shell of polar lipids and apoproteins. Lipoproteins are classified according to their densities. Low density lipoprotein, the major carrier of cholesterol in blood, has a diameter of 22 nm and a mass of about 3×10^6 daltons (Da). LDL is composed of globular particles, with lipid constituting about 75 per cent of the weight and protein (apoprotein 8) the remainder. Cholesterol esters (about 1500 molecules) are located at the core, which is surrounded by a more polar layer of phospholipids and free cholesterol. The shell of LDL contains a single copy of apoprotein B-100, a very large protein (514 kDa). The major functions of LDL are to transport cholesterol to peripheral tissues and to regulate de novo cholesterol synthesis at these sites.

The major site of cholesterol biosynthesis is the liver. The mode of control in the liver has also been discussed: dietary cholesterol (possibly oxysterols,) reduces the activity and amount of HMG-CoA reductase, the enzyme catalyzing the committed step of cholesterol biosynthesis. In some tissues, such as adrenal glands, spleen, lung, and kidney, biosynthesis contributes only a relatively small proportion of the total tissue cholesterol, with the bulk being derived by uptake from LDL in the blood. Investigation upon the interaction of plasma LDL with specific receptors on the surface of some nonhepatic cells has led to a new understanding of the mechanisms of cellular regulation of cholesterol uptake, storage and biosynthesis in peripheral tissues.

5.3 Extraction and Isolation of Sterols

To analyze the sterols in various biological tissues, sterols are first extracted from these tissue with organic solvents. Now let us learn more details about the extraction of sterols.

The choice of an extraction technique is often determined by the nature of the source and the amount of information the investigator chooses to obtain concerning the forms of sterols present as free sterols, glycosylated, and/or esterified sterols through the 3-hydroxy group. Different extraction procedures may vary dramatically depending on the extraction efficiency required for different classes of sterols. Irrespective of the nature of the source, one of four methods of sample preparation is usually employed. Samples can be extracted directly, with little or no preparation; after drying and powdering; after homogenization of the fresh materials; or after freeze-drying or fresh-freezing, followed by powdering, sonication, or homogenization. Extraction procedures vary. The analyst may simply mix the prepared material with the extraction solvent (the most frequently used solvents include mixtures of chloroform and methanol, or chloromethane and acetone) for a short time (0.5–1 h) and separate the organic solvent phase from the aqueous phases and debris by centrifugation. Another common procedure is extraction from a homogenized material by means of a re fluxing solvent in a Soxhlet apparatus for 18 hours or with a boiling solvent for 1 hour. To obtain a total lipid extraction, saponification under basic (*i.e.*, in 10 per cent KOH in 95 per cent ethanol) or acidic conditions is usually conducted prior to the organic solvent extraction. Most of the glycosylated sterols and some esterified sterols cannot be easily extracted into organic solvents without the hydrolysis step.

Many oxysterols contain functional groups (*e.g.*, epoxides and ketones) that may be sensitive to high concentration of acids or bases. Epoxides may undergo nucleophilic attack by strong bases (*e.g.*, NaOH and KOH), followed by ring opening. Moreover, treatment with strong acids can result in ring opening tc form to alcohols, alkenes, and ketones. The hydrolysis of cholesterol epoxides under mildly acidic conditions has been studied. Modified procedures are available for the isolation of steroidal epoxides from tissues and cultured cells by a saponification and/or extraction. Ketones are known to form enolates under the influence of strong bases, which may then form condensation products of higher molecular weight. To circumvent these potential problems, procedures using different extraction techniques are sometimes preferred to a saponification followed by extraction. Mild methods for the removal of the ester function without ketone enolization include extraction by means of sodium or potassium carbonate in heated aqueous solutions of methanol or ethanol. The addition of tetrahydrofuran to these mixtures has been found to significantly increase the solubility of the more polar oxysterols.

There are not many studies on the efficiency of various extraction methods. Most extraction procedures were designed to compare the extraction of lipids from cells or tissues of a single source and have been applied subsequently to plant and animal of various types. The errors in the quantitative analysis of sterols, which are probably introduced in the extraction steps, could be eliminated by using in situ labeling of key sterols. Sterols labeled with deuterium and C have been used to monitor the extraction recovery in human plasma oxysterol analysis.

For the isolation and purification of sterols chromatographic methods are being extensively used.

Chromatographic methods with an organic phase involve the binding of a substrate to the surface of a stationary polar phase through hydrogen bonding and dipole-dipole interaction. A solvent gradient with increasing polarity is used to elute the substrate from the stationary phase. The order of substrate movement will be alkyl > ketone > hindered alcohol > unhindered alcohol. The elution profile is routinely monitored by GC or TLC. Reversed phase column chromatography involves the use of lipophilic dextran (Sephadex LH-20 or Lipidex 5000), and elution with nonpolar solvents like a mixture of methanol and hexane). The substrate could be separated in order of increasing polarity.

Argentation column chromatography is a very powerful chromatographic method in the separation of different alkene isomers. The argentation stationary phase is typically made by mixing $AgNO_3$ solution (10 g of $AgNO_3$) in 10 mL of water) and silica gel or aluminium oxide (90 g of stationary phase in 200 mL of acetone). After acetone has. been evaporated at a moderate temperature (< 35°C) under vacuum, the resulting argentation stationary phase is dried under vacuum at room temperature until a constant weight is obtained. Usually argentation chromatography is suitable only for the separation of sterol acetates. For free sterol, on-column decomposition of substrate was observed. Compared to the conditions of normal adsorption column chromatography, a slightly polar solvent system is usually needed to elute the substrate from the column. For a preliminary cleanup step, good normal silica gel column chromatography running under the gravity will give good results. Column chromatography with fine silica gel and/or aluminum oxide (230–400 mesh) running under medium pressure (10–100 psi, MPLC) could be used in the separation of individual sterols and sterol subclasses.

Thin-layer chromatography is primarily used as an analytical method to detect com- pounds. Also, small samples (< 20 mg) can be separated by means of preparative TLC. For an analytical TLC plate, the thickness of the stationary phase (usually silica gel) is less than 0.25 mm; for preparative TLC a much thicker stationary phase (> 0.5 mm) is required. No more than 1 mg of substrate should be loaded on one-centimeter analytical plates, and 2 mg on preparative plates. Otherwise overloading will prevent the achievement of a good separation. Non-destructive detection methods (such as iodine and UV detection) are usually used to locate individual compounds. The silica gel containing pure individual compounds is scraped from the plates and eluted with polar solvents.

High performance liquid chromatography, adsorption or reversed phase, is becoming the most commonly used technique for the separation of individual sterols from subclass fractions. In adsorption HPLC, microspheres of silica (silicic acid such as μ-Poracil) are employed. As in adsorption, open column, and thin-layer chromatography, this method is characterized by hydrogen bonding and other electronic attractions between the sterol and stationary phase. The small diameter (3–5 μm) and high porosity of the microspheres create a very large effective surface area and allow the simultaneous analysis of many more theoretical plates than the other systems can handle. TLC solvent systems can be used directly for adsorption HPLC. In general, the solvents that give best separations are binary (or trinary) systems of mostly low to moderate polarity solvents with a small amount of a strongly polar solvent such as hexane/benzene (9 : 1), dichloromethane-*n*-hexane-ethyl acetate (94 : 5 : 1).

In reversed phase HPLC (RP-HPLC) systems, alkyl groups C_8 or C_{18} have been chemically bonded to microspheres of silica. These small diameter (2–5 μm), highly porous particles give the stationary phases very large effective surface areas for the interaction of the sterols with the bonded alkyl groups. This large area, in turn, makes possible rapid equilibration of the sterol between the stationary phase and an appropriate mobile phase. yielding columns with as many as 10^4 theoretical plates. A variety of polar mobile phases have been employed in RP-HPLC [*e.g.*, 100 per cent acetonitrile, acetonitrile-water mixtures (0–20 per cent), and methanol- water mixtures (2–20 per cent water)]. Samples are usually injected onto the analytical column at levels of 1 to 100 μg per component (up to 1.0 mg total) for routine separations. Retention volumes of eluted sterols are usually expressed relative to cholesterol.

Gas-liquid chromatography is most frequently used to quantitate sterols in extracts, subfraction mixtures, and isolated sterol fractions. Selectivity is provided by the gas-liquid partitioning. In general, three different detectors have been employed for quantitation: flame ionization detectors (FID), electron capture detectors (ECD), and mass detectors (MD). However, FID is by far the most commonly employed because it is relatively insensitive to temperature changes during analysis and to minor structural differences in sterols and related steroids and because it has a large linear mass range of response. For quantitative analysis it is necessary to determine the linear range or response to a sterol standard (*e.g.*, cholesterol or β-cholestanol) for each FID, and accompanying chromatographic system. For sample analysis, dried fractions are routinely weighed and dissolved in a known amount of solvent to give a mass-to-volume ratio within the linear range of the detector, where upon the samples injected onto the column.

Quantification of sterols in extracts, subclass fractions, or isolated sterol fractions by HPLC is somewhat limited. The selectivity provided by either adsorption or reversed phase chromatography is in many cases greater than that provided by GLC. Most frequently, UV detectors are employed and are set at end adsorption wavelengths. These detectors have limited sensitivity, with monochromatic detectors being most sensitive. Additionally, UV detectors cannot be considered universal sterol detectors. Most sterols differ significantly in their UV absorption properties, even in end absorption regions. Thus, it would not be possible to select a single wavelength for the quantitation of complex sterol mixtures. Therefore, HPLC coupled to variable wavelength detectors or multidiode detectors can be used to quantitate specific sterols in a mixture if the compound in question has a unique absorption spectrum relative to other members of the mixture.

5.4 Waxes

The term 'waxes' is commonly used to refer to the mixtures of long chain apolar compounds found on the surface of plants and animals. Or, a wax is the ester of long chain acid and a long chain alcohol.

Wax is the collective term for a series of natural or synthetically produced substances that normally possess the following properties: kneadable at 20°C, brittle to solid, coarse to finely crystalline, translucent to opaque, relatively low viscosity even slightly above the melting point, not tending to stinginess, consistency and solubility depending on the temperature and capable of being polished by slight pressure.

The collective properties of wax as just defined clearly distinguish waxes from other articles of commerce. Chemically, waxes constitute a large array of different chemical classes, including hydrocarbons, wax esters, sterol esters, ketones, aldehydes, alcohols, and sterols. The chain length of these components may vary from C_2, as in the acetate of along chain ester, to C_{62} in the case of some hydrocarbons.

Waxes can be classified according to their origins as naturally occurring or synthetic. The naturally occurring waxes can be subclassitied into animal, vegetable, and mineral waxes. Beeswax, spermaceti, wool grease, and lanolin are important animal waxes.

Beeswax, wool grease, and lanolin are by-products of other industries. The vegetable waxes include carnauba wax, the so-called queen of waxes, ouricouri (another palm wax), and candelilla. These three waxes account for the major proportion of the consumption of vegetable waxes. The mineral waxes are further classified into the petroleum waxes, ozokerite, and montan. Based on their chemical structure, waxes represent a very broad spectrum of chemical types from polyethylene, polymers of ethylene oxide, derivatives of montan wax, alkyl esters of monocarboxylic acids, alkyl esters of hydroxy acids, polyhydric alcohol esters of hydroxy acids, Fisher- Trop-sch waxes, and hydrogenated waxes, to long chain amide waxes.

The ancient Egyptians used beeswax to make writing tablets and models, and waxes are now described as man's first plastic. Indeed, the plastic property of waxes and cold-flow yield values allow manual working at room temperature, corresponding to the practices of the Egyptians. The melting points of waxes usually vary within the range of 40 to 120°C.

Waxes dissolve in fat solvents, and their solubility is dependent on temperature. They can also wet and disperse pigments, and they can be emulsified with water, which makes them useful in the furniture, pharmaceutical, and food industries. Their combustibility, associated with a low ash content, is important in candle manufacture and solid fuel preparation. Waxes also find application in industry as lubricants and insulators, where their properties as natural plastics, their high flash points, and their high dilelectric constants are advantageous.

The physical and technical properties of waxes depend more on molecular structure than on molecular size and chemical constitution. The chemical components of waxes range from hydrocarbons, esters, ketones, aldehydes, and alcohols to acids, mostly as aliphatic long chain molecules. The hydrocarbons in petroleum waxes are mainly alkanes, though some unsaturated and branched chain compounds are found. The common esters are those of saturated acids with 12 to 28 carbon atoms combining with saturated alcohols of similar chain length. Primary alcohols, acids, and esters have been characterized and have been found to contain an even straight chain of carbon atoms. By contrast, most ketones, secondary alcohols, and hydrocarbons have odd numbers of carbon atoms. The chemical constitution of waxes varies in great degree depending on the origin of the material. A high proportion of cholesterol and lanosterol is found in wool wax. Commercial waxes are characterized by a number of properties. These properties are used in wax grading.

Color and odor are determined by comparison with standard samples in a molten state. In the National Petroleum Association scale, the palest color is rated 0, while amber colors are rated 8. Refined waxes are usually free from taste, this property being especially important in products such as candelilla when it is used in chewing gum. Melting and softening points are important physical properties. The melting points can be determined by the capillary tube method or the drop point method. The softening point of a wax is the temperature at which the solid wax begins to soften. The penetration property measures the depth to which a needle with a definite top load penetrates the wax sample.

Shrinkage and flash point are two other frequently measured physical properties of waxes. The flash point is the temperature at which a nash will occur if a small flame is passed over the surface of the sample. In the liquid state, a molten wax shrinks uniformly until the temperature approaches the solidification point. This property is measured as the percentage shrinkage of the volume.

5.5 The Chemical Analysis of Natural Waxes

Knowledge of the chemical anlaysis of natural waxes is essential for understanding wax biosynthesis, manufacture and application.

The chemical composition of synthetic waxes are constant and depend on the manufacturing process, the natural waxes are much more complicated in chemical composition. In general, natural waxes are isolated by chemical extraction, separated by chromatographic methods, and analyzed by means of mass spectrometry (MS); both gas-chromatographic (GC) and high performance liquid chromatographic (HPLC) techniques are used. The following discussion on chemical analysis is based on an understanding of the general principles of the chemical extraction, chromatography, and mass spectrometry. There are numerous textbooks detailing these principles.

Natural waxes are mixtures of long chain apolar compounds found on the surface of plants and animals. However internal lipids also exist most organisms. In earlier times, the plant or animal tissue was dried, whereupon the total lipid material could be extracted with hexane or chloroform by means of a Soxhlet extractor. The time of exposure to the organic solvent, particularly chloroform, is kept short to minimize or avoid the extraction of internal lipids. Because processors are interested in the surface waxes, it became routine to harvest them by a dipping procedure. For plants this was usually done in the cold, but occasionally at the boiling point of light petroleum, and by swabbing to remove surface lipids. Chloroform which has been widely used is now known to be toxic, dichloromethane can be substituted. After removal of the solvent under vacuum, the residue can be weighed. Alternatively, the efficiency of the extraction can be determined by adding a known quantity of a standard wax component (not present naturally in the sample) and performing a quantification based on this component following column chromatography.

The extract of surface lipids contains hydrocarbons, as well as long chain alcohols, aldehydes, and ketones, short chain acid esters of the long chain alcohols, fatty acids, sterols and sterol esters, and oxygenated forms of these compounds. In most cases it is necessary to separate the lipid extract into lipid classes prior to the identification of components. Separation of waxes into their component classes is first achieved by column chromatography. The extract residue is redissolved in the least polar solvent possible, usually hexane or light petroleum, and transferred to the chromatographic column. When the residue is not soluble in hexane or light petroleum, a hot solution or a more polar solvent, like chloroform or dichloromethane, may be used to load the column. By gradually increasing the polarity of the eluting solvent, it is possible to obtain hydrocarbons, esters, aldehydes and ketones, triglycerides, alcohols, hydroxydiketones, sterols, and fatty acids separately from the column. Most separations have been achieved on alumina or silica gel. However, Sephade LH20 was used to separate the alkanes from Green River Shale. Linde 5 Å sieve can remove the n-alkanes to provide concentrated branched and alicyclic hydrocarbons. Additionally, silver nitrate can be impregnated into alumina or silica gel columns or thin-layer chromatography (TLC) plates for separating components according to the degree of unsaturation.

As the means of further identifying lipids become more sophisticated, it is possible to obtain a sufficient quantity of the separated wax components by TLC. One of the major advantages of TLC is that it can be modified very easily, and minor changes to the system have allowed major changes in separation to be achieved. Most components of wax esters can be partially or completely separated by TLC on 250 µm silica gel G plates developed in hexane-diethyl ether or benzene-hexane.

If TLC is used, the components must be visualized, and the methods employed can be either destructive or nondestructive. The commonly used destructive method is to spray TLC plates with sulfuric or molybdic acid in ethanol and heat them. This technique is very sensitive, but it destroys the compounds and does not work well with free acids. Iodine vapors will cause a colored band to appear, particularly with unsaturated compounds, and is widely used to both locate and quantify the lipids. Since the iodine can evaporate from the plate readily after removal from iodine chamber, the components usually remain unchanged. Iodine vapor is one of the ideal visualization media in the isolation of lipid classes from TLC plates. Commercial TLC plates with fluorescent indicators are available as well, and bands can be visualized under UV light. However, if it is necessary to use solvents more polar than diethyl ether to extract polar components from the matrix, the fluorescent indicators may also be extracted, and these additives will interfere with subsequent analyses.

To isolate lipid classes from TLC plates after a nondestructive method of visualization, the silica gel can be scraped into a champagne funnel and eluted with an appropriate solvent. Or, the gel can be scraped into a test tube and the apolar lipid extracted with diethyl ether by vortexing, centrifuging, and decanting off the ether. Polar lipids are extracted in the same manner, using a more polar solvent such as chloroform and/or methanol. High performance liquid chromatography has been used in the separation and analysis of natural waxes, but its application was halted by the lack of a suitable detector, since most wax components have no useful ultraviolet chromophore. Application of the ultraviolet detector was limited to wave-length around 210 nm. Some components with isolated double bonds and carbonyl group (*e.g.*, esters, aldehydes, ketones) can be detected at this wavelength. Hamilton and coworkers have examined an alternative detection system, an infrared detector at 5.74 µm, which allowed the hydrocarbon components to be detected. While the sensitivity of this detector could not match that of ultraviolet detection, it has merit for use in the preparative mode, where it is feasible to allow the whole output from the column to through the detector. The third useful detector for HPLC is mass spectrometry. The coupling of HPLC and MS makes this mode of chromatography a very important analytical technique.

Table 5.1: TLC Separation of Wax Components

Component	*Solvent Systems*							
	A	*B*	*C*	*D*	*E*	*F*	*G*	*H*
Hydrocarbon	0.95	0.96	0.95	0.85	0.83	0.95	0.85	
Squalene							0.80	
Trialkylglyceryl ethers	0.90							
Steryl esters	0.90					0.95	0.57	
Wax esters	0.90	0.82	0.84	0.71	0.65	0.91	0.75	
β-Diketones		0. 75	0.54					
Monoketones					0.53			
Fatty acid methyl esters	0.65				0.47	0.75		
Aldehydes	0.55	0.65		0.47				0.66
Triterpenyl acetates							0.53	
Secondary alcohols				0.36				
Triacylglycerols	0.35				0.61	0.37		
Free fatty acids	0.18	0.00	0.00			0.35	0.20	
Triterpenols								0.22
Primary alcohols	0.15	0.14	0.16	0.09	0.15	0.21		0.19
Sterols	010				0.16	0.10	0.12	
Hydroxy-β-diketones	0.09	0.04						
Triterpenoid acid					0.05			

When individual classes of waxes have been isolated, the identity of each must be determined. Because of the complex composition of these materials, combined analytical approaches (*e.g.*, GC-MS) have been used to analyze individual wax classes. Mass spectrometry is a major analytical method for the analysis of this class of compounds. With the electron impact-mass spectrometry (El-MS), the wax molecules tend to give cleavage fragments rather than parent ions. Thus soft (chemical) ionization (CI), and fast atom bombardment (FAB), have been frequently used to give additional information for wax analysis.

In GC-MS analysis, the hydrocarbon fraction and many components of the wax ester fraction can be analyzed directly, while long chain alcohols, the aldehydes, and fatty acids are often analyzed as their acetate esters of alcohols, dimethylhydrazones of aldehydes, and methyl ethers of fatty acids. The analysis of wax esters after hydrolysis and derivativation will provide additional information on high molecular weight esters. For example, the chain branching of a certain component might be primarily examined with respect to its unusual retention time on GC analysis, then determined by converting to the corresponding hydrocarbon through the reduction of its iodide intermediate with $LiAlD_4$ (the functional group end is labeled by the deuterium atom). A similar method is to convert the alcohol of the target component to an alkyl chloride via methanesulfonyl chloride. This method labels functional end with a chlorine atom, and its mass spectra are easily interpreted because of the chlorine isotopes. As mentioned above, unsaturated hydrocarbons can be separated from saturated hydrocarbons and unsaturated isomers by column chromatography or TLC with sliver nitrate silica gel or alumina gel media. The position and number of double bonds affect volatility of the hydrocarbons, thereby altering their retention in GC and HPLC analysis. The location of a double bond is based on the mass spectra of their derivatives using either positive or negative chemical ionization.

Epicuticular waxes (from the outermost layer of plant and insect cuticles) comprise very long chain nonpolar lipid molecules that are soluble in organic solvents. In many cases this lipid layer may contain proteins and pigments, and great variability in molecular architecture is possible, depending on the chemical composition of the wax and on environmental factors.

A variety of waxes can be found in the cuticle. On the outer surface of plants these intracuticular waxes entrap cutin, which is an insoluble lipid polymer of hydroxy and epoxy fatty acids. In underlying layers, associated with the suberin matrix, another cutinlike lipid polymer containing aliphatic and aromatic components is found. In some instances internal non-suberin waxes, which are stored in plant seeds, are the major energy reserves rather than triacylglycerides. In insects, intracuticular waxes are the major constituents of the inner epicuticular layer.

A variety of aliphatic lipid classes occur in epicuticular waxes. These include hydrocarbons, alcohols, esters, ketones, aldehydes, and free fatty acids of numerous types. Frequently, a series of 10 carbon atom homologs, occurs, while chain lengths of 10 to 35 carbon atoms are most often found. However, fatty acids and hydrocarbons with fewer than 20 carbon atoms are known, as are esters with more than 60 carbon atoms. Other minor lipids such as terpenoids, flavonoids, and sterols also occur in epicuticular waxes. The composition and quantity of epicuticular wax varies widely from one species to another and from one organ, tissue, or cell type to another. In insects, wax composition depends on the stage of .the life cycle, age, sex, and the external environment.

In waxes, the biosynthesis of long chain carbon skeletons is accomplished by a basic condensation-elongation mechanism. Elongases are enzyme complexes that repetitively condense short activated carbon chains to an activated primer and prepare the growing chain for the next addition. The coordinated action of two such soluble complexes is plastid results in the synthesis of the 16 and 18 carbon acyl chains characterizing plant membranes. Each condensation introduces a β-keto group into the elongating chain. This keto group is normally removed by a series of three reactions: a β-keto reduction, a β-hydroxy dehydration, and an enol reduction.

Variations of the foregoing basic biosynthetic mechanism occur, giving rise to compounds classified as polyketides. Their modified acyl chains can be recognized by the presence of keto groups, hydroxy groups, or double bonds that were not removed before the next condensation took place. It is well established that the very long carbon skeletons of the wax lipids are synthesized by a condensation-elongation mechanism. The primary elongated products in the form of free fatty acids are often minor components of epicuticular waxes. Most of them, however, serve as substrates for the associated enzyme systems discussed. The total length attained during elongation is reflected by the chain lengths of the members of the various wax classes. Normal, branched, and unsaturated hydrocarbons and fatty acids are prominent components of plant waxes, while insect waxes usually lack long chain free fatty acids.

Chapter 6
Chemical Composition of Vegetable Oils

6.1 The Temperate Zone Plant Products

Some temperate zone plants accumulate higher levels of lauric acid in seed oils. Therefore, a priori, it appeared in the early 1980s that lauric acid oils could conceivably be produced in temperate zone agricultural systems.

For industries in the United States, a temperate zone crop source of lauric acid would help address national import/export imbalances; perhaps stabilize world price fluctuations for lauric acid oils; and conceivably provide an oil with higher lauric acid content and the associated savings in processing costs.

When Calgene, Inc., of Davis, California, initiated a project in 1985 to engineer *Brassica napus* canola into a lauric oil producer, there was a long list of technical unknowns. There was no reliable system for putting genes into canola and getting back normal plants. The available data suggested that a gene from a monocotyledonous plant like coconut might not function correctly when transferred to a dicotyledonous plant like canola. The general opinion was that production of lauric acid in nonseed tissues might be detrimental or even lethal to a canola plant. Indeed, even as late as 1990, it was thought that lauric acid might "gum up" triacylglycerol synthesis in canola. Certainly in 1985 there was insufficient understanding of how to limit foreign gene expression to just the developing seed of a transgenic canola plant. Also there was absolutely no experience, on which to base assumptions that a transgenic oil or simply the process of generating transgenic canola would not somehow compromise the agronomic productivity of canola. The perhaps most worrisome aspect of all was that no one knew, despite numerous tries, the mechanistic basis of laurate accumulation in coconut or oil palm kernel, let alone the wild species of *Cuphea glutinosa* or *Umbellularia californica.*

This list of technical challenges provides a framework to understand the steps and areas of technical expertise required to generate a useful new crop type containing a genetically engineered oil.

In the example of lauric acid canola, scientists were able to demonstrate the existence of a unique enzyme comprising lauroyl-acyl carrier protein (ACP) thioesterase in embryos of seed from *Umbellularia californica* that could conceivably account for the production of lauric acid. Moreover, that specific enzyme activity appeared to be missing in similar extracts from canola seed. Encouraged by that finding, a team of biochemists purified the *U. californica* lauroyl-ACP thioesterase protein sufficiently to obtain, by means of an automated protein sequencer, a partial amino acid sequence. This amino acid sequence allowed the design of synthetic DNA primers, which were then used with templates made by molecular biologists from messenger RNA isolated from developing seed of *U. californica* to generate gene- specific DNA probes. The DNA probes were subsequently used to identify lauroyl-ACP thioesterase complementary DNA (cDNA) clones from a cDNA bank that was constructed using, again messenger RNA (mRNA) from developing seed of *U. californica.*

Independently in the same laboratory, another group of molecular biologists isolated a canola gene specifically expressed at high levels in developing embryos during the normal period of canola seed development when storage lipids are formed. This natural *B. napus* gene was dissected down to the elements necessary to encode proper gene expression timing and tissue-specific localization within the canola plant. These "promoter" or genetic expression elements were then combined with the central portion of the *U. californica* lauroyl-ACP thioesterase cDNA close corresponding to the open reading frame encoded by the original naturally occurring messenger RNA. This synthetic gene was then combined with a second gene (the selectable marker gene) in a specific manner relative to other DNA signal sequences in a unique microbe known as a disarmed *Agrobacterium tumefaciens*.

Plant cell biologists then took the genetically engineered strain of *A. tumefaciens* and cultivated it for a brief time with sectioned hypocotyl tissues from germinated *B. napus* plantlets. This now-routine method of gene transfer into canola had to be developed while the biochemists were studying the lauroyl-ACP thioesterase enzyme in *U. californica* seed extracts and while molecular biologists were identifying the gene expression control elements from embryo-specifically expressed genes in seed of *B. napus*. After the treatment of *B. napus* hypocotyl sections with *A. tumefaciens*, the bacteria were completely removed and the plant tissues cultured on a series of different growth media in different containers, to regenerate complete *B. napus* canola plants. During this process of regeneration, most of the plantlets were deliberately killed by addition of in antibiotic. The selectable marker gene that was linked to the synthetic lauroyl-ACP thioesterase gene in the *A. tumefaciens* strain was a synthetic gene designed to detoxify the antibiotic used in these experiments. Thus, the only plants that survived the antibiotic treatment were transgenic plantlets that had received the kanamycin resistance gene (and almost always the lauroyl-ACP thioesterase gene as well). This is how the cell biologist can selectively produce 30 to 300 different transgenic plants without having to sort through thousands of plants that only might be transgenic. Each transgenic plant coming out of this process is potentially different and is generally regarded as a different event. As it turns out, each event tends to be unique in one or several features, most notably, in this case, with respect to how much lauric acid accumulates in the seed oil.

Progeny from each different event can be grown up and examined for the lauric acid content in the seed oil, and for other traits as well. As mentioned above, the transgenes show inheritance like regular canola genes, so gene segregation is possible, especially when a transgenic canola is crossed with another line.

It becomes the plant breeder's job to genetically fix the transgenes in a given canola line as well as to select for other traits such that the lauric acid content is reliably constant from generation to generation and agronomic characters such as seed yield are also preserved. Practically speaking, of course, this means multiple seasons of field testing and selections. In addition, just like other canolas, lines optimally adapted to different environments have to be tested and selected in different environments. The importance of this phase to the eventual success of developing a new crop type with a novel oil composition should not be underestimated.

In summary, then, development of laurate canola required basic biochemistry research, applied protein purification, gene cloning molecular biology, embryo-specific gene expression molecular biology, cell biology, and breeding, with many of these tasks and much of the technology development carried out in parallel. As well as an agronomically acceptable lauric acid canola, there had to be development of an infrastructure at the farm level to produce and harvest a new canola without contamination by regular canola seed. And as we shall see, another set of expertises to recognize and take advantage of special properties that may be present in a new oil also had to be established. Clearly, the development of a new crop type by directed genetic modification of a seed oil is not a small undertaking.

6.2 Modification of Seed Storage Lipids

One of the first considerations in thinking of specific ways to modify seed storage lipids is feasibility. Certainly there are some limits to what is practically possible. For example, seed storage lipids serve an important role in the life cycle of a crop plant. Germinating seeds presumably require the energy stored in the seed lipids and/or the young seedlings require access to those seed lipids in the cotyledons. If a lipid has been modified into particular structures that interfere with the ability of germinating seeds or young seedlings to tap that energy, or if the oil contains fatty acids that are difficult to metabolize, there may be a limit to the quantities of modified lipids of those kinds that one can achieve in a seed oil and still have a viable crop variety.

Moreover, the synthesis and incorporation of certain unusual fatty acids into storage lipids during seed development may also result in those fatty acids becoming incorporated into structural lipids that are essential for normal seed function. For example, the castor bean endosperm contains very high levels of ricinoleic fatty acid in the storage triacylglycerols, with very little ricinoleic acid in structural lipids. Clearly, as the castor bean lineage evolved synthetic mechanisms for ricinoleic fatty acid in seed oils, it also evolved mechanisms to either prevent incorporation of ricinoleic acid into structural lipids or to clear that fatty acid from structural lipid molecules. If one chooses to engineer soybean to produce ricinoleic acid in seeds, it may be found that soybean lacks the ability to maintain the integrity of the seed structural lipids and the seed is not fertile.

It should perhaps be noted that lipid biosynthesis occurs in all plant cells and is essential to support growth. Whereas the triacylglycerols comprising vegetable oils found in seeds are neutral lipids used as a means to store energy and fixed carbon, polar lipids found in all cells-including those in seeds-serve important structural functions and are thus often referred to as structural lipids. Both structural lipids and seed triacylglycerols contain glycerol-bound fatty acids, and thus it is not surprising that their respective synthesis pathways share many common steps. Re-engineering triacylglycerol biosynthesis must not compromise the synthesis of the structural lipids necessary to support growth and viability.

If some desirable lipid compositions are indeed incompatible with viable seed, there may be a unique opportunity in engineering sucoil types into either oil palm or avocado mesocarp tissues. These are oil-rich tissues whose natural fate is to rot away after fruit dehiscence. Since plant progeny do not directly depend on the viability of these tissues, they might be engineered to "self-destruct" by making, for example, very solid fats rich in long chain saturated fatty acids. Engineered oil palm fruits could be harvested by standard means and the palm kernels, (the oil palm seed) would be normal-assuming that the technology employed worked to ensure that the changed fatty acid composition was limited to just the mesocarp.

Another type of limitation that is of commercial interest is the ultimate amount of oil obtainable from a crop. While focused on traits like disease resistance and plant stature, much of the oil seed crop breeding community measures success by how much seed is harvested per acre. As crop yields go up, one expects the cost of food oils to come down. At a different level, breeders and genetic engineers alike are interested in increasing the oil content on a per-seed basis. Soybeans typically contain only 20 to 25 per cent oil by weight, whereas canola seed is typically 40 to 45 per cent. Peanuts can be 50 per cent oil, while the cacao bean is 60 per cent cocoa butter. It seems physiologically possible, therefore, to increase the seed oil content in soybean and canola with correspondingly dramatic effects on the cost of producing vegetable oils from these major crops.

Even though the scope of possible lipids modification projects must be very widely based on the natural variation in seed lipids found among plants from around the world, there are factors, other than natural ones that practically limit what will be done in the near term.

Genetic engineering of crop plants requires a battery of expertises: plant physiology, enzymology, molecular biology, gene transfer cell biology, prototype evaluation, and plant breeding. Much of an engineering project entails sequential applications of each area of expertise to the eventual goal. Coupled to the obvious costs of having these expertises in place and having the time needed to go from concept to practice is a consideration of the technical risk entailed when one is embarking on a new project. Financial modeling and analyses of the eventual value and return versus development costs and risk may argue against ever starting certain projects.

Part of the eventual value may hinge on the production system that allows the value to be ensured and protected. Since genetically engineered genes behave like other genes once present in a plant, they are necessarily transmitted by pollen. Vegetable oils are often zygotic characters. That is, the chemical composition is determined by both parents. Thus if a transgenic canola crop is grown immediately adjacent to normal canola, an interchange of pollen will decrease the purity of both crops. So, novel oilseed crops may have to be grown in carefully managed districts. Of course, there are some precedents for producing "specialty" oils, including the previously mentioned low-linolenic flax, high-oleic sunflower, and erucic-containing industrial rapeseed, These identity-preserved production practices typically add costs. Moreover, seed from crops with different types of oil cannot be stored together or crushed together; this restriction adds more complexity to production. These especial production practices need to be considered when the development of a new food oil is being planned. In the case of canola, where the precedent of identity-preserved production of high erucic oils already exists, it appears that the incremental costs are likely less than 5¢/lb.

Regulatory and political issues should be anticipated as well. Since genetic engineering is a new technology, products are receiving additional scrutiny by the public and by regulatory agencies to ensure that environmental and food safety issues are as fully reviewed as feasible. The nature of the issues varies considerably according to the nature of the altered traits. For example, different issues are raised by herbicide resistance and the potential introduction of a new food oil rich in myristic acid. Some segments of society may be less willing to try foods containing ingredients based on biotechnology. Religious issues for some novel oils may need addressing (*e.g.*, kosher and halal definitions). These market realities may affect what genetic engineering projects make sense.

Although plant lipid biosynthesis research has blossomed dramatically in the past 10 years, fueled by interest in transgenic plant applications, there is still much to be learned. Even though plant oils exist in nature that are greater than 85 per cent laurate, there may not be enough known to justify the laying out of a scientific strategy to convert soybean into a producer of an oil with 85 per cent lauric acid. On the other hand, if a technical route to some financially attractive target appears to be straightforward, an organization should consider the competitive aspects carefully. Perhaps a competitor elected to gamble at an earlier stage when the technical route was less clear and that organization may have gained an insurmountable lead to a commanding patent position.

Obviously, embarking on a long-term project with technical risk and complexity requires not only an evaluation of the utility and marketplace need of the final product, but also an extensive competitive analysis. The technology underlying the genetic engineering of oil composition may provide a clear basis for the patentability of the product of the research. Thus, since whoever is first may very well be in a position to block similar approaches by others, a consideration of which groups are after the same goal and where they are in the process is in order. In addition, for vegetable oils, there may be more than one means to the same end, and these alternatives should be anticipated as well as possible. For example, one could consider the genetic engineering of soybean with a *C. lanceolata* gene to produce an oil rich in capric acid. Alternatively, another group may choose more conventional means of domesticating *C. lanceolata* into an economically feasible crop. Either or neither approach may succeed; but if both succeed, the eventual marketplace value to each for the financial and time investment may be less.

In the case of structured oils and fats, there are of course chemically and enzymatically based synthetic methods that utilize glycerol and fatty acids sourced from animal fats as well as vegetable oils. Cost of raw materials and synthetic capacities are factors to consider when this approach is compared to the genetic engineering of crop plants. In the latter case, volumes are limited only to the number of acres that can be planted; in the former case, capital-intensive hard assets may be required in the form of manufacturing plants.

Of course, the possibility of competitive products becoming available and concerns about the profitability and financial return on investment versus risk are critical concepts for the private sector entities considering the development of new food lipids via the genetic engineering of crop plants. On the other hand, certain targets of lipids modification that will not pass these private sector hurdles for profitability versus risk may nonetheless have value and importance to the human community. For example, a cottonseed oil lacking malvalic acid may not be able to command much of a premium over regular cottonseed oil as long as cheap soybean oil (which naturally lacks malvalic acid) is also available. This might be a lipids modification project that could appropriately be undertaken by public sector research groups like U.S. Department of Agriculture, since the benefit could be spread among a large number of independent cooperative ginning entities throughout the cotton-growing delta of the Mississippi River. Similarly, improving the nutritive value of lipids in rice may not be attractive to private sector companies for many reasons, including the long and perhaps risky research phases that would be necessary, the structure (or lack of structure) in how rice planting seed is sold, the prevalence of small (subsistence) family farms in specific growing regions of Asia, and the resistance of farmers to paying premiums. But clearly, if the technology can be used to improve traits like carotenoid or vitamin content even slightly in a basic commodity foodstuff, the benefits would be tremendous to society at large.

The important issue to address regarding public sector initiative research on tailored vegetable oils is the coordinated recognition of technical feasibility along with marketplace-defined needs (*e.g.*, shelf life or processing traits like colors or free fatty acid content) and consumer common interest objectives (*e.g.*, levels of atherogenic saturated fatty acids or essential dietary fatty acids). How this process sorts out common-good objectives and provides for the necessary funding and coordination of properly qualified laboratories, plant breeding organizations, and the equally necessary product development and introduction functions is a challenge for everyone working in areas relating to foods and health.

6.3 The Temperate Zone Oilseed Crops

The temperate zone oilseed crops tend to have storage lipids rich in unsaturated fatty acids, and this liquid oils that are in themselves not suitable for products like shortenings and margarine. This shortcoming is typically addressed by a postharvest treatment of partial hydrogenation of vegetable oils. Partial hydrogenation does carry some incremental cost and generally results in the introduction of trans-unsaturated fatty acids, which may be undesirable in foods for health reasons. Of the tropical tree oils, only oil palm is produced on a scale large enough to provide a "fat" fraction rich in palmitic acid as a hardstock. Illipe, shea, and sal are tropical trees with seeds rich in the C18 fatty acid stearate, which contributes more solids to a fat; however, such tropical fats are limited in availability and generally too expensive to use as hardstock ingredients. Inspection of the pathway for biosynthesis of fatty acids reveals that both palmitic and stearic fatty acids are precursors to the unsaturated fatty acids making up the bulk of the oil found in seeds from temperate zone crops. Not surprisingly, then, it has been possible to engineer the pathway so that fewer unsaturated fatty acids are made and more of the precursor palmitic and stearic acids accumulate in the triacylglycerols.

Numerous thioesterase enzymes have been identified by cloning cDNAs from plant mRNAs that share DNA homology with the

California bay tree thioesterase cDNA described above. Every plant so far examined appears to have an enzyme with high activity on oleoyl-ACP as a substrate, and this enzyme presumably is the basic thioesterase in the fatty acid biosynthesis pathway. The oleoyl-ACP thioesterase typically has less but significant activity on both palmitoyl-ACP and stearoyl-ACP substrates. Another class of thioesterase enzymes has been found in many plants, however, with enhanced activity on palmitoyl-ACP substrates; genetically engineered overexpression of such enzymes in seed of transgenic canola plants results in seed oils enriched in palmitic acid. On the other hand, thioesterase enzymes with low activity on palmitoyl-ACP substrates and relatively high activity on steroyl-ACP appear to be rare. When researchers looked at cDNA clones made from mRNA from developing seed of mangosteen fruit (mangosteen seed oil typically contains 40–50 per cent stearic acid), a clone was identified that corresponds to an enzyme with enhanced levels of stearoyl-ACP activity relative to other thioesterases. Genetically engineered over expression of that mangosteen enzyme in seed of transgenic canola plants results in a seed oil enriched in stearic acid, up to 30 per cent in some seed.

Alternatively, one can consider engineering a seed to have less desaturase activity so that fewer molecules of the saturated precursor stearoyl-ACP are converted to oleoyl-ACP. This has been demonstrated in transgenic canola and soybean by suppressing levels of the stearoyl-ACP desaturase enzyme. Suppression of an enzyme can be achieved by genetic engineering methods of either antisense or cosuppression. Each allows for tissue-specific suppression. In this particular application, tissue-specific suppression is important because the stearoyl-ACP desaturase is an essential enzyme in leaves and other tissues for plant viability.

Yield is a very important trait in reducing the cost of vegetable oils. Theoretically, there are several arguments for the feasibility of raising seed oil content in most crops. The coco bean has 60 per cent oil by weight, so the physiological limit for crops like soybean and canola may be at least this high. Individual seed of rapeseed can typically vary from 35 to 50 per cent; so even within existing germ plasm, the current average oil content from canola of 42 per cent will likely be raised by straightforward breeding selections in the coming years. Historically, soybean protein content has been the most important component of the bean, with the 20 per cent of the seed that is oil a valuable by-product. Clearly, there is room to increase oil content in soybean by some means; however, increased oil will be balanced against maintaining value in the meal component for animal feed uses.

Despite the evident technical premise and motivation for increasing oil content, exact scientific strategies are still developing. Acetyl-CoA carboxylase is an enzyme activity often considered to be a rate-limiting step for fatty acid biosynthesis. The data for this assumption seem clearest in animal cells and fairly convincing in the bacterium *E. coli*, but somewhat less clear in higher plants. Transgenic modifications of levels of this enzyme activity have been achieved with slight and possibly significant increases in oil content; however, experiments continue. The development of other strategies to increase fatty acid biosynthesis at the expense of nondigestible fiber and other less desirable components of seed represent an area of increasing research.

The availability of gene suppression technologies allows one to think of decreasing certain constituents of vegetable oils. An early and far-reaching application is evident in the use of cosuppression in canola and soybean to dramatically reduce the levels of polyunsaturated fatty acids. Other potential projects might be, for example, the suppression of enzymes directly responsible for formation of cyclopropene fatty acids such as malvalic acid in cottonseed oil or achieving a decrease in the high palmitic acid content in palm oil by suppressing the level of palmitoyl-ACP thioesterase activity in the oil palm mesocarp so that more fatty acids can be elongated and desaturated to oleic acid.

For the most part we have discussed modification of the kinds of fatty acid found in triacylglycerols. However, vegetable oils can also contain, other important lipid components. These include the antioxidants of tocopherols (*e.g.*, vitamin E) and of mixed carotenoids (including β-carotene and lutein), plant sterols such as sitosterol (which may have cholesterol-lowering properties in the human diet), and economically important by-products of vegetable oil processing such as lecithin. As the enzymatic bases for the synthesis of these compounds and the genetic bases for regulation of amounts of these compounds become better understood, there should be growing prospects for the genetic engineering of levels of these compounds as well as flavor components in vegetable oils.

6.4 Coconut Oil Production

The world's major coconut oil producer is the Philippines, where thousands of small farmers with small production acreage accounted for 43 per cent of production and 65 per cent of total exports in 1992–1993. Because the Philippine coconut industry has suffered from chronic underinvestment, production and exports have decreased and are expected to continue to do so for the foreseeable future. Estimates are that as many as one-third of the 300 million coconut trees in the Philippines will become nonproductive over the next 10 years. Production in these secondary countries, spurred by government investment, is expected to increase. Coconut production in any given year is heavily influenced by weather factors, including rainfall amounts and the impact of typhoons.

The second major source of lauric oils is palm kernel. Malaysia is the largest producer of palm kernel oil (PKO), accounting for 54 per cent of the world's production in 1992–1993. Unlike coconut production in the Philippines, palm (along with palm Kernel) is produced by large sophisticated plantations and enjoys major government and private investments that are allowing it to forward-integrate into end-use derivatives. As a result, palm kernel oil production is projected to increase significantly through the year 2002, with yields of 2175 metric tons as new plantations become productive.

In the world of foods, where, typically the whole triglyceride is utilized as one of the components, lauric oils can be replaced–to some degree–by the nonlauric oils, especially as specialty fat blends are developed that mimic the physical properties of these lauric fats through a series of processing steps including special hydrogenation and fractionation. In the industrial area, however, lauric acid derived molecules invest unique properties into the surface activity of the resulting compound and formulation, and they allow for little or no substitution of the fatty acid moiety without a direct effect on performance. This is especially true in the soap, detergent, and personal care markets, where the C12 moiety provides unequaled detergency and mildness. Because the perennial nature of lauric oilseeds prevents quick response to changes in supply and demand, and much of the demand side is inelastic owing to the unique functionality of the oil, lauric oil prices are typically much more volatile than prices for nonlauric oils. Rising incomes and standards of living in Southeast Asia and Eastern Europe are expected to result in an increase in consumption of detergent and personal care products and, therefore, increased demand for lauric oils.

6.5 The Application of Transgenic Oilseed Products

The strategy developed for the application of transgenic oilseed products is the use of laurage canola as an alternative for coconut and palm kernel fats in food systems.

Although these markets are smaller in volume than the industrial applications of laurics, the functional nature of the uses would allow the recapture of the research investment, assuming that at least equal functionality was shown in existing applications.

The main function of lauric oils is to provide desirable structure and mouthfeel characteristics in a variety of foods. Both coconut and palm kernel provide relatively high solids at room temperature; but a steep melting curve at about body temperature gives food products based on these oils a good melt-away sensation in the mouth. The solids profile of a given fat is typically, determined by dilatometric methods, although techniques using pulsed NMR spectrometry allow for similar determinations of the solid/liquid ratio in a given fat as a function of temperature. Usually, the solids fat index (SFI) is determined for food fats at a select number of temperatures. The plots of solids versus temperature have been related over the years to the actual performance of given fat systems in specific food formulas. The portion of the curve below 25°C (77°F) relates to the "hardness" of the fat, specifically the resistance to deformation it provides to the base food, while the solids present between the temperatures of 25°C and 30°C (86°F) represent the amount of solids typically found at ambient temperatures and are a measure of the degree of heat resistance provided to the food product by the fat system. Solids that occur above 37°C (or about body temperature, 98.6°F) are perceived as "waxiness" in the mouth, although low levels of solids are acceptable. A rapid rate of solids decrease between ambient temperatures and body temperature provides a desirable cooling sensation in the mouth and is deemed a positive for confectionery fat systems. A major use of these fats is in confectionery coatings, where the lauric fats provide an effect similar to that of cocoa butter in chocolate, at a much reduced cost.

The major fat used in the manufacture of chocolate is cocoa butter. It is also one of the most expensive ingredients used in the formula, and because so much of it must be used to achieve the desired functional effects, it contributes the highest cost of goods for the total formulation. Aside from a significant contribution to the final flavor of the chocolate, cocoa butter has some physical properties that make it unique among the triglycerides found in nature. First of all, it is composed of a very few sets of highly structured triglycerides.

These triglycerides are essentially composed of only three fatty acids: palmitic (C16 : 0, P), stearic (C18 : 0, S), and oleic (C18 : 1: 0). Practically all the oleic acid occurs esterified at the *sn*-2 position of the various triglycerides, with the *sn*-1 and *sn*-3 positions containing the esterified palmitic or stearic acids. This specificity results in the presence of three predominant structured triglycerides: POP, POS, and SOS. These three triglyceride types make up over 80 per cent of the triglycerides found in cocoa butter. These specific triglycerides, which resemble each other very closely, provide cocoa butter with its unique solids profile and desirable melting characteristics.

Going back to the basic issue, however, cocoa butter is expensive, and supply limited. Over the years, various researchers have spent considerable amounts of time developing fats that could serve as alternatives to cocoa butter by providing properties similar to its melting profile and solids content at various temperatures. Several classes of alternatives have been developed, ranging from cocoa butter "equivalents" produced from selected blends or fractions of natural fats high in specific triglyceride contents that are miscible in all proportions to cocoa butter, to fats that contain totally different triglyceride distributions but mimic, in many ways, the melting behavior of cocoa butter. An example of the production of a cocoa butter "equivalent" would involve the purification and fractionation of a series of different naturally occurring fats to obtain the proper proportion of the desired triglycerides having the desired structure, at the appropriate levels. The majority of the POP portion required is obtained from palm midfraction.

The reconstruction of a cocoa butter equivalent from the isolation of specific triglycerides, through fractionation from a variety of source oils and subsequent blending, obviously is not an inexpensive process, especially since the oils used for feedstocks are sourced from tropical plants that are not grown and cultivated in the most efficient manner and in them- selves are often supply-constrained. These "equivalents" are commercially viable only when the price of cocoa butter is high enough to justify their high costs. When cocoa butter is relatively inexpensive, they are not used.

Table 6.1: Triglyceride Composition of Sal, Kokum, Shea, and Illipe Fats

Triglyceride[a]	*Composition (wt %)*			
	Sal	*Kokum*	*Shea*	*Illipe*
POO	3	–	2	–
POSt	11	5	5	35
StOSt	42	72	40	45
StOO	16	15	27	3
StOL	1	–	6	–
StOA	13	–	2	4
OOO	3	2	5	–
POP	1	0	0	7
Others	10	6	13	6

[a] A: Arachidate; O: Oleate; P: Palmitate; St: Stearate.

Another approach to mimicking the functional properties of cocoa butter is through the use of special fractions of a combination of domestic and tropical (nonlauric) fats that have been selectively hydrogenated. The fats often used include soy, canola, cotton, and palm. The manufacture of chocolate-flavored coatings using these final blends of fat in place of cocoa butter is well known, and these coatings have a real place in the market in certain applications where rate of melting of the fat in the mouth and ultimate flavor release are 'not key factors in consumer acceptance of the product. They do have the drawback, however, of some amount of "waxiness" that can be detected in the mouth, since a fraction of their triglycerides does melt above body temperature. This effect is minimized when chocolate-flavored coatings made with these cocoa butter replacers are used on baked goods, where the crumb structure of the substrate aids in the mastication of the coating and helps minimize the perception of waxiness in the mouth.

Table 6.2: Fatty Acid Composition (Weight %) of Processed Palm Kernel and Coconut Oils

Oil	*C12*	*C14*	*C16*	*C18*	*C18 : 1*	*C18 : 2*
Palm kernel oil	47.0	17.0	8.5	3.0	11.0	2.0
Hydrogenated at 35°C	47.0	17.0	8.5	13.0	7.0	–
Hydrogenated at 40°C	47.0	17.0	8.5	19.0	1.0	–
Fractionated	55.0	21.0	8.5	2.0	7.0	1.0
Hydro/Fractionated	55.0	21.0	8.5	10.0	–	–
Interesterified/Hydrogenated	47.0	17.0	8.5	19.0	1.0	–
Coconut oil	47.5	17.5	8.5	2.5	7.0	1.5
Hydrogenated	47.5	17.5	8.5	10.0	1.0	–

The most popular approach for the formulation of chocolate-flavored coatings is the use of cocoa butter substitutes (CBSs), based on lauric fats. These coatings are widely used in both the chocolate/confectionery and baking industries. The principal base fats used in their manufacture are palm kernel oil and coconut oil, with the former being the more widely used. These fats contain triglycerides that are high in the esters of lauric and myristic acid. Triglycerides of these compositions form crystals that are relatively small and uniform, and have a sharp melting point around body temperature. The set point of these fats is also, typically, very sharp, and they provide a coating base that has significant solids at room temperature. With their high levels of lauric and myristic acids, however, it is obvious that a significant percentage of these fatty acids will occur on all three of the available positions, in a random manner. Because these triglycerides are of vastly different compositions and are nonstructured, coatings based on them are not compatible with cocoa butter; hence the ultimate coatings can tolerate only minimal levels of cocoa butter containing ingredients, such as cocoa powder, and the flavor impact of these coatings is usually degraded by these restrictions. They do have better melting properties than the cocoa butter replacers discussed above, however, with the noticeable absence of any lingering waxy mouth feel.

With the production of the first crop of laurate canola, we were faced with a totally new type of triglyceride: one that was high in lauric acid esters, though not as high as coconut or palm kernel oils, but one in which all the C12 : 0 and C14 : 0 occurred on the *sn*-1 and *sn*-3 positions, and the *sn*-2 position was occupied solely by C18 : x fatty acids, where $x = 1, 2$, and/or 3. Selective hydrogenation of this oil, then, would allow us to closely control the melting properties of the resulting triglyceride and to manipulate the solids profile of the final fat system. When these new fats were utilized in typical formulated food systems, it was noted that the systems offered flavor release that was far superior to that obtainable with the corresponding tropical laurics. Although this statement is based on anecdotal data, a great deal of effort is under way to quantify the flavor release properties of these new structured fats versus the random lauric fats typically used. It could be immediately demonstrated, however, that this new source of a lauric fat used in confectionery was highly compatible with cocoa butter–a trait not encountered with the tropical laurics–hence could be used with cocoa butter sources that lead to higher flavor in the finished product. These obvious differences in how the genetically engineered laurate canola performed with respect to existing laurics were explored at some length and are discussed later.

6.6 Saturate Levels in Liquid Oils

Saturate levels in liquid oils have become a significant issue for consumers because of reports linking saturates to coronary heart disease. Saturates have been shown to lower the levels of high density lipoprotein (HDL) cholesterol. The FDA now requires that food labels list saturate levels. These concerns have already had a major effect of the U.S. salad and cooking oil segment. Consumption of saturated tropical fats has decreased tremendously. Consumer concern over saturates is projected to continue to play a key role in this market segment, and food companies will continue to search for products that address this concern.

Since canola oil contains the lowest level of saturates of any of the commonly used food oils, it seemed a natural base from which to launch a variety of approaches that would lead, ultimately, to the desired product. As it turned out, the research was not trivial. A combination of mutagenesis and the use of two different genes led to a canola oil having a saturates level between 4.0 and 4.5 per cent, but with questionable agronomic performance. The presence of relatively high levels of polyunsaturated C18 fatty acids also contributed negative performance characteristics when the oil was used for frying purposes. Several different companies are continuing to work in this project area using conventional plant breeding to lower the saturates. In most cases, these companies are working to combine this attribute with low linolenic and/or high oleic traits to offer an oil that is low in saturates with improved stability under high temperature uses. None of these groups has yet developed an oil with less than 8 per cent saturates.

The presence of *trans* fatty acids in the diet has become a major issue for consumers. In 1993 margarine sales began to decline as a number of studies indicated a potentially adverse impact of the *trans* fatty acids found in partially hydrogenated vegetable oils used in the manufacture of most margarines and spreads. These studies suggested that *trans* fatty acids contributed to coronary heart disease by raising levels of LDL cholesterol and lowering levels of HDL cholesterol. Although there is still considerable debate in the medical community about the true health impacts of *trans* fatty acids, they have become a source of concern for many consumers and have affected buying habits. At the same time that U.S. consumers became concerned over the trans issue, butter manufacturers began to aggressively compete with margarine manufacturers for market share through price reductions and increased marketing and promotional activities. The net result is that margarine sales have declined since 1993, falling by more than 6 per cent from 1993 to 1994.

An opportunity was seen to exist for the development of a suitable oil containing sufficient saturates in its natural makeup to eliminate the need for hydrogenation. Hydrogenation, normally used to increase the saturated fatty acid levels in a triglyceride system, is the primary producer of *trans* fatty acids when polyunsaturated systems are reduced to mono-unsaturated systems. Thus, elimination of hydrogenation would result in the elimination or minimizing of *trans* fatty acid occurrence in food systems. Key to assessing the opportunity is understanding the needs of the margarine and spreads industries.

Margarines and spreads are prepared by blending fats and oils with other ingredients, including water, milk, edible proteins, salt, flavorings, and colorings. Spreads are differentiated from margarines by their lower fat content. By FDA regulation, a product must contain at least 80 percent fat to be labeled margarine; products with lower fat levels must be labeled as spreads. As consumers have become more concerned about overall fat consumption, spreads have become increasingly popular. The key to the formulation of both margarines and spreads is to provide a solid, spreadable fat with appropriate melting characteristics. This is achieved by using partially hydrogenated vegetable oils, which have had a portion of their unsaturated fatty acids converted to both saturated fatty acids and less saturated, *trans* fatty acids. Saturated and *trans* fatty acids have higher melting points than unsaturates and provide the requisite functional properties for the finished product. The specific oils used and the degree to which they are hydrogenated vary as a function of specific marketing objectives for the particular product.

In Europe, some margarines contain an oil portion consisting of a liquid vegetable oil blended with one that has been fully hydrogenated–a formulation ploy seen increasingly in the United States. Unlike partially hydrogenated oils, fully hydrogenated oils do not contain any *trans* fatty acids. This allows for the manufacture of margarines that are *trans*-free but also contain higher levels of saturates than those containing partially hydrogenated oils. European manufacturers also use more tropical oils in the manufacture of margarines and spreads than firms in the United States, since European consumers are not as concerned as U.S. consumers about tropical fats.

Several companies are actively pursuing the development of seed oils that contain levels of saturated fatty acids high enough to permit the elimination of needs for hydrogenation, with the concomitant production of *trans* fatty acids. As an integral part of this higher production of functional, saturated fatty acids, it is also necessary to have the desired fatty acids in a low polyunsaturated fatty acid background, to assure the ultimate stability of the finished formulation. To compete in the large markets involved under these two sets of conditions of composition, one must also strive for a resulting plant with agronomic vigor and oil yields comparable to those of existing sources of commodity oils.

Chapter 7

Extraction, Cleaning and Storage of Oils and Fats

7.1 Introduction

For several, thousand years, fats and oils recovered from oil-bearing seeds, fruits and fatty animal tissues have been used for food cosmetics, lubricants and lighting fluids. Processing seeds into edible oils can be broken into four sets of operations; recovery, refining, conversion and stabilization.

Oil recovery is often referred to as extraction of crushing when plant sources are being processed and rendering in the case of processing animal tissues. Oil extraction involves either pressing the oil-bearing material to separate crude oil from the solids high in protein or washing flaked or modestly pressed material with solvent, almost always hexane. The defatted solids are known as cake after pressing and as meal after solvent extraction. The oil, "crude oil" because it contains undesirable components, such as pigments, phosphatides, free fatty acids, and undesirable flavors and odors, must be refined to remove these contaminants and produce high quality edible oils. Refined oils consist primarily (> 99 per cent) of triglycerides and can be "converted," usually by hydrogenation; but winterizing, fractional crystallization, and interesterfication should also be considered to be conversion processes because they achieve properties different from those of the original oil, such as converting liquid oil into semisolid or solid fats. Plasticizing, tempering, and stehling are operations designed to stabilize crystal-oil mixtures used for shortenings and margarines.

All oil recovery processes are designed to obtain oil triglycerides as free as possible from undesirable impurities, to obtain as high a yield as possible consistent with economics of the process, and to produce cake, meal, or flour (finely ground meal) usually high in protein content, of maximum value. Three general types of process are used to crush oilseeds: hard pressing, prepress solvent extraction, and direct solvent extraction. The extraction process of choice depends primarily on the oil content of the material, the amount of allowable residual oil in the meal, the amount of protein denaturation allowed, the amount of available investment capital, and local environmental laws concerning emissions of volatile organic compounds (VOCs).

The oldest oil recovery method is hard pressing, where the seed is pressed, usually after various pretreatments to enhance oil recovery, to squeeze the oil from the solids known as cake. In the early years, lever presses and screw-operated presses, often driven by oxen or other work animals, were used; then, during the Industrial Revolution, batch hydraulic presses were introduced, which evolved at the turn of the century into continuous screw presses connected to line shafts driven by steam engines. Today, continuous screw pressing and direct, solvent extraction have become the preferred processing methods because the oil is more completely recovered. For a long time, the rule of thumb was that materials containing more than 30 per cent oil require pressing, either hard pressing or prepressing, prior to solvent extraction. Hard pressing involves squeezing as much oil as possible; in prepress solvent extraction, only part of the oil is squeezed out before the partially deoiled material is subjected to more complete extraction with solvent. The recent adoption of the expander has largely done away with this rule, and even materials high in oil content can be solvent-extracted today with no or little prior oil extraction. Direct solvent extraction, without any prior pressing or expanding, has long been the most widely practiced method: the oil is more completely recovered, and the oil and meal are economically recovered, undamaged by heat.

Table 7.1: Major Edible Fats and Oils in the United States and Methods of Processing

Source	*U.S. Oil Consumption (million pounds)*	*Oil Content (%)*	*Prevalent Method of Recovery*
Soybean	13,000	19	Direct solvent extraction
Corn (germ)	1,300	40	Wet or dry milling and prepress solvent extraction
Tallow (edible tissue)	1,265	70–95	Wet or dry rendering
Canola	1,246	42	Prepress solvent extraction
Coconut (dried copra)	1,069	66	Hard pressing
Cottonseed	965	19	Hard pressing or prepressing or direct solvent extraction
Lard (edible tissue)	910	70–95	Wet or dry rendering
Palm	331	47	Hard pressing
Palm kernel	309	48	Hard pressing
Sunflower	295	40	Prepress solvent extraction
Peanut (shelled)	201	47	Hard pressing or prepress solvent extraction

Storage for extended periods at moisture contents exceeding critical moisture levels will damage oilseeds, reducing the recoveries of oil and protein, and diminishing the quality of the oil (notably color and refining loss). Seeds at harvest are alive and respire, converting seed mass to CO_2 and other metabolites, albeit at a low rate when the moisture content is below the critical moisture level. The critical moisture level for safe storage varies with the seed species: usually, the higher the oil content, the lower the critical moisture value. At moisture contents exceeding the critical moisture level, respiration rate increases, and the seed can even germinate, becoming subject to fungal attack. Respiration and germination liberate heat and, when aeration is insufficient, further accelerate these reactions. Under extreme conditions, the seed may become scorched or even catch fire (especially cottonseed). Modern facilities employ temperature-monitoring systems to alert elevator and storage operators when seed temperatures exceed critical set points. Then, the seed is moved to another bin to disperse hot spots and/or aerated. The percentage of seed that is heat damaged is often a factor in the U.S. grades and standards (*e.g.*, soybeans).

Overdrying can increase seed fragility, leading to excessive breakage during handling, storing, and processing. In the case of soybeans, the cotyledon is prone to splitting into halves when the hull becomes separated from the cotyledon during conveying and transporting, a problem that becomes worse when the beans are overdried. "Splits" are undesirable because they are difficult to separate from foreign matter, and the oil deteriorates at a faster rate. Oil from soybean splits is higher in free fatty acids, phosphatides, iron, and peroxides as a result of activation of catabolic enzymes. Oils from field-and storage-damaged seed usually are poor in flavor. For these reasons, damaged kernels and splits are also factors in the U.S. grades and standards for soybeans.

When oilseeds are received at the crushing plant, samples are often taken for analysis or moisture, foreign matter, damaged seed, oil, protein, and free fatty acid contents. Shipments arriving with similar values for these analyses may be segregated based on actions that are required to minimize further degradation. Seed containing more than the critical moisture content are immediately processed or dried. Seed with excessive foreign matter is transferred to scalping operations to reduce the level of contaminants. Foreign matter usually contains more moisture than the seed and tends to become concentrated at certain locations, during placement into storage bins. Removing foreign matter reduces bulk moisture content and improves storability. Usually, foreign matter is also a factor in grading systems.

Both inside and outside storage systems are used. Bins are constructed of both metal and concrete. Cottonseed is often stored in special buildings having roof slopes the same as the angle of repose for the seed. Some oilseeds are stored outside on concrete pads, and the piles may be covered with tarpaulins and/or equipped with air-distribution systems.

Any type of handling, including normal harvesting, conveying, and transporting, increases seed damage. Nature has given seeds a high degree of subcellular organization. Seed, endosperm is composed of many cells containing oil and storage protein, which supply energy, nitrogen reserves, and other metabolites to support the growth of the embryo during germination. Storage protein (the greatest proportion of seed nitrogen) is concentrated in discrete bodies known as "protein bodies"; lipids (predominantly triglycerides) are stored in organelles called spherozomes. Phospholipids are largely associated with pseudomembranes around protein bodies and spherozomes. Enzymes and cellular metabolites are present in the cytoplasm. Some of these enzymes, notably lipase, hydrolyze triglycerides; this increases free fatty acids, which must be removed by refining. Phospholipase activity may render some phospholipids nonhydratable and difficult to remove from the oil by means of refining procedures. Lipoxygenases, unique to soybeans and other legumes, oxidize linoleic and linolenic acids, causing painty, green, beany flavors in protein; these enzymes may also reduce the oxidative stability of the oil.

In intact seeds, these enzymes are kept away from the oil by natural compartmentalization within the cell. However, damage by weather, harvesting, and/or handling bruises the seed, breaking cell walls and membranes and allowing the oil and enzymes to come into contact with one another. This series of events accelerates adverse reactions, whose rates are dependent on temperature, moisture, and extent of damage.

The first step in processing oilseeds usually involves cleaning the seed. Unless removed by cleaning operations, foreign matter will reduce oil and protein contents and increase wear and damage to expensive processing equipment; in addition, it may adversely affect oil quality (especially color). Magnets are placed in chutes just ahead of processing equipment to prevent damage by tramp iron. Vibrating and/or shaker screens with or without aspiration are used to remove stems, pods, leaves, splits, broken grain, dirt, and extraneous seeds.

7.2 The Screw Press Principle and Dehulling

A screw press is basically a continuous screw auger designed to accept feed material and subject it to gradually increasing pressure as it is conveyed through the barrel cage.

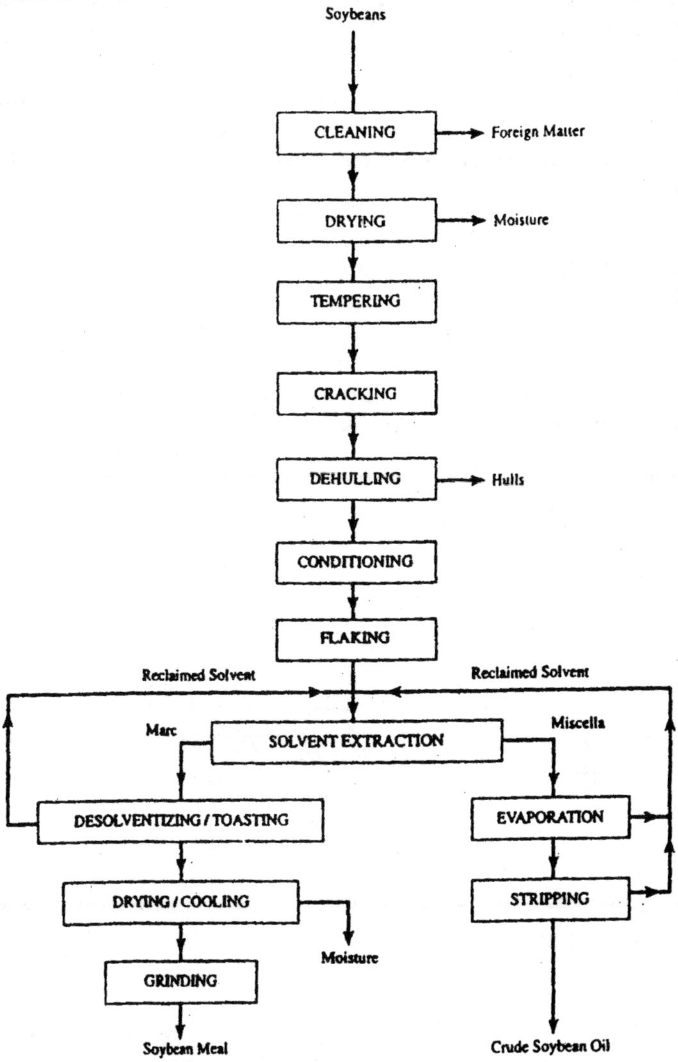

Figure 7.1: Steps in Processing Soybeans into Meal and Edible Oil

The barrel is composed of bars surrounding the screw and oriented parallel to the screw axis. The bars are separated by spacers of decreasing size, which allow the oil to drain. A plug of compressed oil-lean solids, the "cake," forms at the discharge end. Increases in pressure down the length of the barrel are achieved by increasing the root diameter of the screw, decreasing the pitch of the screw flights, and controlling the opening for the discharging cake by means of a choke. This design causes fresh material to be rammed against the plug.

Screw presses comprise three sections: feeding, ramming, and plugging. The meats or flakes fall into a rapidly rotating feed screw, which introduces them into the pressing cage to expel entrapped air and squeeze out the easily removed oil. In some screw presses, the feeding section is a separate vertical screw, or it may be mounted directly to the ramming screw. In either case, the feeding screw turns faster than the ramming screw. Knife bars with projecting nibs are

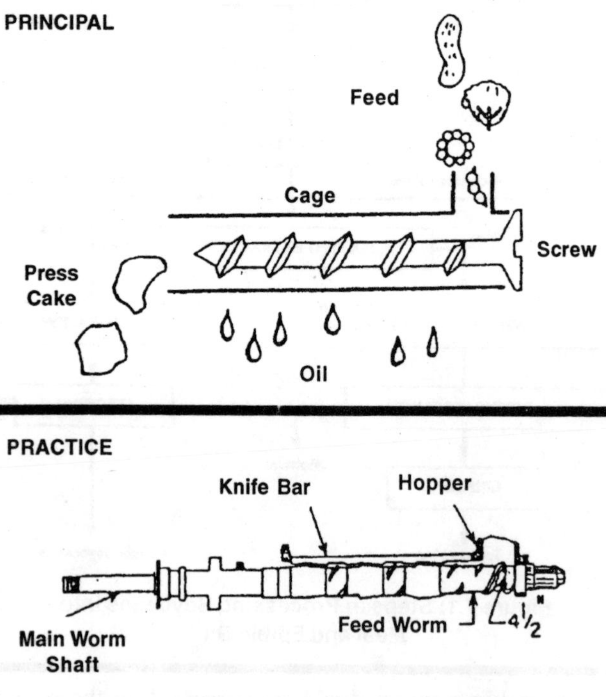

Figure 7.2: The Screw Pressing Principle

clamped between the half-cages to prevent cake slippage and rotation. Maximum pressure is developed in the ramming section as partially deoiled cake is rammed against the deoiled plug. The plug section provides the resistance against which the ramming occurs. The friction in the barrel generates heat that must be removed to achieve low residual oil. Some screw presses recycle cooled pressed oil over the cage to remove excess heat, while others use water-cooled shafts and bar cages.

In hard pressing, screw presses are choked to put maximum pressure on flaked, cooked seed while maintaining cake discharge. The oil is pumped or flows by gravity to a basin, to allow settling of cellular debris (foots) that was pressed with the oil. The cake is then ground into meal.

There is interest in hard screw pressing oilseeds to produce organically grown vegetable oil or to comply with local laws (*e.g.*, in California) that prevent construction of new solvent plants. A small amount of soybeans is hard-screw-pressed where the meal has particularly high value (*e.g.*, rumen bypass dairy cattle feed). Screw pressing of copra, peanuts, sesame, and cocoa butter also remains popular in developing countries, where investment capital is limited.

So far we have studied about screw pressing principle. Now let us try to understand some details about dehulling process.

Usually, it is desirable to remove the hull or seed coat that surrounds the oilseed "meat." Hulls always contain much less oil and protein than do meats. Removing the hull reduces the amount of material that must be handled, extracted, and desolventized, thus increasing downstream plant capacity. The protein content of the meal is raised by removing the hull. Oftentimes, as with corn and sunflower seed, the hull contains waxes on the outer surface as a natural protective mechanism of the seed. These waxes must be removed during latter oil refining steps, often by winterization or cold centrifugation; otherwise, the wax may become insoluble at cold temperatures, making the oil unattractively cloudy and causing separation in emulsions such as mayonnaise. Removing the hulls can alleviate these wax problems. On the other hand, hulls can be helpful at times, for example, by providing fiber to allow easier hard pressing or to enhance solvent drainage, and thus they are not always removed.

Dehulling must be done carefully, to ensure that the meat is not broken into too many small pieces, which would be difficult to separate from the hull.

7.3 Rate of Oil Extraction

In the extraction of oil different methods are used. Prepress solvent extraction, and direct solvent extraction are two popular methods. Now let us learn some informations about the extraction of oil, and also about the rate of oil extraction.

In prepress solvent extraction, part of the oil that is easily removed is pressed out, as described above. The press is choked, developing less pressure. Consequently, less oil is extracted and throughput is increased. Usually, the oil content of prepress cake is 15 to 18 per cent, and the partially deoiled cake is then extracted with solvent. The cake may be broken into pieces and even flaked to increase bulk density and extractor capacity, and to speed extraction. The remaining steps are the same as for direct solvent extraction, described next.

Cell walls are impermeable to oil and nearly so to extraction solvents. Consequently, cell walls and membranes must be "distorted" or ruptured to get the oil out, regardless of whether the process chosen is solvent extracting or screw pressing. This requirement often calls for a reduction in seed size. Small pieces transfer heat and moisture more readily during conditioning or cooking. But excessive size reduction means less mechanical distortion during flaking, however, since flaking mills distort larger particles more than smaller particles. Of course, particles must be small enough to pass into the nip between rolls. Heating prior to flaking reduces oil viscosity, inactivates enzymes, coagulates protein, ruptures some cell walls and membranes, and makes the seed particle plastic for subsequent flaking or pressing. Proper plastic texture is necessary to produce thin, non fragile flakes with minimum fines and maximum cell distortion. The flaked material must have tenacious, thin structure, with porosity that allows transport of the oil or *miscella* (solvent-oil mixture).

It has long been recognized that flake thickness influences the rate of oil extraction. Oilseeds are conditioned prior to flaking by means of vertical stack cookers or rotary steam tube driers to heat the seed to 70 to 80°C over 20 to 30 minutes, while maintaining 10.5

to 11.5 per cent moisture. This treatment makes the meats soft and pliable enough to be flaked with smooth-surfaced roller mills to low thickness, 0.25 mm (0.010–0.012 in.) without producing excessive fines, which adversely affect other operations.

Extraction has been linked to cleaning paint from a brush. As such, key to getting the brush clean is getting good contact and penetration of the solvent; then there must be enough clean solvent to dissolve more solute, and enough time and heat to quickly dissolve more solute. However, oilseeds extraction does not follow the single mechanism of leaching, as the brush example implies and others have often assumed in oilseeds extraction. Instead, oilseeds extraction involves a combination of leaching, diffusion, and dialysis, which results in an ever decreasing rate of extraction as the relative importance of each mechanism changes during the course of extraction. For flakes, the larger proportion of readily extractable oil is derived from ruptured cells, especially near the surface. The transfer of oil from distorted interior cells probably is governed by capillary flow, and the rate of oil transfer is partly dependent on viscosity of the miscella. A portion of the slowly extracted oil is contained within intact undistorted cells and must be transferred by osmosis. This transfer has been shown to be very slow. Presumably, the process of extruding flaked meats, known as *expanding,* shifts the relative importance toward leaching because nearly all the cells are ruptured and the collet structure is quite porous.

Another portion of the slowly extracted oil relates to slowly soluble extractable materials such as phosphatides, free fatty acids, nonsaponifiables, and pigments, which contribute to refining loss. The best quality oil, high in triglyceride content, is extracted first, while with more exhaustive extraction, poorer quality oil is extracted. Thus, at low residual oil levels, the proportion of free fatty acids and phosphatides extracted is greater, as is the refining loss. However, current industry practice is to strive for the most complete extraction possible. Typically, residual oil contents range 0.5 to 1.0 per cent.

Flake thickness and solvent temperature have profound effects on extraction rate, and empirical; relationships to extraction time have been observed. While these factors are easily controlled by adept operators, general lack of understanding and appreciation often exist in practice. Moisture content of the flakes is another factor affecting rate of solvent extraction. In most cases, 9 to 11 per cent moisture is ideal. Hexane and water are immiscible, and higher moisture contents interfere with the penetration of hexane. Lower moisture levels reduce the structural strength of the flakes, leading to the production of additional fines.

To reduce the amount of solvent used in extractors, countercurrent flow of the solvent to the flakes is used. That is, the freshest flakes contact the oldest solvent and progress through the process until nearly oil-free flakes contact fresh solvent. Flakes enter the extractor through a plug vapor seal that allows the material to enter while keeping hexane vapors from escaping. The extractor is an enclosed vessel designed to wash, extract, and drain flakes.

Two principal types of extractor have been employed over the years: immersion extractors and percolation extractors. An *immersion extractor* immerses and soaks the material in solvent (an industrial example is the Hildebrandt U-tube extractor, and a laboratory example is the Soxhlet extractor). Generally, more solvent usage is required by immersion extractors. Few immersion extractors remain; percolation extractors now dominate. In a *percolation extractor,* the solvent percolates by gravity through a bed of material (a laboratory example of a percolation extractor is the Goldfisch). The solvent flows over the surface of the particles and diffuses through the material during its downward, circuitous travel.

Most modern extractors are of the percolation type. The extraction principles employed by most extractors are the same, but there are different methods of achieving counter current flow of solvent to flakes. The shallow bed, chain extractor is one of today's popular extractors and resembles a full-loop conveyor. Flakes fed into an inlet hopper are conveyed down the first leg of the loop, where they are washed with moderately dilute miscella to extract surface oil and. penetrate the cells. As the flake bed moves into the bottom horizontal section, full miscella is recycled through the bed for filtering, and then to a liquid cyclone for final fines removal and, finally, to the evaporation system. Flakes are conveyed counterclockwise, through a progressively more dilute series of miscella washes until a final wash with fresh solvent is used in the top horizontal section of the loop. The latter half of the top loop is for drainage, after which the marc (solvent-laden spent flakes) passes to the desolventizer. The following advantages are claimed: the shallow bed (usually < 1 m) promotes drainage and, thus, low solvent carryover to the meal desolventizer and uniform contact of the flakes with solvent, since the bed is turned over while going up the right vertical leg.

Today, hexane is the solvent of choice, although many other solvents were used during development of solvent extraction technologies. Alternative extraction solvents were extensively reviewed in the early 1980s, and few practical developments have occurred since that time. The primary disadvantage of hexane is its flammability; in addition, its availability and price structure are tied to petroleum prices.

Flakes are conveyed to the extractor, where they are contacted with a countercurrent, flow of hexane and extracted for 30 to 60 minutes. Generally, less than 1 per cent residual oil in the extracted material is achieved, and the amount is lower for soybeans (about 0.5 per cent). The miscella contains 22 to 30 per cent oil, and the solvent is separated from the oil by distillation and stripping columns. The basic process is to heat the miscella under vacuum to evaporate the solvent. This step is usually done in two stages of evaporators. The first-stage evaporator concentrates the oil to about 90 per cent and uses reclaimed heat from heated solvent vapors from meal desolventization. Steam is used to heat the second-stage evaporator, where the oil is concentrated to better than 99 per cent. Most of the remaining solvent is removed in a disk-and- doughnut stripping column where evaporation is promoted by means of heat, vacuum (450–500 mmHg absolute pressure), and steam sparging. Crude oil leaves the stripping column with less than 0.15 per cent moisture and hexane. Trading specifications require the oil to have a flash point greater than 250°C, which is equivalent to no more than 800 ppm of hexane.

The marc generally contains 30 to 32 per cent solvent holdup, which must be recovered and recycled. Heat must be used to evaporate the solvent holdup from the meal. Live steam is also injected to aid heat transfer and to provide moisture vapor to strip the solvent. Regardless of the type of extractor used, the extracted flakes ("spent flakes") must be drained of the solvent held up by the material. The solvent that will not drain is referred to as solvent holdup, and as indicated earlier, the solvent-laden flakes are called marc. Solvent holdup should be minimized because this solvent must be removed by evaporation with heat. Greater solvent holdup increases the energy required for desolventizing the meal.

Toasting is often needed for feed meals to efficiently denature trypsin inhibitors and the enzyme urease (soybeans), to bind gossypol to protein (cottonseed), and to improve protein digestibility. Of course, none of these objectives can be obtained without considerable denaturation of protein and the accompanying loss of water solubility by the protein. However, depending on the method used, meals with great differences in protein solubilities or dispersibilities can be produced.

The first two employ live steam injection through nozzles within the sweep arms to evaporate the majority of the solvent. Meal advances down through the trays, and a series of gates and floats serves to control the levels in each tray. The lower four trays are essentially toasting/drying sections, where the meal is held at a minimum temperature of 100°C, and the meal is dried to a value suitable for dryers that follow the DT. Drying at normal DT conditions to less than 17 per cent moisture is detrimental to available lysine. However, meal should not leave the DT at more than 22 per cent moisture, for this would result in prohibitive drying energy requirements.

In recent years, the Schumacher-type desolventizer/toaster/dryer/cooler has become widely accepted. This device consists of four trays: the top tray is for predesolventizing; the second for desolventizing-toasting with injection of steam through its perforated bottom (achieving countercurrent use of steam relative to solvent evaporation); the third for drying, with hot air blown through its perforated bottom; and the fourth tray is for cooling, by blowing cold air through its perforated bottom.

The flash desolventizer and the vapor desolventizer were developed to reduce protein denaturation and produce highly soluble protein food ingredients (*e.g.*, protein isolates) from soybeans. Integrating these systems with cooking systems produces edible protein flours having a broad spectrum of protein dispersibility characteristics. The system includes a desolventizing tube, a flake separator, a circulating blower, and a vapor heater. These units are arranged in a closed loop in which hexane vapor is superheated under pressure and continuously circulated. Solvent-laden flakes, usually from dehulled soybeans, are fed into the system and conveyed by the high velocity circulating vapor stream. The turbulent

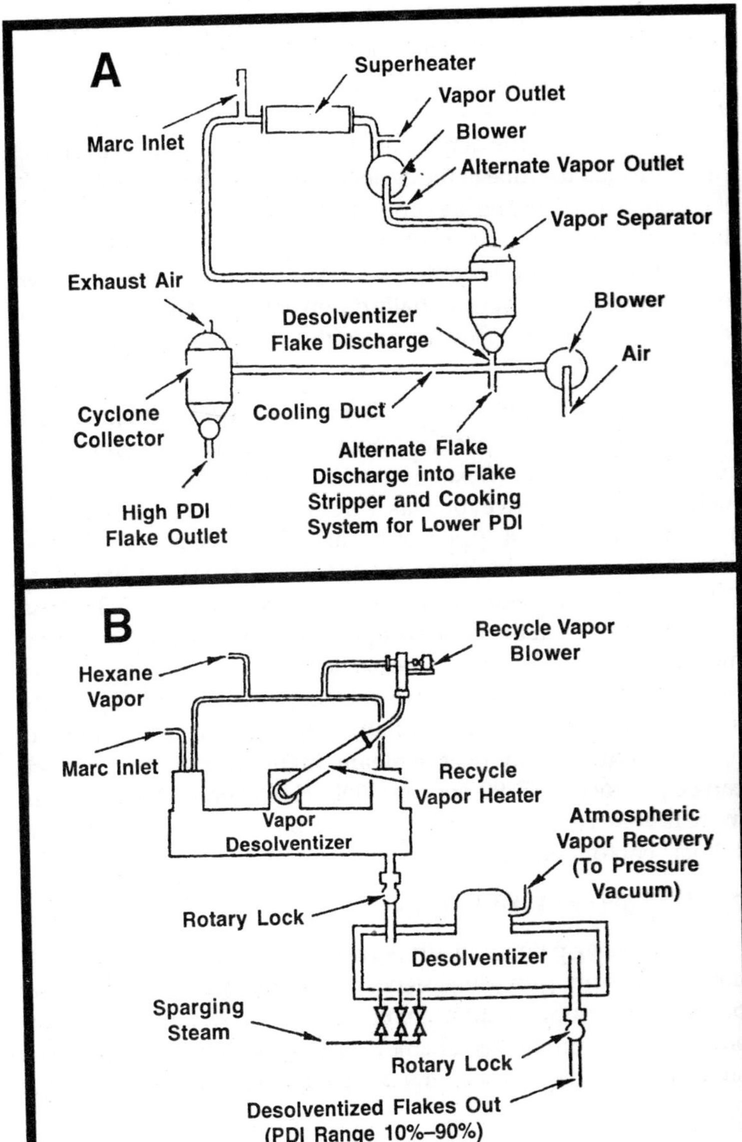

Figure 7.3: Setups for (A) Flash Desolventizing and (B) Vapor Desolventizing

superheated vapor flow (157–166°C) elevates the temperature of the flakes to 77 to 88°C, well above the boiling point of hexane (65°C) in less than 3 seconds. Because the flakes enter the flash desolventizer at low moisture for a very short period and no steam is injected into the vapor stream, little denaturation of protein occurs. As the flakes travel through the tube to the cyclone separator, the greatest portion of the entrained hexane is evaporated. At this point, if care is taken during conditioning, the protein dispersibility index (PDI) of the meal protein will be within 2 to 5 per cent of the untreated seed (native protein). The substantially desolventized flakes are removed from the system through a cyclone with a vapor-tight, rotary airlock and go to deodorizers.

Vapor desolventizing is similar to flash desolventizing in that superheated hexane vapor furnishes the required heat energy. Flakes are contacted with hot hexane vapor in a horizontal drum equipped with an agitator /conveyor. Flakes from either system usually enter a deodorizer to be stripped of hexane traces using only indirect heat. A slow moving agitator gently tumbles the spent flakes. The PDI may be further reduced by up to 10 percentage units. The final PDI is controlled at the flake stripper. Sparge steam may be used to minimize solvent loss and produce low PDI products (50– 65 per cent PDI). If only indirect steam is used, medium range PDI products are produced (60–75 per cent PDI). If the stripper is bypassed or operated without any heat or steam, highly dispersible products can be produced (75–90 per cent PDI). However, as PDI increases, more hexane remains with the flakes as they exit the system: 0.5 to 1.2 per cent for high PDI products.

7.4 Important Edible Oils

Oil-bearing palms and olives are two examples of commercially important fruits providing important edible oils. Palm fruit can provide two distinctly different fats, one from the fleshy mesocarp and the other from the seed kernels. Olives and palms are processed slightly differently. However, only palm oil production system is being discussed.

Palm trees are perennials grown on plantations particularly in Malaysia. It takes 3 years for the plantings to mature sufficiently to bear fruit, and they produce for about 25 years. Palm fruits are hand harvested and immediately transported to the mill, where they are

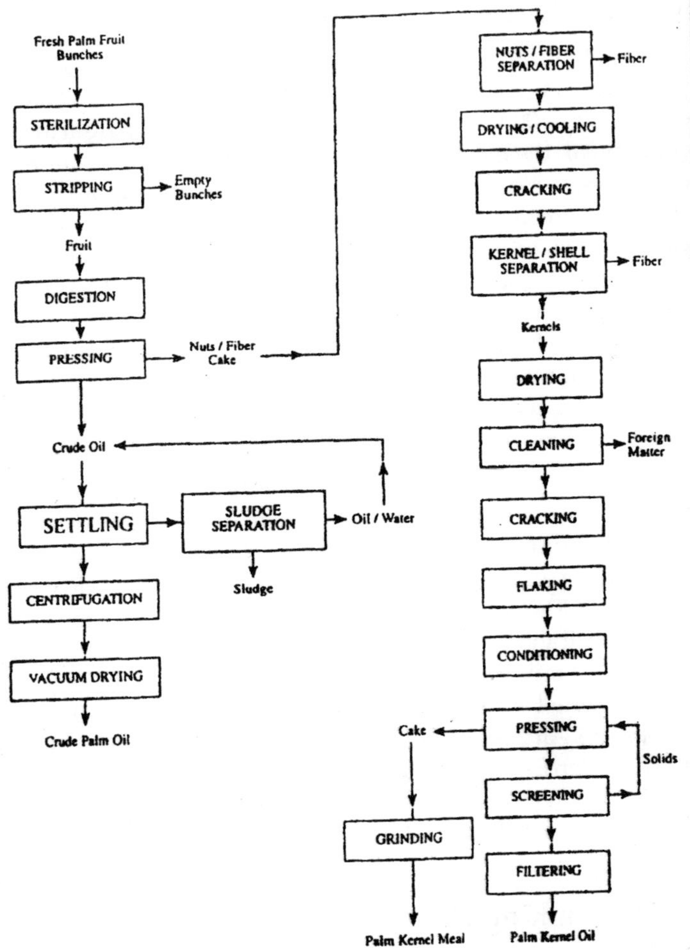

Figure 7.4: Steps in Processing Palm Fruits

quickly "sterilized." Fruits are processed within hours after harvesting because the oil begins to degrade immediately after harvesting. The fruits are sterilized by heating under steam pressure (145°C) for 1 hour to inactivate the enzyme lipase, which otherwise would quickly hydrolyze the oil and increase refining loss.

Sterilization also aids in stripping the fruit from bunch stalks and preconditioning the material for subsequent steps.

The sterilized fruits are stripped from bunch stalks with drum-type strippers. The material is then sent to a digester, where it is reheated (95–100°C for 20 min) to loosen the pericarp from the nuts and to break the oil cells. The material is conveyed to continuous screw presses similar to, but not quite the same as, those used for oilseeds, to extract oil from the fruit flesh, but not the kernel. The liquid extract, or *press liquor,* from the screw press contains about two-thirds oil, one-quarter water, and one-tenth solids. The press cake contains fruit flesh fiber and nuts. Water must be added to the press liquor to facilitate satisfactory settling of solids after screening, an operation referred to as *clarification.* Oil is skimmed off the top and passed to a centrifuge (clarifier) and then to a vacuum dryer. The crude oil is cooled and placed in storage.

The press cake is conveyed by means of a breaking conveyor to an aspirator (a vertical air column), where the nuts fall to a rotating polishing drum at the bottom, and the fruit fiber is blown to a cyclone, where it is separated from discharge air. The fiber is used to fuel the steam boiler. The nuts are conditioned by drying to loosen the kernels from the shell and by cooling to harden the shell. The nuts are cracked in an impact mill into two or more pieces. The shells are separated from the kernel with winnowing columns and by hydrocloning or clay bathing. All three of these operations separate shells from kernels based on density differences. The kernels are dried and screw-pressed or solvent-extracted to produce palm kernel oil and meal. For every 10 tons of palm oil produced, one ton of palm kernel oil is produced.

7.5 Degumming and Alkali Refining

Degumming is a water-washing process for the removal of phosphatides. Degumming may be conducted either as a separate operation or simultaneously with neutralisation.

Hydration makes phosphatides insoluble in oil, and they precipitate, yielding an oil that is unattractive because of unsightly sludge or "gums." Phosphatides can degrade and cause dark colors when the oil is heated as in the later deodorization step. All soybean oil in the export trade is degummed. Phosphatides are also surfactants and, if present in frying oils, can cause dangerous

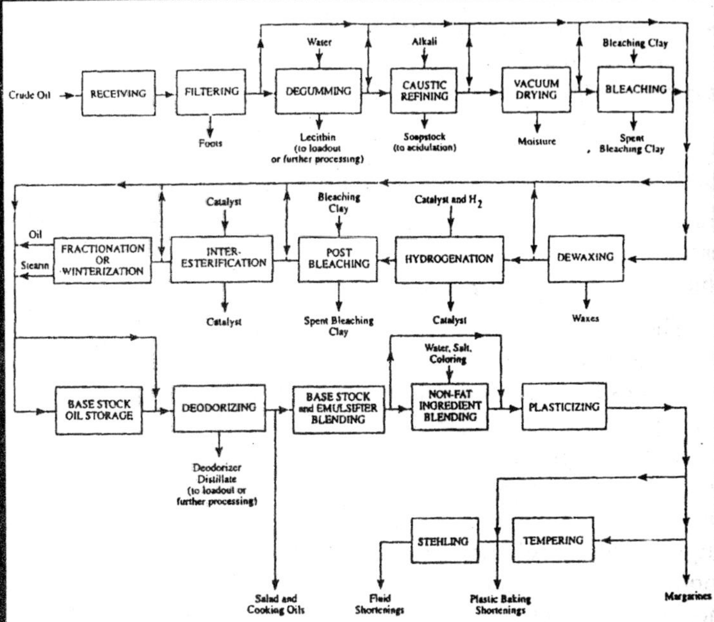

Figure 7.5: Steps in Refining and Converting Edible Oils

foaming. (When hot oil foams up and spills over the rim of a cooking vessel, it may burn the user; if it contacts a flame, it will catch fire.) Phosphatides also known as lecithin, are important food emulsifiers and, in the case of soybean oil, often times become economical to recover.

The gums are rendered insoluble in oil by hydrating them with 1 to 3 per cent water. As a general rule, the amount of water used should be equivalent to the hydratable phosphatide content of the oil. If single-bleached lecithin is to be recovered, hydrogen peroxide may be added to the water; if double-bleached lecithin is to be recovered, benzoyl peroxide may also be added. Alternatively, the degummed lecithin alone may be treated with hydrogen peroxide or both. The mixture is intensively mixed and then agitated for 30 to 60 minutes at 60 to 80°C in a slow mixing vessel (hydration tank) to allow the phosphatides to become fully hydrated and the hydrated micelles (gum particles) to coalesce. Hydration of the phosphatides is not instantaneous, and adequate time must be allowed. Higher temperatures solubilize more phosphatides, and lower temperatures increase oil viscosity; both extremes reduce the efficacy of degumming. The gums, being more dense than oil, can be removed by settling or filtering; more often, however, they are centrifuged out. The "wet" degummed oil is either dried (as described later) or immediately neutralized. Usually, about 90 per cent of the phosphatides are removed by this process. The gums typically contain 25 per cent water and 75 per cent oil-soluble substances (of which one-third is neutral oil).

There are both hydratable and nonhydratable phosphatides. Of the 1 to 3 per cent phosphatides in soybean oil, 0.2 to 0.8 per cent are generally regarded nonhydratable. The phosphatides are composed of phosphatidylcholine, phosphatidylinositol, phosphatidylethanolamine, and phosphatidic acid. The first two are always hydratable; but, the latter two can complex with divalent metal ions, rendering them nonhydratable. "Acid degumming" and "superdegumming" make more of the phosphatides hydratable. Nonhydratable phosphatides remain oil soluble. The nonhydratable phosphatides are believed to be calcium and magnesium salts of phosphatidylethanolamine and, especially, phosphatidic acid that arise from the enzymatic action of phospholipases when the cellular structure of the seed is damaged. Nonhydratable phosphatides are particularly problematic in soybeans.

Acid degumming is an improvement over conventional degumming described above and has become the usual practice in the U.S. soybean industry. A small amount (0.05–0.2 per cent) of concentrated phosphoric acid (75 per cent) is added to warm oil (70°C) followed by stirring for 5 to 30 minutes and degumming as described in connection with conventional degumming. Longer mixing times are often substituted for lower reaction temperatures. Phosphoric acid is added to make the phosphatides more hydratable by binding calcium and magnesium ions before adding water. Phosphoric acid pretreatment also partially removes chlorophyll from the oil.

Phosphatide content varies widely in vegetable oils but is the highest in crude soybean oil. Soybean oil is the only oil that is regularly degummed. The use of expanders in preparing soybeans for extraction almost doubles the usual phosphatide content of soybean oil. Only about half the phosphatide content of soybeans is extracted with hexane when soybeans are prepared by flaking alone. Sometimes, the degumming operation is conducted at the mill so that the gums may be added back to the meal. Gums contribute digestible energy to livestock. The available U.S. supply of soy lecithin is about twice the volume that can be economically sold. The gums for lecithin production are dried and may be further purified and/or bleached. Soybean lecithin is a mixture of about 40 per cent phosphatides (16 per cent phosphatidylcholine, 14 per cent phosphatidylethanolamine, and 10 per cent phosphatidylinositol), 35 per cent oil, 17 per cent phytoglycolipids, 7 per cent carbohydrate, and 1 per cent moisture.

Recently superdegumming processes have been developed in which more of the phosphatides are rendered hydratable. A strong solution of citric acid is added to warm oil (70°C), and the mixture is stirred and cooled to 25°C to precondition the gums. Then water is added with stirring for an additional 3 hours to hydrate the gums. This process causes the phosphatides to form liquid phospholipid crystals, which are easily removed during centrifugation.

Another variation of degumming, dry degumming, is occasionally applied to oils relatively low in phosphatide content, such as palm, coconut, and peanut oils. The oil is treated with concentrated acids to agglomerate the gums, which are then separated from the oil by being adsorbed to bleaching earth during subsequent steps of bleaching and filtration.

Now let us learn about alkali refining of oils. This process is generally known as neutralization. The term "neutralization" comes from the processing that is applied to an oil to eliminate the acidity emanating from the presence of free fatty acids.

Some use the term "refining" to refer only to neutralization. Neutralization is the most important operation in refining edible oils. An improperly neutralized oil will present problems in subsequent refining steps of bleaching and deodorizing, and in conversion operations of hydrogenation and interesterfication.

Neutralization is achieved by reacting the free fatty acid with caustic soda (sodium hydroxide) to form a soap referred to as soapstock. "Saponification" is the term that describes reactions between glycerides and sodium hydroxide to also form soaps.

Table 7.2: Properties of Some Crude and Refined, Bleached, Deodorized (RBD) Oils

Properties	*Oils*									
	Soybean		*Cottonseed*		*Canola*		*Palm*		*Sunflower Seed*	
	Crude	*RBD*	*Crude*	*RBD*	*Crude*	*RBD*	*Crude*	*RBD*	*Crude*	*RBD*
Triglycerides (%)	95–97	> 99	NA	> 99	NA	99	NA	> 99	NA	> 99
Phosphatides (%)	1.5–2.5	0.003–0.045	0.7–0.9	NA	2.7–3.5	NA	0.006–0.013	0.012	0.5–1.0	NA
Unsaponifiable matter (%)	1.6	0.3	NA	NA	0.5–1.2	NA	NA	NA	< 1.3	NA
Plant sterols	0.33	0.13	0.37	NA	NA	NA	0.036–0.062	0.011–0.016	NA	NA
Tocopherols	0.15–0.21	0.11–0.18	0.11	0.06	0.06	NA	0.06–0.10	0.04–0.06	0.05	NA
Hydrocarbons (squalene)	0.014	0.01	NA	NA	NA	NA	0.02–0.05	NA	NA	NA
Free fatty acids (%)	0.3–0.7	< 0.05	0.9–37	< 0.05	0.4–1.0	< 0.05	2.0–5.0	< 0.10	0.8–2.4	< 0.05
Trace metals										
Iron (ppm)	1–3	0.1–0.3	NA	NA	1.5	< 0.1	5–10	0.12	NA	NA
Copper (ppm)	0.03–0.05	0.02–0.06	NA	NA	0.10	< 0.01	0.05	0.05	NA	NA

Neutralization must be done correctly or some of the glycerides will be saponified, resulting in an increase of refining loss. The oil, low in acid value, is termed neutral oil. Removal of the soapstock must also be carefully done to prevent high losses of entrained neutral oil, a second means of increasing refining loss.

Soapstock is a coproduct of refineries in that it can be acidulated with sulfuric acid to produce a salable product. Once reacidified, the fatty acids (95 per cent fatty acids) will separate in settling basins as 35 to 40 per cent free liquid, so-called acid oil, from an emulsified layer (high in phosphatides) and a water layer. Most acid oil is used as a high energy ingredient in livestock feed, but when market prices are attractive, it is sold to fatty acid producers, who distill it to produce feedstocks for various oleochemicals (*e.g.*, surfactants).

In the case of cottonseed oil, proper and timely neutralization is important to achieve adequate removal of gossypol and oil that is low in red color. For reasons that are not clear, gossypol is adsorbed onto soapstock particles even though gossypol is unsaponifiable.

The amount and strength of sodium hydroxide used depend on the amount of free fatty acids present in the oil. Nearly all oils other than soybean and rapeseed oils are simultaneously degummed with neutralization. Free fatty acids form water-soluble sodium soaps, and any phosphatides become hydrated and water insoluble. The amount of sodium hydroxide used is termed "treat." The proper treat produces adequately refined oil with the lowest refining loss. Excessive treat can saponify triglycerides and reduce the yield of refined oil. The proper treat is determined by titrating the oil to determine the free fatty acid content and using industry tables, such as those published in the *Official Methods and Recommended Practices of the American Oil Chemists' Society.*

Proper neutralization is dependent on proper amount of sodium hydroxide, proper mixing, proper temperature, adequate contact time, and efficient separation. As in acid degumming, some oils are preconditioned with phosphoric acid. That is, prior to neutralizing the oil is treated with 0.02 to 0.5 per cent phosphoric acid at 60 to 90°C for 15 to 30 minutes, making the phosphatides less soluble in the oil and easier to remove. The proper amount of caustic is proportionately metered into the warm oil stream with good mixing and sent to retention, or dwell, mixers (5–10 min mixing time). The emulsion is then thermally shocked by heating to about 75°C to

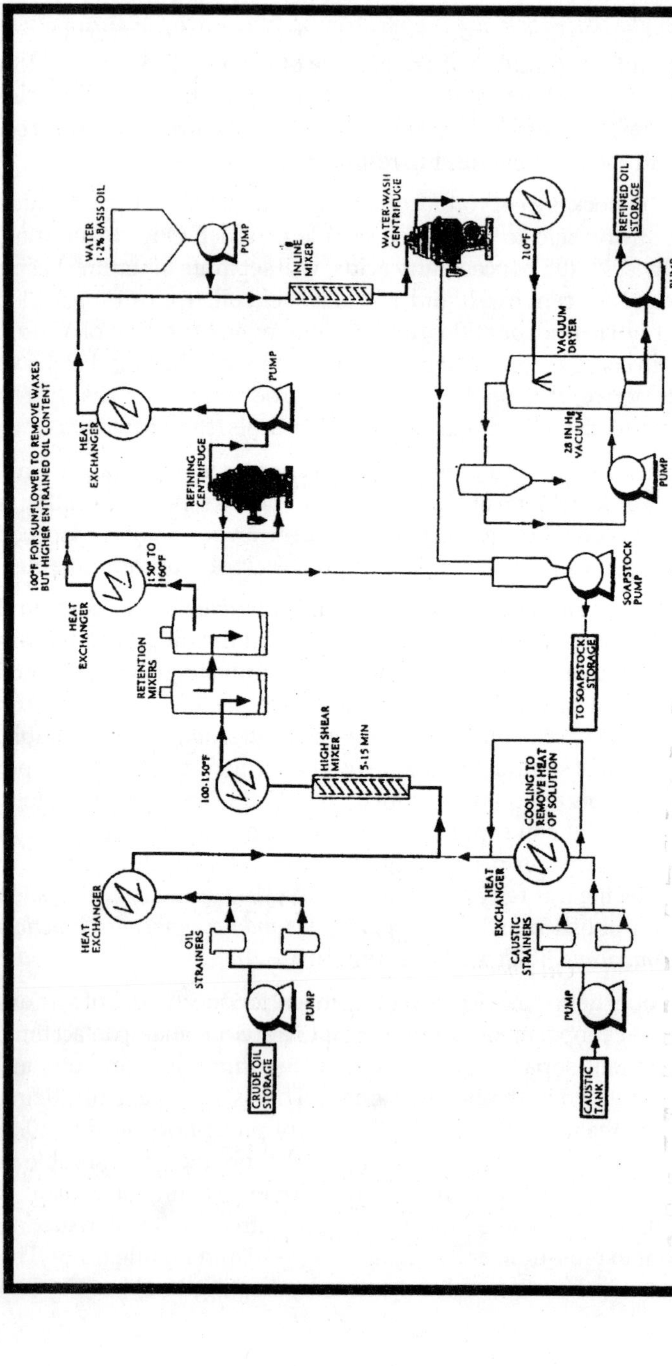

Figure 7.6: Process Flow Sheet for Alkali Refining

break out the soapstock. Soapstock is removed from the oil by means of continuous, disk-type centrifuges. Refined oil is then washed with soft water (10–20 per cent) at 90°C and recentrifuged to complete the removal of 90 per cent of the soap.

Alkali refining or neutralizing in the presence of hexane is known as miscella refining. In the case of cottonseed, it is desirable to carry out alkali refining as quickly as possible after extraction at the plant; otherwise, gossypol may become fixed in the oil, hence, unremovable. Also, carrying out alkali refining in the presence of hexane reduces the viscosity of the oil phase and increases the density difference between the miscella phase and the water/soap phase. This improves the separation efficiency and reduces refining loss. Usually the oil content of the miscella is concentrated to 40 to 60 per cent oil. The oil is mixed with sodium hydroxide with high shear mixers, sometimes using high pressure piston pumps with, homogenizing valves. The mixture is heated to 65°C to melt the soapstock and then cooled to 45°C, and the aqueous and oil phases are separated by centrifuging. Water washing is not required in miscella refining. The neutral oil miscella must then be evaporated and the oil stripped, dried, and cooled. Miscella refining produces oil with better color. The soapstock is usually added back to the marc and contributes digestible energy to the meal.

The oil then is marketed as "once-refined oil." Once-refined oil is re-refined like any other oil, but much less caustic is required because most of the free fatty acid content has already been neutralized. Although any oil can be subjected to miscella refining, the additional capital investment in the safety features required for centrifuges that work with hexane is justified only for cottonseed.

7.6 Dewaxing and Deodorization

Waxes can be removed by cooling the oil to 6 to 8°C, and filtering or centrifuging at these cold temperatures, a process similar to winterization.

Waxes can harm the appearance of bottled oil by causing unsightly cloudiness or sediments. Corn, rice bran, safflower, sesame, and sunflower seed oils are notorious for problematic high wax contents (02–3.0 per cent) and must undergo dewaxing; occasionally canola oil also has wax problems.

Sometimes, dewaxing is accomplished simultaneously with removal of the gums and/or soapstock by carrying out the centrifugation at cool temperatures. Sunflower seed oil is often predewaxed (from 1500 ppm to 400 ppm) by cooling to 25°C for 24 hours and then degumming with a centrifuge at this temperature. Alternatively, sunflower seed oil may be simultaneously dewaxed and alkali-refined. After neutralization, as already described, the oil-soapstock mixture is cooled to 5 to 8°C and held there for 4 to 5 hours under gentle mixing; then the oil is mixed with 4 to 6 per cent of water heated to 18°C. The soapy water phase wets and causes the small wax crystals to form a heavy suspension in soapy water. The soapy suspension is centrifuged to produce a wax/soapstock fraction and a refined, dewaxed oil. Sometimes, sodium lauryl sulfate is added to help wet the crystals.

The final step in refining fats and oils is deodorization. Oils that are converted by a variety of processes are done so before deodorizing. The primary objective of deodorization is to remove compounds responsible for undesirable odors and flavors, such as residual free fatty acids (especially low molecular weight fatty acids), aldehydes, ketones, and alcohols. Deodorization also removes peroxide decomposition products: freshly deodorized oil should have a peroxide value of zero and free fatty acid content of less than 0.03 per cent. These compounds are more volatile than triglycerides and are preferentially removed. Lard is one fat that is often not deodorized, since the better grades have mild, unobjectionable flavors. Some losses of monoglycerides, sterols, sterol esters, tocopherols, and other natural antioxidants also result. The tocopherols are potent natural antioxidants and contribute significantly to the greater oxidative stability of crude oil compared to deodorized oil.

Deodorization is essentially steam distillation performed at high temperatures (180– 270°C) and under high vacuum (3–8 mmHg absolute pressure). Steam is sparged to carry away the volatiles and to provide agitation. Vacuum is usually provided by three to five stages of steam ejectors connected in series. Because deodorization is a mass transfer operation, deodorizers are designed to provide large surface areas and shallow oil depths. Only minor amounts of triglycerides are lost. The concentration of materials to be removed is in the range of 0.1 to 1 per cent for most oils, and the original

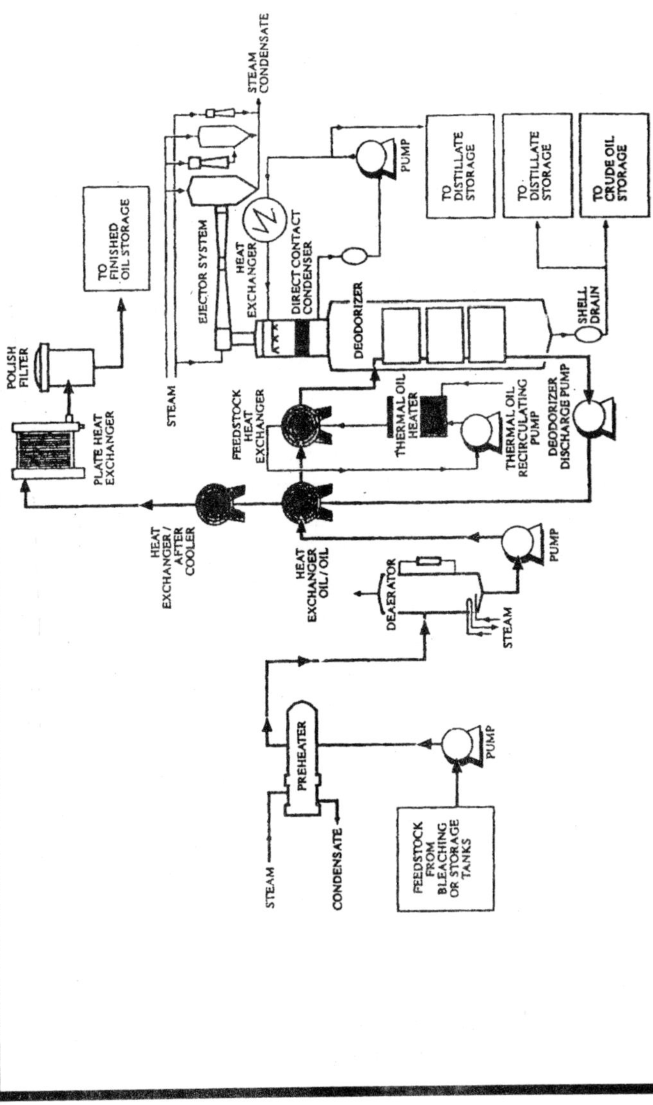

Figure 7.7: Process Flow Sheet for Deodorization

values should be reduced by over 99 per cent. The usual loss of oil weight is 0.2 to 0.8 per cent. The amount of stripping steam ranges from 10 to 50 kg of steam per 100 kg of oil.

Factors that affect the efficacy of deodorization are the vapor pressures of the materials to be removed, the product flow rate, the intimacy of steam mixing with the oil, the absolute pressure achieved during deodorization, the temperature of deodorization (which controls the vapor pressure of the materials being removed), the sparge steam rate, and the time of deodorization.

Usually steam alone is not used to heat the oil because very high steam pressures would be required to heat the oil sufficiently. Rather, a eutectic mixture of diphenyl and diphenyl oxide, known by the trade name Dowtherm A, is used. This product has a boiling point of 258°C and at 304°C generates a pressure of only 110 kPa.

Deodorization may be conducted in batch, semibatch, or continuous vessels. The type of process largely depends on the volume and number of different products being processed. Batch deodorizers have cycle time of 6 to 8 hours. Continuous deodorizers are most suitable when a limited number of products are manufactured in very large volume. There are five stages in deodorization: deaeration, heating, deodorization/steam stripping, heat recovery/cooling, and final cooling. Stripping steam is provided through sparging rings and airlift pumps. After the deodorized oil has cooled, a small amount (0.005–0.01 per cent) of citric acid is added to chelate metal cations, which otherwise would promote oxidation and reduce shelf life. Crude oils usually have greater oxidative stability than refined oils. Indeed, many processors do not expose deodorized fats and oils to the atmosphere, but discharge oil into tanks blanketed with nitrogen and fill bottled oil under nitrogen.

Deodorization also removes any residual pesticides and hexane. Some pigments, such as β-carotene, are destroyed by the high heat in deodorization, and thus, the yellow and sometimes red colors are reduced. Deodorized oils have improved flavor, odor, and color. The high temperatures used in deodorization cause limited geometric isomerization. Although deodorization removes most peroxides, it cannot reclaim rancid oxidized oils.

Deodorizer distillate is condensed, recovered, and sold at higher prices per pound than the oil itself. Soybean deodorizer distillate typically contains 12.3 per cent tocopherols and 21.9 per cent sterols. Deodorizer distillate may be further processed into valuable fractions rich in tocopherols (vitamin E), which are in high demand by the food and pharmaceutical industries. The sterols may also be purified and sold into the pharmaceutical industry for manufacturing various synthetic hormones.

7.7 Physical Refining and Conversion of Oil

Physical refining is also known as steam refining. These terms are applied to the removal of the free fatty acids from the oil rather than reacting them with alkali, as well as to the removal of the compounds normally targeted by deodorization. Physical refining combines neutralization and deodorization into one operation. Physical refining is always preceded by degumming and bleaching steps. The major advantage of physical refining is that the yield of oil is improved because there is none of the neutral oil loss that accompanies the production of soapstock. This process also affords the possibility of recovering fatty acids for the oleochemical industry without acidulating soapstock and attendant wastewater production. The equipment used for physical refining is similar to deodorization, but with additional steam sparging trays.

Not all oils are suitable for physical refining. The oil must be low in phosphorus content (most comes from phosphatides). The exact upper limit for phosphorus in oil suitable for physical refining is a matter of long-standing controversy, but it may be less than 5 ppm. Higher levels result in dark colored oils. Physical refining is also more appropriate for oils higher in free fatty acid content because the higher yield justifies the cost of the process. Most regard soybean oil as being unsuitable for physical refining, or at best difficult to refine, while palm, rice bran and coconut oils and animal fats are well suited to physical refining.

Often times, it is desirable or necessary to convert or transform liquid refined oils high in unsaturation into more saturated forms; there may be a requirement for greater oxidative stability or altered physical forms of plastic and solid fats, or for less saturated forms to provide greater physical stability at cold temperatures. The term "conversion" is often restricted to mean only hydrogenation, a process by which hydrogen is added to highly reactive, unsaturated, double carbon-carbon bonds. However, winterization, fractional

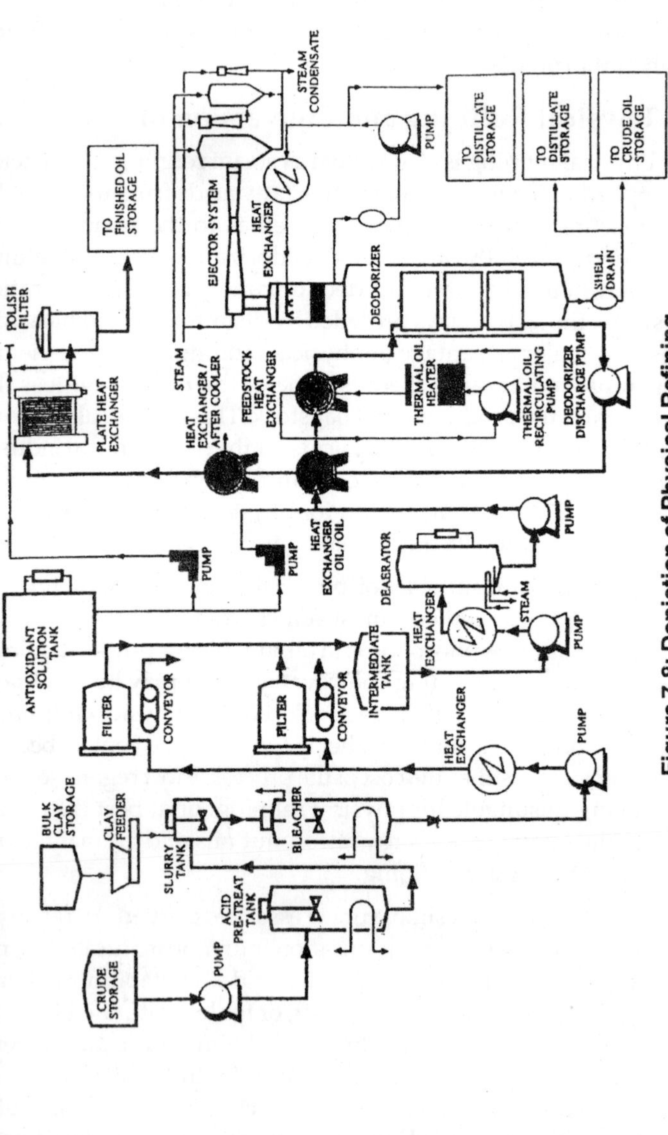

Figure 7.8: Depiction of Physical Refining

crystallization, and interesterification, particularly directed interesterification, should also be regarded as modes of conversion, since these processes can also significantly change oxidative stability, physical stability, and functional properties of fats and oils, and their fractions. Stehling, votating, and tempering might also be considered to be conversion processes, in as much as they are applied to stabilize fluid and plastic fats.

7.8 Hydrogenation Reaction Mechanism

The hydrogenation process was first applied to whale oils in 1903. This process was used earlier to hydrogenate chemical feedstocks.

In 1909 hydrogenation was patented for use in a lard substitute, shortening produced from cottonseed oil. Hydrogenation is used for two purposes: to improve oxidative stability (by hydrogenating some of the double bonds to saturated ones) and to convert liquid oils or soft fats into plastic or hard fats (facilitating uses for which less saturated forms are unsatisfactory). Liquid oils are converted into shortenings and margarine fats; oils prone to rapid oxidation, such as soybean oil, are partially hydrogenated for use as salad and frying oils.

Unsaturated double bonds are converted to saturated bonds by addition of hydrogen (H_2). The reaction between the liquid oil and H_2 gas is accelerated by using a suitable solid catalyst; thus the reaction is "heterogeneous," involving three phases. Hydrogenation is exothermic, and heats about 1.7°C per unit drop in iodine value (IV).

For successful hydrogenation, many of the crude oil impurities must be removed, and refining operations must be correctly carried out beforehand. Many of the contaminants (soaps, gums, sulfur, magnesium, potassium, chromium, zinc, and mercury) can "poison" the catalyst, reducing its activity. Canola and rapeseed oils are particularly notorious for high levels of natural sulfur content, which can cause problems in hydrogenation.

Both batch and continuous processes are used. For hydrogenation to occur, gaseous H_2 liquid oil, and solid catalyst must be brought together at a suitable temperature. Thus, hydrogenation is a mass transfer issue, and mass transport of reactants is the rate-limiting factor. Hydrogen is first dispersed as pubbles; then it must dissolve in the bulk oil, diffuse to the catalyst particle, and from there diffuse to the catalyst surface. Triglycerides also must diffuse to the catalyst surface, receive the hydrogen, and then diffuse out into the bulk oil. Therefore, a higher rate of reaction is achieved by raising the temperature and increasing agitation to disperse the gas in bubbles as small as possible. Small bubbles have more surface area for transfer per unit of mass for diffusion into the bulk oil. Agitation also keeps the film thicknesses small. Increasing gas flow rate helps to keep dissolved H_2 at high levels. The reaction is carried out in pressurized factors because increased pressure increases the saturation concentration of H_2 hence, the driving force for dissolution. Other factors that influence the reaction are the quantity and activity of the catalyst.

Pressurized reaction vessels are also used to contain the H_2, which is highly explosive and must be prevented from possible ignition in the presence of air (oxygen). The reaction in the pressure vessel is normally conducted at 250 to 300°C over 40 to 60 minutes. The catalyst is mixed with a small part of the oil at room temperature. The bulk of the oil is pumped into the reactor, a vacuum is established, deaerating the oil, and the oil is heated. Finally, the oil,

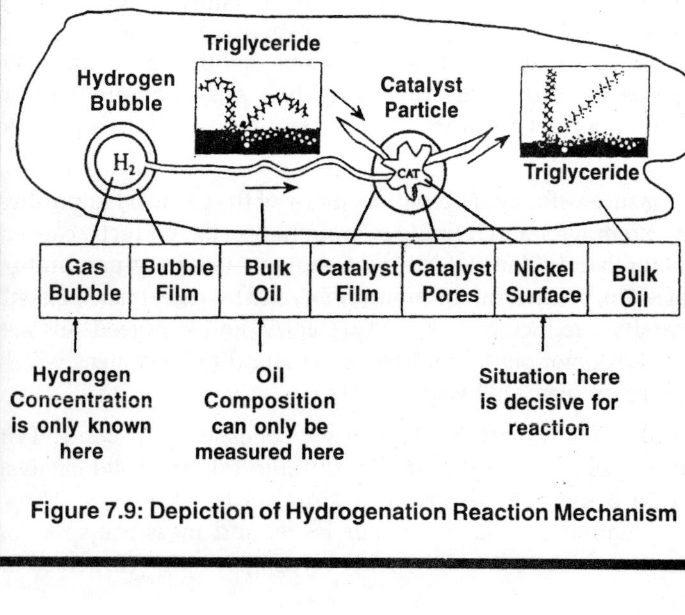

Figure 7.9: Depiction of Hydrogenation Reaction Mechanism

catalyst slurry is added. The extent of reaction is followed by monitoring the refractive index (RI) of the oil, which is directly related to IV.

A catalyst increases the rate of reaction without being consumed in the reaction. Small amounts of catalyst are effective; in the case of hydrogenation, the usual amount is in the range of 0.01 to 0.02 per cent of the weight of oil. Reduced nickel is the most widely used catalyst by the vegetable oils industry; but copper, platinum, palladium, and ruthenium are also effective. Nickel is the catalyst of choice because of its good activity, selectivity, filterability, reusability, and economical use. Copper has been occasionally used, but it must be completely removed to prevent accelerating oxidation. Since catalytic activity is a surface phenomenon, inducing pore formation increases surface area and thus, activity. However, the size and shapes of pores in the catalyst are important: pores must be at least 10 nm in diameter to accommodate the transfer of triglycerides into them.

Catalyst promoters are substances that enhance the activity of catalysts without having catalytic activity themselves. It is believed that the promoter function is structural; that is, these substances somehow permit a larger number of active sites on the catalyst particle. For this reason, hydrogenation catalysts are commonly supported on siliceous materials, aluminum oxide, chromium oxide, cobalt oxide, or copper oxide.

Catalysts are expensive, but they can be reused numerous times. Hydrogenation catalysts can also catalyze oxidation. For these reasons, it is important to efficiently remove the catalyst from the hydrogenated oil and then recover it, which is accomplished by filtration. While surface area per unit weight increases as the size of the catalyst particle decreases, so does the relative difficulty of removing the catalyst by filtration. For that reason, the catalyst is often incorporated onto a support, such as 17 to 25 per cent nickel fixed onto kieselguhr. The catalyst is removed with leaf filters, and the filtered, hydrogenated oil is bleached to assure removal of all catalyst before deodorizing. This process, though termed "postbleach," is carried out in the same manner as normal bleaching.

Chapter 8

Phenolic and Enzymatic Antioxidants

8.1 Introduction

The most commonly used primary antioxidants in foods are synthetic compounds. Antioxidants can be broadly classified by mechanism of action as primary antioxidants and secondary antioxidants.

Primary antioxidants are mono-or polyhydroxy phenols with various ring substitutions. Substitution with electron-donating groups ortho and para to the hydroxyl group of phenol increases the antioxidant activity of the compound by an inductive effect. These hindered phenolic antioxidants decrease the reactivity of the hydroxyl group by increasing its electron density, substitution with butyl or ethyl groups para to the hydroxyl enhances the antioxidant activity. Because of steric hindrance, however, the presence of longer chain or branched alkyl groups at the para positions can decrease antioxidant effectiveness. Substitutions of branched alkyl groups at ortho positions enhance the phenolic antioxidant's ability to form stable resonance structures and further reduce the antioxidant radical's ability to participate in propagation reactions.

The most commonly used primary antioxidants in foods are synthetic compounds. Examples of important primary phenolic antioxidants include butylated hydroxyanisole (BHA), butylated hydroxytoluene (BHT), propyl gallate (PG), and tertiary butylhydroquinone (TBHQ). However, a few natural components of food also act as primary antioxidants and are commonly added to foods. Tocopherols are the most commonly used natural primary antioxidants. Carotenoids are another group of natural compounds that have primary antioxidant activity, although the mechanism differs from the phenolics.

Secondary, preventive, or type 2, antioxidants act through numerous possible mechanisms. These antioxidants slow the rate of oxidation by several different actions, but they do not convert free radicals to more stable products. Secondary antioxidants can chelate prooxidant metals and deactivate them, replenish hydrogen to primary antioxidants, decompose hydroperoxides to non-radical species, deactivate singlet oxygen, absorb ultraviolet radiation, or act as oxygen scavengers. These antioxidants are often referred to as synergists because they promote the antioxidant activity of type I antioxidants. Citric acid, ascorbic acid, ascorbyl palmitate, lecithin, and tartaric acid are good examples of synergists. Some of the more important types of secondary antioxidant mechanism are discussed in the following sections.

Consumers are concerned about the safety of their food and about potential effects of synthetic additives on their health. Despite the superior efficacy, low cost, and high stability of synthetic antioxidants in foods, the suspicion that these compounds may act to promote carcinogenicity has led to a decrease in their use. A trend toward the use of "natural" food additives in the food industry has been apparent for quite some time–a result of consumer demand. Some natural preservatives exist inherently in foods; others can be added to the product or can arise as a result of processing or cooking. Natural food antioxidants such as citric acid and ascorbic acid are used widely in the food industry. Recent research has focused on isolation and identification of effective antioxidants of natural origin.

Synthetic antioxidants are intentionally added to foods to inhibit lipid oxidation. Synthetic antioxidants approved for use in food include BHA, BHT, PO (also octyl and dodecyl gallate), ethoxyquin, ascorbyl palmitate, and TBHQ. The synthesis of novel antioxidants for food use is limited by rising costs of research and development, costs associated with safety assessment, and the time required to obtain regulatory approval of additives. These restrictions, as well as growing consumer preference for natural food additives, has led industry to emphasize natural materials as a source of novel antioxidants.

Phenolic compounds represent some of the oldest and most frequently used antioxidants in foods. The differences in antioxidant activity of the phenolic antioxidants are due to variations in structure that directly influence physical properties. Phenols in which the aromatic ring contains alkyl groups (hindered phenols) like extremely effective antioxidants. Hindered phenols are also effective antimicrobials in foods. A food product's characteristics ultimately determine the selection of the phenolic antioxidant. BHA and BHT are fairly heat stable and are used in heat-processed foods. PG decomposes at 148°C and is inappropriate for high temperature processing. Therefore, heat-stable TBHQ is useful in frying applications. BHA and BHT are strongly lipophilic and are used extensively in oil-in-water emulsions. BHA and BHT are also typically used together in mixtures, acting synergistically.

Titles 9 and 21 of the U.S. Code of Federal Regulations (CFR) govern the use of antioxidants in meat and poultry products and in foods, respectively. Phenolic antioxidants are effective at low concentration and are often used at levels less than 0.01 per cent. The level of phenolic antioxidants permitted in the United States varies according to the product.

Commercial antioxidant preparations are available in solid and liquid blends. Liquid blends are convenient because the antioxidant is solubilized for easy addition during processing. Solvents include vegetable oils, propylene glycol, glyceryl monooleate, ethanol, and acetylated monoglycerides. Antioxidant preparations typically contain mixtures of phenolic antioxidants, a synergist and a solvent system. Some of the most important synthetic antioxidants are discussed in the following sections.

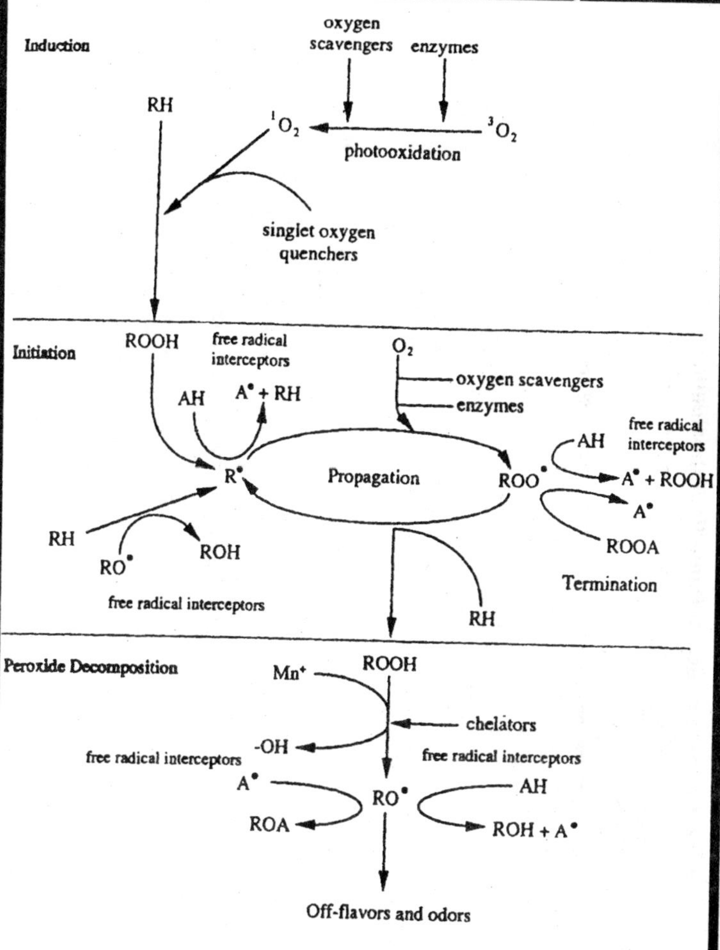

Figure 8.1: Overview of Lipid Oxidation and the Interaction of Antioxidants

8.2 Characteristics of Phenolic Antioxidants

Characteristics of phenolic antioxidants are described in Table 8.1.

Table 8.1: Characteristics of Phenolic Antioxidants

Compound	Molecular Weight	Appearance	Boiling Point (ºC)	Melting Point (ºC)	Solubility	Functionality	21CFR (9 CFR)
Butylated hydroxyanisol (BHA)	180.25	Waxy solid	264–270	48–63	Insoluble in water. Slightly soluble in glycerol and mineral oil. Soluble in fats, alcohol, propylene glycol, petroleum, ether; glyceryl monooleate; paraffin.	Good carry-through in baked and fried products. Volatile, distillable. Most effective in animal fats. Used in packaging materials. Synergistic with other antioxidants. Slight phenol odor	172.110 182.3169 172.615 172.515 165.175 166.110 164.110 (381.147) (318.7)
Butylated hydroxytoluene (BHT)	220.356	White crystals	265	70	Insoluble in water, glycerol, and propylene glycol. Slightly soluble in mineral oil. Soluble in fats, paraffin, glycerylmonooleate, alcohol, petroleum, ether, most organic solvents	Less expensive than BHA. Synergistic with BHA and other antioxidants. Volatile, distillable. Less carry-through in baked and fried products than BHA. More sterically hindered than BHA. Most effective in animal fats.	171.115 182.3173 172.615 137.350 165.175 166.110 164.110 (381.147)

Contd...

Table 8.1–Contd...

Compound	Molecular Weight	Appearance	Boiling Point (ºC)	Melting Point (ºC)	Solubility	Functionality	21CFR (9 CFR)
						Slight phenol odor. Poor carry-through properties. Synergistic with other antioxidants.	(318.7)
Propyl gallate (PG)	212.20	White crystals	Decomposes above 148	150	Slightly soluble in water, fats, mineral oil, glyceryl monooleate. Soluble in alcohol, glycerol, propylene glycol	Decomposes at frying temperatures. Stability of octyl and dodecyl forms greater than propyl. Discolors in presence of metals; always used in combination with a chelator. Less soluble in fats than BHA and BHT. More effective in vegetable oils than BHA and BHT Poor carry-through in baking, but good carry-through in frying. Synergistic with other antioxidants.	184.1660 172.615 165.175 166.110 164.110 (318.7) (381.47) 171.185 165.175 166.110 164.110 (318.7)

Contd...

Table 8.1–Contd...

Compound	Molecular Weight	Appearance	Boiling Point (ºC)	Melting Point (ºC)	Solubility	Functionality	21CFR (9 CFR)
Tertiary butyl-hydroquinone (TBHQ)	166.22	White to tan crystals	300	126.5–128.5	Slightly soluble in water. Moderately soluble in fats, propylene glycol, glyceryl monooleate. Soluble in alcohol	Excellent antioxidant in vegetable oils. Does not discolor in presence of metals. Little odor Poor carry-through in baking and frying.	(381.147)
Trihydroxybuty rophenone (THBP)	196	Tan powder	–	149–153	Slightly soluble in water. Moderately soluble in fats. Soluble in alcohol, propylene glycol, paraffin	Synergistic with other antioxidants. Used in packaging materials. Browns in presence of metals	172.190
Ethoxyquin	217.31	Yellow liquid	123–125		Insoluble in water. Soluble in most organic sol. vents	Used extensively in animal rations. Effective in pigment retention	172.140

8.3 The Antioxidative Mechanism

Extensive research has been dedicated to identification of antioxidative mechanism. First of all we will learn about natural antioxidants.

Ascorbic acid and tocopherols are the most important commercial natural antioxidants. Other sources of natural antioxidants include carotenoids, fiavonoids, amino acids, proteins, protein hydrolysates, Maillard reaction products, phospholipids, and sterols. Numerous naturally occurring phenolic antioxidants have been identified in plant sources and vegetable extracts. Enzymes also play important roles as antioxidants. Processing of foods can induce the formation of antioxidants. Maillard reaction products, protein hydrolysates, fermentation products, and nitrosyl compounds from curing have been reported to possess antioxidant activity. Natural antioxidants allow food processors to produce stable products with "clean" labels that tout all-natural ingredients. However, these products can have several drawbacks, including high usage levels, undesirable flavor and/or color contributions, and lack of stability due to low antioxidant efficiency.

The safety of natural antioxidants should not be taken for granted. Cautions must be heeded, since numerous natural products are potential carcinogens, mutagens, or teratogens, and the safety of many natural compounds with antioxidant activity has not been established. A case in point is nordihydroguaiaretic acid (NDGA), which was used extensively as an antioxidant earlier in this century. NDGA is a natural constituent of the creosote bush, which was removed from GRAS status when unfavorable toxicological results were reported. Regardless of the politics surrounding the issue of safety and "natural" additives, natural antioxidant products are commercially important and desired by the consumer. The main advantage of substances naturally present in foods is that the burden of proof of safety may be less rigorous than that required for synthetic products. No safety testing is required if the antioxidant is a natural constituent of GRAS ingredients. In addition, some natural antioxidants derived from spices, herbs, and Maillard reactions can be listed as navorants rather than antioxidants, a technical distinction that serves to exempt the substances from safety testing requirements.

Tocopherols and tocotrienols (Figure 8.2) comprise the group of chromanol homologs that possess vitamin E activity in the diet. They are natural monophenolic compounds with varying antioxidant activities. Eight naturally occurring homologs are included in the vitamin E family. They are fat-soluble 6-hydroxy chroman compounds. The α-, β-, γ-, and δ-tocopherols are characterized by a saturated side chain consisting of three isoprenoid

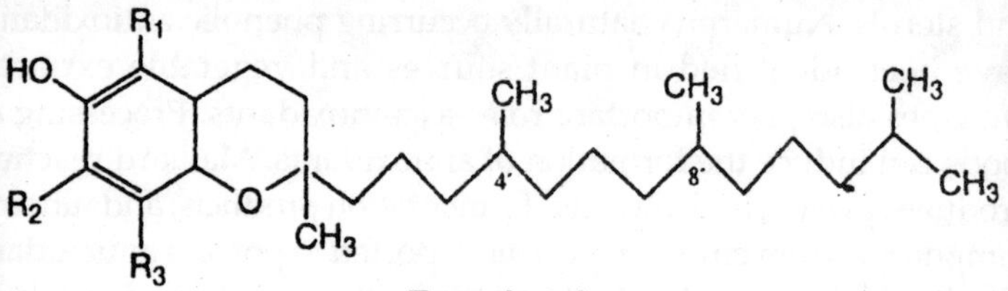

Tocopherols

Trivial Name	*Chemical Name*	*Abbreviation*	*Ring Position* R_1	R_2	R_3
α-Tocopherol	5,7,8-Trimethyltocol	α-T	CH_3	CH_3	CH_3
β-Tocopherol	5,8-Dimethyltocol	β-T	CH_3	H	CH_3
γ-Tocopherol	7,8-Dimethyltocol	γ-T	H	CH_3	CH_3
δ-Tocopherol	8-Methyltocol	δ-T	H	H	CH_3

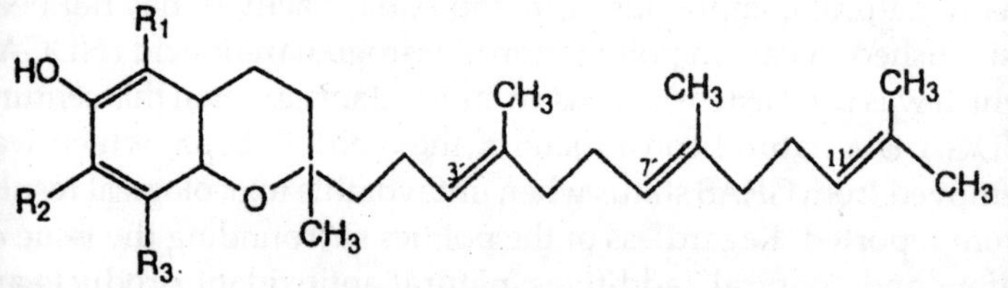

Tocotrienols

Trivial Name	*Chemical Name*	*Abbreviation*	*Ring Position* R_1	R_2	R_3
α-Tocopherol	5,7,8-Trimethyltocotrienol	α-T	CH_3	CH_3	CH_3
β-Tocopherol	5,8-Dimethyltocotrienol	β-T	CH_3	H	CH_3
γ-Tocopherol	7,8-Dimethyltocotrienol	γ-T	H	CH_3	CH_3
δ-Tocopherol	8-Methyltocotrienol	δ-T	H	H	CH_3

Figure 8.2: Structures of Tocopherols and Tocotrienols

units. The corresponding tocotrienols (α-, β-, γ-, and δ-) have double bonds at the 3, 7, and 11 positions of the isoprenoid side chain. Only *RRR*-isomers are found naturally. Synthetic α-tocopherol (all-*rac*-α-tocopherol) consists of eight stereoisomers found in equal amounts in the synthetic mixture. Biologically, *RRR*-α-tocopherol is the most active vitamin E homolog. Tocopherols and tocotrienols are widely distributed in the plant kingdom, with vegetable oils providing the most concentrated source of vitamin E. Tocotrienols are less common but are present in palm oil, rice bran oil, cereals, and legumes. Palm oil has a unique vitamin E profile, providing tocotrienols in higher concentrations than other food sources. Tocopherols and tocotrienols are retained throughout the edible oil refining process, although there is some loss during the deodorization step. Worldwide, the main commercial source of natural tocopherols is in the soybean oil refining industry. Commercial natural antioxidant preparations prepared from soybean oil typically consist of greater than 80 per cent γ- and δ-tocopherol. Synthetic tocopherols are commercially available and vary in isomeric form. Tocopherols provide a useful natural antioxidant source for foods marketed under an "all-natural" label. They are permitted in foods according to GMP regulations. Natural tocopherols are limited to 0.03 per cent (300 pp in animal fats). Because of the high vitamin E content of most vegetable oils, addition of tocopherols can lead to prooxidant activity.

The antioxidative mechanism of α-tocopherol is well understood. α-Tocopherol donates a hydrogen to a peroxy radical resulting in a α-tocopheryl semiquinone radical. This radical may further donate another hydrogen to produce methyl-tocopherylquinone or react with another α-tocopheryl semiquinone radical to produce an α-tocopherol dimer.

The antioxidant activity of the tocols and tocotrienols increases from α through δ. Tocopherol and α-tocotrienol are fully substituted benzoquinone derivatives and are the most effective antioxidants. α-Tocopherol is highly reactive with peroxy radicals and prevents them from participating in propagation reactions. The α-tocopheroxyl radical is stable because a resonance structure forms on the benzoquinone ring. A comprehensive review of the fundamental antioxidant chemistry of tocopherols and tocotrienols has been compiled by Kamal-Eldin and Appelqvist.

L-Ascrobic acid or Vitamin C is widespread (Figure 8.3) in nature as component of plant tissues and is produced synthetically in large quantities. Ascorbic acid is attractive as an antioxidant because it has GRAS status with no usage limits, is a natural or nature-identical product, and is highly recognized as an antioxidant nutrient by the

Ascrobic Acid and Related Compounds

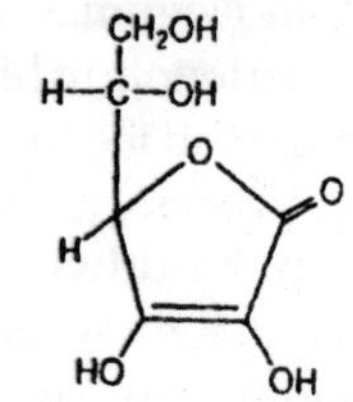

L-Ascrobic Acid

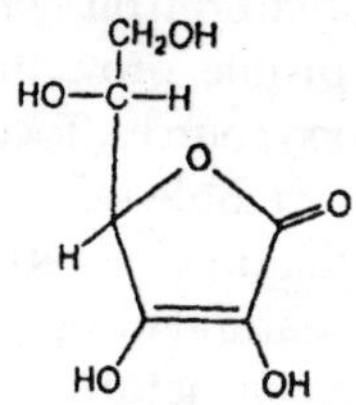

Isoascrobic Acid
Erythorbic Acid
D-Araboascrobic Acid

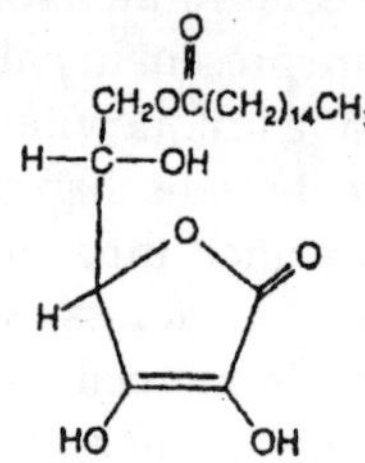

Ascrobyl Palmitate

Carotenoids

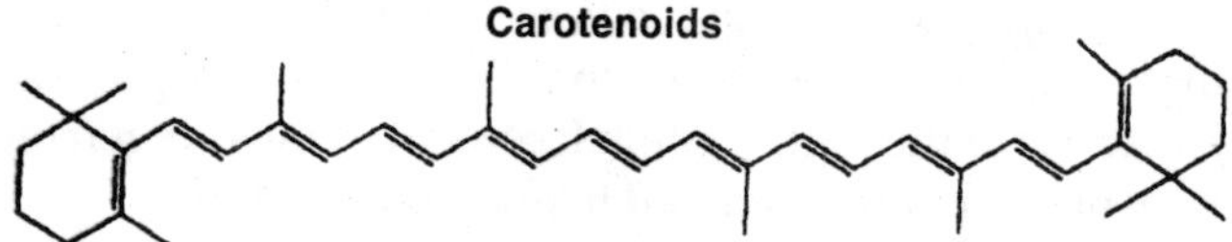

β-carotene
β,β-carotene

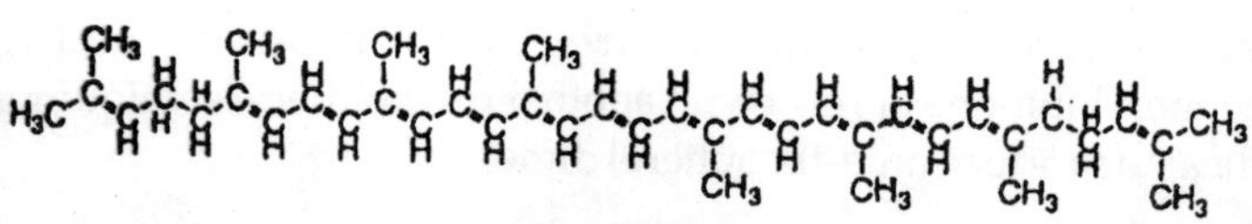

Lycopene
ψ,ψ-carotene

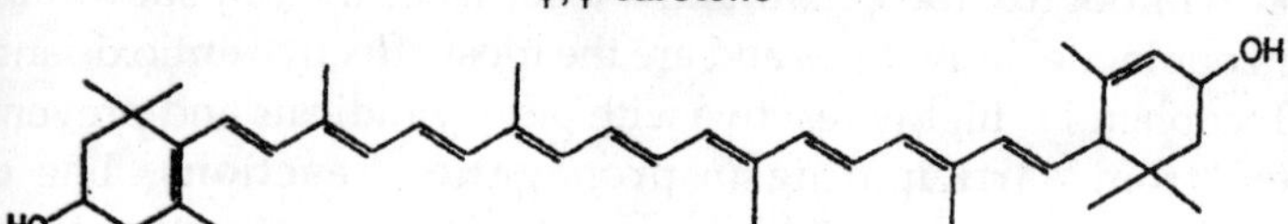

Lutein
β,β-carotene-3,3'-diol

Figure 8.3: Structures of Ascorbic Acid and Related Compounds and Carotenoids

consumer. In some food products it is also a flavorant and acidulant. In foods that are heat-treated, however, ascorbic acid can participate in nonenzymatic browning and may be degraded through reductone reactions.

Ascorbic acid acts as a primary or a secondary antioxidant. *In vivo*, ascorbic acid donates hydrogen atoms as a primary antioxidant. Ascorbic acid is also capable of scavenging radicals directly by converting hydroperoxides into stable products. Ascorbic acid is an important antioxidant in plant tissue and is essential for the prevention of oxidative cellular damage by hydrogen peroxide. In foods, ascorbic acid is a secondary antioxidant with multiple functions. Ascorbic acid can scavenge oxygen, shift the redox potential of food systems to the reducing range, act synergistically with chelators, and regenerate primary antioxidants. The vitamin is commonly used as a synergist to donate hydrogen to primary antioxidants such as tocopherol.

Now let us learn about carotenoids.

Carotenoids (Figure 8.3) are yellow, orange, and red lipid-soluble pigments that are ubiquitous in green plants and in fruits and vegetables. They are 40-carbon isoprenoids or tetraterpenes with varying structural characteristics. The two classes of carotenoids are carotenes and xanthophylls. Carotenes (*e.g.*, β-carotene and lycopene) are polyene hydrocarbons that vary in degree of unsaturation. Xanthophylls (*e.g.*, astaxanthin and canthaxanthin) are synthesized from carotenes by hydroxylation and epoxidation reactions and therefore, contain oxygen groups. About 10 per cent of the 600 or so identified carotenoids have the biological activity of vitamin A and are referred to as provitamin A compounds. β-Carotene is the most abundant of the provitamin A carotenoids found in food and also has the highest vitamin A activity. β-Carotene is a polyene, synthesized from eight isoprene units. It has an intense orange-red color and, as a food additive, is used primarily as a colorant for oils and fats. Carotenoids can act as primary antioxidants by trapping free radicals or as secondary antioxidants by quenching singlet oxygen. Carotenoids are typically secondary antioxidants in foods. However, in the absence of singlet oxygen (low oxygen partial pressure), carotenoids may also prevent oxidation by trapping free radicals and acting as chain-breaking antioxidants.

Singlet oxygen is unstable and can react with lipids to produce free radicals. In the presence of β-carotene, singlet oxygen will preferentially transfer energy to β-carotene to produce triplet state β-carotene. The transfer of energy from singlet oxygen to carotenoid takes place through an exchange electron transfer mechanism.

We are studying natural antioxidants. Numerous amines, amino acids, peptides and protein hydrolysates, have antioxidant activity.

A comprehensive list of all the proteinaceous antioxidant sources is not presented here. Amines have been shown to possess antioxidant activity. Recently, spermine and spermidine isolated from fish sources were used to inhibit fish oil oxidation. Numerous amines such as hypoxanthine and xanthine can be readily isolated from marine sources. Amino acids have chelating abilities, but also exhibit antioxidant activity when used alone. Glycine, methionine histidine, tryptophan, proline, and lysine are effective antioxidants in oil.

Proteins and protein hydrolysates possess antioxidative factors. Iron-binding proteins such as ferritin and transferritin have antioxidant function. Histidine-containing peptides such as carnosine. as well as synthetic peptides and peptides obtained from protein hydrolysis, possess antioxidant activity. Glutathione tripeptide (γ-Glu-Cys-Gly, GSH) has antioxidant activity, which is mediated by the sulfhydryl group of cysteine and glutathione peroxidase. The sulfhydryl group on one GSH will react with the sulfhydryl group of another GSH to produce oxidized glutathione (GSSG) with a disulfide bond. Glutathione/glutathione peroxidase was effective in preventing lipid oxidation in a minced mackerel system. In addition, GSH can form disulfides with proteins or other thiols. It is capable of acting as a free radical scavenger by forming a thiyl radical. GSH also donates electrons for the reduction of dehydroascorbate to ascorbic acid in the presence of glutathione dehydrogenase. Ascorbic acid is then returned to its active, free radical scavenging form. The enzyme has been purified from wheat flour, which has a relatively high activity.

The good example of natural process-induced oxidation inhibitors are Maillard reaction production (MRPs).

MRPs are formed during cooking of low moisture foods at temperatures above 80°C. They are produced from the reaction of amines and reducing sugars. Lipids, vitamins, and other food constituents also participate in Maillard reactions. MRPs are presumed to be safe because they occur naturally as products in cooked foods. They are often used as bases for flavorants and gravies, The brown pigment associated with MRPs can be an advantage in cooked products where brown color is desirable. Use of MRPs as a source of antioxidants in foods has been intensively studied, but out understanding of the compounds responsible for the antioxidant activity is incomplete and the mechanisms of action are unknown. Identification of the compounds responsible for antioxidant activity has proved difficult because of the complexity of the Maillard reaction, the vast number and variety of MRPs, and the diversity of the model systems that can be studied. MRPs have been shown to have antioxidant activity in model systems as well as in some fat-containing foods. Lingnert and Eriksson reused processing parameters and the Maillard reaction to prevent oxidation in cookies, and Sato *et al.* inhibited warmed-over flavor in cooked beef with MRPs.

Because of the conflicting views present in the literature, it is difficult to state conclusively which of the numerous MRPs are actually responsible for antioxidant activity. It is even more difficult to attempt to describe the mechanism of action of these suspected antioxidants. Investigators have shown a correlation between colorless, low molecular weight, intermediate MRPs and antioxidant activity. Evans *et al.* showed a correlation between antioxidant activity and reductone levels in MRP mixtures. In contrast, Kirigaya *et al.* showed that antioxidant activity was proportional to the color intensity of MRP and proposed that nondialyzable, high molecular weight melanoidins, which inhibit hydroperoxide and carbonyl compound formation, were responsible for antioxidant activity. By completely oxidizing the reductones in the MRP solution with DPI butanol, these investigators also demonstrated that reductones contributed little antioxidant activity in their model system. Other researchers found a direct relationship between color intensity of Maillard reaction solutions and antioxidant activity. Yamaguchi *et al.* also showed increases in antioxidant activity with increases in melanoidin formation.

Theories on the mechanism of antioxidant activity of MRPs conflict as well. Kawakishi *et al.* hypothesized that the protective effects of melanoidins against autoxidation were likely to depend on their ability to chelate metals. Amadori compounds may behave like reductones, which inhibit autoxidation. Eichner believes that MRP intermediates may scavenge oxygen. These conflicting reports likely reflect the different reaction conditions used in the experiments and the multiple antioxidative functions exhibited by MRPs with different mechanisms of action. Because of the overwhelming complexity of even model systems of the Maillard reaction it would be imprudent to discount any of these theories as to the nature and activity of antioxidant MRPs.

8.4 Sources of Phenolic Compounds

Numerous plants have been identified as sources of phenolic compounds with antioxidant activity. The list of natural antioxidants is growing as a result of the amount of research that is being conducted to isolate and identify these compounds in plants. A comprehensive listing of all the sources of antioxidant compounds identified in plant materials is beyond the scope of this chapter. Recent reviews by Pratt and Pratt and Hudson contain detailed information on numerous phenolic compounds. Some recent research findings are given below. The diversity of plant sources and compounds considered to be antioxidants is evident even from this brief description.

For example, flavonoid derivatives have been recently identified as potent antioxidants found in apples and chrysanthemum. Antioxidant activity has been correlated to flavonoids found in rice, buckwheat, barley, and malt. Other examples of antioxidants include tannins from bark, lignans from papua mace, and capsaicin from peppers. Compounds with antioxidant activity isolated from high alpine plant species, marine sources, and wood smoke have also been characterized.

Tea extracts are a source of natural antioxidants. Tea catechins have potent antioxidant activity. Extracts of green and black tea contain epicatechin, epicatechin gallate, epigallocatechin, epigallocatechin gallate, and gallocatechins. During the fermentation process required to produce black tea, catechins are oxidized to produce flavins and flavin gallates that have antioxidant activity. Tea antioxidants have been patented for use in several food products.

Several antioxidative compounds have been isolated and identified from sesame seed oil. The compounds are lignanphenols such as pinoresinol sesaminol, sesamol (and its dimer), and sesamolinol. Sesame seed oil has a much longer shelf life than many edible oils because of the presence of these phenolic compounds.

Soybeans have several antioxidative constituents that include soy proteins. Soy and other vegetable hydrolysates contain phenolic compounds with antioxidant activity. Fermented soy products contain isoflavones and genistein, which inhibit oxidation. Diets containing soy provide genestin, whi.ch is a phytoestrogen. Dietary estrogens are structurally similar to endogenous estrogens and mimic their action by binding to the estrogen receptor. Dietary estrogen can produce biological effects similar to those of the endogenous estrogen. The role of diatary estrogens such as genistein in preventing or enhancing the progression of such chronic diseases as breast cancer, heart disease, or bone loss has not bee clarified at this time. However, the interest in the biological effects of dietary estrogens is great both in the research community and among consumers.

Spices and herbs have been used as flavorants for thousands of years. The antioxidant activity of spices and herbs is thought to be primarily due to the presence of phenolic compounds and especially phenolic acids and flavonoids. The strong flavor of spices and herbs precludes their use in many food products. Researchers have tried to identify and isolate specific antioxidant components of spices and herbs that do not contribute undesirable flavor or color to foods. To date, only rosemary and sage (*Perilla* plants) are commercially available as flavorless, odorless, and colorless antioxidant extracts. Thyme, fenugreek, and turmeric possess antioxidative components and may provide sources for commercial products.

Rosemary antioxidants (Figure 10.4) have been used in processed foods for decades. Currently, concentrated rosemary extracts are available that do not impart flavor or color to foods. The use of these products is increasing significantly with the rising consumer demand for natural food additives. These extracts are substantially more expensive than synthetic alternatives and require higher usage levels because of lower efficiencies. The diterpene phenolics carnosol and carnosic acid have been shown to be major antioxidant constituents of rosemary extracts. These compounds are as active as α-tocopherol in bulk oils. Carnosic acid was even more potent than BHA and BHT in soybean oil.

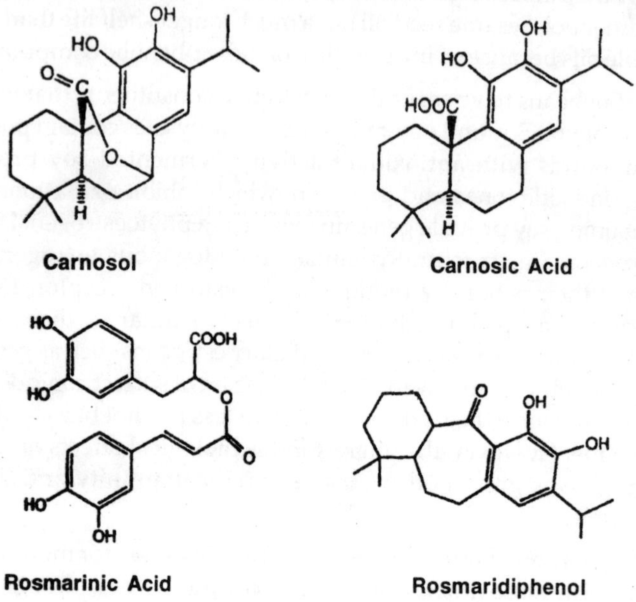

Figure 8.4: Structures of Rosemary Antioxidants

Antioxidant nutrients including vitamin C (ascorbic acid), vitamin E, carotenoids (such as β-carotene), and natural phenolic constituents of foods are being touted as free radical scavengers that may act to prevent cancer, heart disease, and cataracts. Some researchers suggest that synthetic antioxidants in the diet may also prevent cancer formation. Free radicals, produced in the body as a natural product of oxidative reactions, can cause oxidation of cell lipids and DNA damage that may lead to serious diseases. Dietary antioxidants are thought to scavenge these free radicals, prevent them from damaging cells and DNA, and possibly reduce oxidized fatty acids or mutagens that lead to heart disease or cancer. Epidemiological studies have shown trends suggesting that antioxidants may be beneficial in disease prevention. However, a cause-and-effect relationship cannot be determined in humans because lifestyle factors interfere with interpretation of research results.

Chapter 9

Triacylglycerol Acylhydrolases

9.1 Sources of Triacylglycerol Acylhydrolases

Triacylglycerol acylhydrolases (lipases) from various organs and tissues of several mammalian species have been investigated but human and other pancreatic lipases are the most thoroughly studied.

Lipases (triacylglycerol acylhydrolases), are enzymes that catalyze the reversible hydrolysis of triacylglycerols under natural conditions. Widely distributed in animals, plants, and microbes, lipases differ from other esterases and are unique in that their activity is greatest against water-insoluble substrates and is enhanced at the substrate (oil)-water interface; that is, they exhibit "interfacial activation." Optimum activities are obtained in systems such as emulsions, where high surface areas of the substrate can be obtained. Lipases are active not only in normal phase emulsions where the substrate is emulsified into an aqueous system (oil-in-water), but they are also active, often more active, in invert (water-in-oil) emulsions and in reverse micelle systems containing an organic solvent solution of the substrate. Furthermore, lipases are exceedingly versatile in that they can also catalyze transesterification reactions and the stereospecific synthesis of esters, and they can act on a broad range of substrates.

Major advances have been made in our understanding of lipolytic enzymes over the past few years through solving the crystal structures of lipases from several sources. Knowledge of their structures has given insight into the mechanism of action, interfacial activation, specificity, and the nature of the active site. The high current interest in lipases, shown by the large volume of recent scientific literature on the topic, is driven by the great potential in a diversity of commercial applications for these enzymes. This chapter gives an overview of lipases and their properties, with a focus on the topics of current research interest such as lipase structure and its implications in interfacial activation and selectivity. The emphasis is on microbial lipases, particularly from fungi, with references to human pancreatic lipase for comparison.

Pancreatic lipases are secreted into the duodenum and active on dietary triacylglycerols.

They are a class of structurally similar 50 kDa glycoproteins that are characterized by their specificity toward triacylglycerols with little or no activity toward phospholipids, activation at the oil (substrate)-water interface and by colipase, and inhibition at micellar concentrations of bile salts. Pancreatic lipase can catalyze the complete breakdown of triacylglycerols to free fatty acids and glycerol. Nonhuman pancreatic lipases may differ from corresponding human lipases; for example, guinea pig pancreatic lipase differs in that it exhibits phospholipase A activity.

Other mammalian lipases have been studied and, while similar, they exhibit some characteristics that differ from those of pancreatic lipase. In addition to pancreatic lipase, fat digestion is aided by a series of lingual, pharyngeal, and gastric lipases that may be responsible for up to 50 per cent of dietary fat breakdown. Highly stable at low pH, these enzymes are activated by bile salts and show a preference for the *sn*-3 position of the substrate. Human lipoprotein lipase, which functions in hydrolyzing triacylglycerols in chylomicrons and very low density lipoproteins, shows many similarities to pancreatic lipase (*e.g.*, high sequence homology, presence of the lid). However, this lipase is not responsive to colipase but instead,. requires apolipoprotein C-II (apoC-II) for activity, functions as a dimer, and is activated by heparin. Hepatic lipase is confined to the liver, where it also is involved in the metabolism of lipoproteins, but it is not activated by apoC-II. Human milk lipase, which functions in the digestion of milk fat ingested by infants, is activated by bile salt.

There is substantial current interest in developing microbial lipases for use in biomedical and industrial applications because of their versatility and availability; in addition, they can be produced less expensively than corresponding mammalian enzymes. Hou and Johnston screened 1229 bacteria, yeasts, actinomycetes, and fungi and found that about 25 per cent were lipase positive. Microbial lipases that have received the greatest attention are inducible, extracellular enzymes having properties that are generally similar to those of human pancreatic lipase, despite differences in detail. At least some of the microbes produce a mixture of extracellular lipases formed from multiple genes, and some lipases vary by degrees of glycosylation. Fungal lipases typically exist as monomers with molecular masses ranging from about 30 to 60 kDa. They vary in specificity, specific activity, temperature stability, and other properties; however, dimeric lipases have been reported.

Plant lipases have not received the same attention as those from other sources, but they have been reviewed recently by Mukherjee and Hills. Oilseed lipases have been of greatest interest among the plant lipases, and those from a variety of plant species show differences in their substrate specificity, pH optima, reactivity toward sulfhydryl reagents, hydrophobicity, and subcellular location. These lipases are relatively specific for the native triacylglycerols of the species from which they were isolated. They are absent from the ungerminated seed and formed during germination.

9.2 Structures of Mammalian and Microbial Lipases

Generally lipases are α/β proteins with a central core of a mixed β sheet containing the catalytic triad composed of ser... His... Asp, and a surface loop restricting access of the substrate to the active site. Now let us study different lipases.

Human pancreatic lipase is folded into two domains, a larger N-terminal domain comprising residues 1 to 335 and a smaller C-terminal domain. The core of the N domain is formed by a nine-stranded, β-pleated sheet in which most of the strands run

Table 9.1: Properties of Some Recently Characterized Extracellular Microbial Lipases

Organism	*Molecular Weight (kDa)*	*pH Optima*	*Optimum Temperature (ºC)*	*Specific Activity (U/mg)*	*Specificity*
Neurospora sp, TT-241	55	6.5	45	8203	No specificity but preferred 1- and 3-positions
Botrytis cinerea	60	6.0	38	2574	Highest activity with oleic acid esters
Pseudomonas fluorescens AK102	33	8–10	55	6200	Nonspecific for fatty acids
Rhizopus niveus 1f04759					
Lipase I	34	6.0–6.5	35	4966	–
Lipase II	30	6.0	40	6198	–
Candida parapsilosis	160	6.5	45	–	High specificity for long chain fatty acids, particularly polyunsaturated fatty acids
Propionibacterium acidi-propionici	6–8	7.0	30	–	Preferred substrates with high saturated fatty acids
Neurospora crassa	54	7.0	30	44	Preferred triglycerides with C_{16} and C_{16} acyl chains
Pythium ultimum #144	270	8.0	30	63	1,3-Specific, preferring substrates with higher unsaturation
Rhizopus delemar ATCC 34612	30.3	8.0–8.5	30	7638	–
Fusarium heterosporium	31	5.5–6.0	45–50	2010	1,3-Specific; preferred triacylglycerols with C_6–C_{12} fatty acids
Penicillium roquefortii	25	6.0–7.0	30	4063	Preferred triacylglycerols with C_4-C_6 fatty acids
Penicillium sp. uzim-4	27	7.0	25	1001	1,3-Specific, discriminates against diglycerides and active at low surface pressures

parallel to one another. Seven α-helical segments of varying length occur in the strand connections, and six of them pack against the two faces of the core sheet. The C domain is formed by two layers of antiparallel sheets, the strands of which are connected by loops of varying length. The N domain contains the active site, a glycosylation site, a Ca^{2+}-binding site, and possibly a heparin-binding site. The active site is buried beneath a short amphipathic α-helical surface loop, termed the "flap" or "lid." Colipase, binds exclusively to the C domain of the protein through hydrophobic interactions and ion pairing, and the binding does not induce a conformational change. Calcium ions activate and reduce the lag phase of human pancreatic lipase, particularly for mixed bile acid-lipase complexes.

The structures of pancreatic lipase from other systems have been recently reported, including guinea pig and horse. Overall, the nonhuman pancreatic lipases are structurally similar to that from humans but may differ in detail. For example, the guinea pig pancreatic lipase does not possess the lid that is typical of other lipases. Also, unlike the pancreatic lipases from humans and pigs, the horse enzyme is not glycosylated.

The crystal structures of several fungal lipases have also been determined: for example, *Rhizomucor miehei, G. candidum, C. rugosa, C. antarctica* and *Humicola lanuginosa*. Although there are no obvious sequence similarities between pancreatic and fungal lipases, except for the Gly-X-Ser-X-Gly consensus sequence in the active site region, there are structural similarities. Fungal lipases are α/β proteins and have similar topologies based on a large central mixed β-pleated sheet pattern, mainly parallel, but the connectivities between strands vary. The two lipases of G. *candidum* have 554 amino acids and share about 85 per cent sequence homology. Also, many fungal lipases possess the serine protease catalytic triad Ser…His…Asp in their active sites; however, glutamic acid is substituted for aspartic acid in the lipases from *G. candidum* and *C. rugosa*. Lipase II gene from *G. candidum* has been cloned, and the results of probing the active site by site-directed mutagenesis are consistent with X-ray crystallography data in that the Ser...His...Glu is the active site.

Fungal lipases also possess a lid that prevents access of the substrate to the active site, form a functional oxyanion hole, and have of an interfacial binding site (Gly-X-Ser-X-Gly), but the motion of the lid differs between fungal and pancreatic lipases. Lid rearrangements by human pancreatic lipase, *C. rugosa* and *G. candidum* involve more than one loop. The lid is closed and covers the active site in *G. candidum* and is open in *C. rugosa;* otherwise lipases from these two sources are very similar in sequence and structure. The three-dimensional structure of the extracellular lipase from *Rhizopus delemar* based on X-ray crystallographic coordinates, with the lid partially open exposing the active site residues Ser, Asp, and His and residues 86 to 92 in the partially closed conformation.

Lipases from several *Pseudomonas* species have been crystallized, and preliminary X-ray crystallographic analyses of such species as *P. cepacia, P. glumae,* and *P. fluorescens* have been conducted. Their molecular masses, which are in the 30 to 35 kDa range, show extensive sequence homology to one another but little to those of other lipases; the common G-X-S-X-G sequence in the active site region is conserved, and activity appears at the substrate-water interface. On the other hand, lipases from *Bacillus* species have molecular weights at about 19 kDa, have the sequence A-X-S-X-G instead of the characteristic sequence, and do not exhibit interfacial activation.

9.3 Interfacial Activation of Lipases

Solving the crystal structures of lipases has given insight into the mechanism of interfacial activation. It is believed that the preferred conformation in aqueous solution is with the lid covering the active site, thus denying access to the substrate. Adsorption of the lipase to the interface involves a conformational change in the enzyme whereby the lid, which covers a cavity containing the active site and is held in place by mostly hydrophobic and some hydrogen bonds, undergoes reorientation. This is accompanied by additional conformational changes that expose the active site and a large hydrophobic site, and allow access of the substrate to the active site. The working hypothesis for the mechanism of interfacial activation is based on the three-dimensional structural analysis of lipases from *Rhizomucor miehei*, human pancreatic lipase, *G. candidum,* and and *C. rugosa*. The two latter lipases, for example, are globular, single-domain proteins built around an 11-stranded mixed β-sheet with domains of 45 Å × 60 Å × 65 Å.

Activation of *R. miehei* lipase involves the movement of a 15 amino acid long lid in a hinge-type, rigid-body motion that transports some of the atoms of a short α helix by over 12 Å. This, combined with another hinge movement, results in the exposure of a hydrophobic area representing 8 per cent of the total molecule surface. In C. *rugosa,* comparison of open and closed conformation indicates that activation of the lipase requires the movement and refolding, including a cis-to-trans isomerization of a proline residue, of a single surface loop to expose a large hydrophobic surface where the substrate likely interacts. Lid reorganization contributes to the formation of a catalytically competent oxyanion hole and creation of a fully functional active site. The scissile fatty acyl chain is bound in a narrow, hydrophobic tunnel, where modeling studies suggest that the substrate must adopt a tuning fork conformation. There is a tryptophan residue at the tip of the lid, in the lipase from *H. lanuginosa* that plays an important role in hydrolytic activity. When Trp is substituted with other amino acids, activity drops substantially and variably depending on the substituent amino acid.

It is difficult to visualize how the hydrophobic substrate molecule, which is buried in the oil surface, gains access to the active site, which is buried within the water-soluble lipase molecule. Blow has suggested that upon activation, a "hydrophobic seal" forms at the interface that allows the substrate to enter the active site without interacting with the bulk water; that is, the enzyme partially withdraws the substrate molecule from the bulk oil with at least some of the acyl chains projecting into the lipid.

Cygler *et al.* have postulated that there are two tetrahedral intermediates in the lipase-catalyzed hydrolysis of esters. Formation of a noncovalent Michaelis complex between the lipase and triacylglycerol is followed by formation of a tetrahedral, hemiacetal intermediate resulting from a nucleophilic attack by the serine O′. The oxyanion resulting from formation of the enzyme-substrate complex is stabilized by the amide groups of the oxyanion hole and α helix following the active site serine. Intermediate formation is followed by cleavage of the substrate ester bond, breakdown of the tetrahedral intermediate to the acyl enzyme, and protonation and dissociation of the diacylglycerol. The serine ester of the acylated enzyme is attacked by an activated water molecule to form a second tetrahedral intermediate, which is cleaved to give rise to the protonated enzyme (serine residue) and fatty acid.

Calcium may stimulate lipase-catalyzed hydrolytic activity by (*a*) binding to the enzyme resulting in a change in conformation, (*b*) facilitating adsorption of the lipase to the substrate-water interface, and/or (*c*) removing from the interface fatty acid products of hydrolysis that may reduce end-product inhibition of the reaction. Activation of human pancreatic lipase by calcium is complex and variable depending on the substrate and presence or absence of bile acids. Calcium effects on microbial lipase activity may be variable depending on the enzyme source and assay conditions. For example, stimulation of *C. rugosa* lipase activity was attributed to the formation of calcium salts of fatty acid products in a normal phase emulsion, with olive oil as the substrate but not with tributyrin; calcium, however, had no effect in an invert emulsion. In a non-emulsion system (*i.e.*, without emulsifier), calcium had no effect on *C. rugosa* lipase activity with olive oil as the substrate but tended to offset the inhibitory effects of bile acid (Mozaffar and Weete, unpublished).

Human pancreatic lipase is inhibited by bile salts, and the inhibition can be overcome by the 10 kDa protein colipase. Bile salt coating of the substrate micelles creates a negatively charged surface that is believed to inhibit adsorption of the bile salt-lipase complex to the interface. Colipase overcomes the inhibitory effect of the bile salts through formation of a 1 : 1 complex with lipase that facilitates adsorption at bile-salt-coated interfaces. Naka and Nakamura found that although the bile salt sodium taurodeoxycholate inhibited pancreatic lipase activity when tributyrin was the substrate, a result that has been widely reported and cited by others, and colipase could reverse the inhibition, the bile salt actually stimulated hydrolytic activity when triolein was the substrate. This was attributed to the fact that triolein is a more natural substrate for the lipase than tributyrin. On the other hand, when sodium taurocholate was added to an emulsion assay of the lipase from *C. rugosa* with olive oil as the substrate, activity was progressively inhibited from 0.1 mM to 0.8 mM concentration of the bile salt. Relatively high activity at the lowest concentration was attributed to the role of the bile salt in the stabilization of oil particles in the emulsion and providing high interfacial area for adsorption of the lipase, and inhibition was due to interaction with the enzyme such that adsorption was reduced.

A variety of substances have been shown to inhibit lipase activity; examples include anionic surfactants, certain proteins, metal ions, boronic acids, phosphorus-containing compounds such as diethyl *p*-nitrophenyl phosphate, phenylmethyl sulfonylfluoride, certain carbamates, β-lactones, and diisopropylfluorophosphate.

9.4 Assay and Modification of Lipases

Lipases are assayed in emulsions where high substrate surface areas can be achieved. Since, in addition to high interfacial area, the chemical and physical environments at the substrate-water interface are important to the adsorption and activity of lipases, it is useful to understand the nature of the different types of emulsion used for lipase assay or in various applications.

Emulsions are mixtures of two immiscible liquids (*e.g.*, oil and water); one of the components is dispersed as very small droplets, or particles, and the mixture is stabilized by a surface-active agent, or surfactant. Emulsions are classified as macro- or microemulsions where the dispersed particles are either greater or smaller than one micrometer in diameter, respectively. Macroemulsions are turbid, milky in color, and thermodynamically unstable (*i.e.*, they will ultimately separate into the two liquid phases). On the other hand, microemulsions are homogeneous and stable.

Depending on the conditions of formation, and particularly the nature of the surfactant, macroemulsions may be normal phase (oil-in-water) or reversed phase (water-in-oil, invert). In the former, the oil is emulsified into the aqueous phase, and the surfactant forms a monolayer film around the dispersed oil droplets, whereby the hydrophobic moiety of the surfactant extends into the oil and the polar moiety is at the droplet surface. In reversed phase emulsions, the aqueous phase is dispersed in the oil with the orientation of the surfactant molecules reversed.

Amphiphilic (surfactant) molecules undergo self-organization into spheroidal particles when dissolved in certain organic solvents such as isooctane with the polar head groups oriented inward and hydrophobic tails outward. These particles are referred to as reverse micelles and are less than 1 μm in diameter. Water can be "solubilized" in the organic solvent by becoming entrapped in the particles at up to several dozens of molecules per molecule of surfactant. Typically, reverse micelles are formed when the molar ratio of water to surfactant is less than 15, which is expressed as water activity, w_0 (*i.e.*, molarity of water to molarity of surfactant) or *R*. Enzymes such as lipases can be entrapped within the aqueous

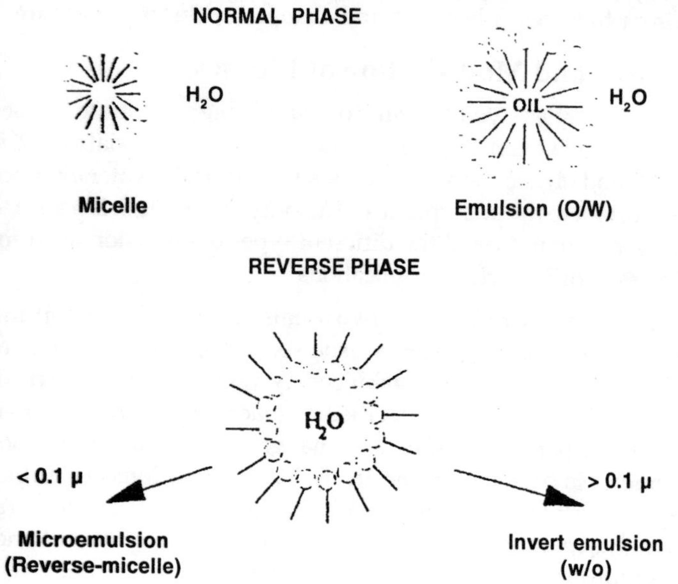

Figure 9.1: Diagrammatic Representation of Globules (Particles) in Various Types of Emulsions. In the ball-and-stick model of amphiphilic (surfactant) molecules, the ball represents the polar head group and the stick the hydrophobic region of the molecule.

phase particles of the invert emulsions or reverse micelles of microemulsions, where they retain their activity. In the latter case, the enzymes are isolated from the organic solvent. Luisi *et al.* and Sanchez-Ferrer and Garcia-Carmona can be consulted for more information on microemulsions.

Early studies with fungal lipases focused on the isolation and characterization of extracellular lipases from various species. Some of the thoroughly studied fungal lipases include those from *C. (cylindracea) rugosa, R. miehei, P. camembertii, H. lanuginosa, C. antarctica B. Rhizopus delemar,* and *G. candidum,* all of which are commercially available. Many of these lipases have relatively high specific activities: 3485 U/mg for *Mucor miehei* lipase A and 7638 U/mg for *Rhizopus delemar* (M). Lipases that have received the most attention are mainly those having relatively high activities or certain properties that make them commercially attractive. Other than additional strains of known lipase producers, there seems to be no pattern among the fungi or yeasts from a taxonomic point of view that would direct future studies on where to find prolific lipase producers or lipases with specific properties.

Lipases have been modified using either chemical or molecular approaches to alter their properties and to identify structure-activity relationships. For example, lipases have been chemically modified with polyethylene glycol to render them more soluble in organic media. Recently, Kodera *et al.* produced amphipathic chain-shaped and copolymer derivatives of lipases from *Pseudomonas fragi* or *P. cepacia* that were soluble in aqueous and hydrophobic media and exhibited catalytic activities for esterification and transesterification reactions, as well as for hydrolysis. The modified lipase showed preference for the *R* isomer of secondary alcohols in esterification reactions.

Now let us learn some details about the uses of lipases

Lipases are currently used, or have the potential for use, in a wide range of applications: in the dairy industry for cheese flavor enhancement, acceleration of cheese ripening, and lipolysis of butterfat and cream; in the oleochemical industry for hydrolysis, glycerolysis, and alcoholysis of fats and oils; and for the synthesis of structured triglycerides, surfactants, ingredients of personal care products, pharmaceuticals, agrochemicals, and polymers.

The Colgate-Emery process, currently used in the steam fat-splitting of triacylglycerols, requires 240 to 260°C and 700 psi, has energy costs, and results in an impure product requiring redistillation to remove impurities and degradation products. Also, this process is not suitable for highly unsaturated triacylglycerols. Lipase-catalyzed reactions offer several benefits over chemical reactions, including stereospecificity, milder reaction conditions (room temperature, atmospheric pressure), cleaner products, and reduced waste materials.

The largest current use of industrial enzymes is in laundry detergents, where they combine environmental friendliness and biodegradability with a low energy requirement and efficiency at low concentrations. The current U.S. market share of enzymatic laundry detergents is approaching 80 per cent, and the U.S. detergent enzyme market is about $140 million. Essentially four types of enzyme are used in detergents: proteases, amylases, lipases, and cellulases. These enzymes perform multiple functions (*e.g.*, stain removal, antiredeposition, whiteness/brightness retention, and fabric softening). Proteases were the first and are the most widely used enzymes in detergent formulations. Lipases are relatively new introductions to detergents, where they attack oily and greasy soils and contribute to making the detergents particularly effective at lower wash temperatures. However, a current limitation is that most lipases are unstable in alkaline conditions in the presence of anionic surfactants used in laundry detergents. However, some lipases may be relatively resistant to certain surfactants.

Chapter 10

Precise Analysis of Lipids

10.1 Introduction

The precise analysis of lipids in foods is important for determining nutritive value, uniformity as well as for promoting an understanding of the effects of fats and oils on food functionality.

Lipids in nature are associated with other molecules via van der Waals interaction (*e.g.*, interaction of several lipid molecules with proteins), electrostatic and hydrogen bonding, mainly between lipids and proteins, and covalent bonding between lipids, carbohydrates, and proteins. Lipids have to be separated and isolated from the complex cellular matrix for further quantitation and analysis.

The general classification of lipids can be explained as follows.

Simple lipids: compounds with two types of structural moiety

[*Glyceryl esters*: esters of glycerol and fatty acids (*e.g.*. triacylglycerols, partial acylglycerols)

Cholesteryl esters: esters of cholesterol and fatty acids

Waxes: the true waxes are esters of long chain alcohols and fatty acids; esters of vitamins A and D are also included

Ceramides: amides of fatty acids with long chain di- or trihydroxy bases containing 12 to 22 carbon atoms in the aliphatic chain (*e.g.*, sphingosine)

Complex lipids: compounds with more than two types of structural moiety

Phospholipids: glycerol esters of fatty acids, phosphoric acid and other groups containing nitrogen

Phosphatidic acid: diacylglycerol esterified to phosphoric acid

Phosphatidylcholine: phosphatidic acid linked to choline, lecithin

Phosphatidylethanolamine

Phosphatidylserine

Phosphatidylinositol

Phosphatidylacylglycerol: more than one glycerol molecule is esterified to phosphoric acid (*e.g.* cardiolipin; diphosphatidyl acylglycerol)

Glycoglycerolipids: 1,2-diacylglycerol joined by a glycosidic linkage through position *sn*-3 to a carbonhydrate moiety (*e.g.*. monogalactosyl diacylglycerol, digalactosyl monoacylglycerol, sulfoquinovosyl diacylglycerol (monoglycosyl diacylglycerol at the position 6 of the monosaccharide moiety is linked by carbon-sulfur bond to a sulfonic acid)]

Gangliosides: glycolipids that are structurally similar to ceramide polyhexoside and also contain 1 to 3 sialic acid residues, most contain an amino sugar in addition to the other sugars

Sphingolipids: derivatives of ceramides

Sphingomyelin: ceramide phosphorylcholine

Cerebroside: ceramide monohexoside (*i.e.*, ceramide linked to a single sugar moiety at the terminal hydroxyl group of the base)

Ceramide dihexoside: linked to a disaccharide

Ceramide polyhexoside: linked to a tri- or oligosaccharide

Cerebroside sulfate: ceramide monohexoside esterified to a sulfate group

Derived lipids: compounds that occur as such or are released from other lipids as a result of hydrolysis (*e.g.*, fatty acids, sterols, fatty alcohols. hydrocarbons; fat-soluble vitamins A, D, E, and K)

Now let us try to understand about lipid extraction.

The existing procedures of lipid extraction from animal or plant tissues usually include several steps:

1. Pretreatment of the sample (which includes drying, size reduction, or acid hydrolysis).
2. Homogenization of the tissue in the presence of a solvent.
3. Separation of liquid (organic and aqueous) and solid phases.
4. Removal of nonlipid contaminants, and
5. Removal of solvent and drying of the extract.

Standard methods of lipid extraction have been established by the Association of Official Analytical Chemists (AOAC) International for materials/tissues of different types. In practical situations, however modification of the assigned method may be necessary in individual cases.

As with any form of chemical analysis, proper sampling and storage of the samples are essential for obtaining valid results. According to Pomeranz and Meloan, an ideal sample should be identical in all intrinsic properties to the bulk of the material from which it is taken. In practice, a sample is satisfactory if the properties under investigation correspond to those of the bulk material within the limits set by the nature of the test. Sample preparation for lipid analysis depends on the type of food and nature of its lipids. Effective analysis calls for a knowledge of the structure, chemistry, and occurrence of principal lipid classes and their constituents. Therefore, it is not possible to devise a single standard method for extraction of lipids of all kinds in different foods. Use of high temperatures should be avoided, and maintaining an inert atmosphere during sample preparation and extraction, if at all possible, may minimize oxidation reactions of unsaturated lipids.

Extraction of lipids should be performed as soon as possible after the removal of tissues from the living organism, to minimize any subsequent changes. Immediate extraction is not always possible, however, and samples usually are stored at very low temperatures in sealed containers, under an inert (nitrogen) atmosphere or on dry ice. Yet the freezing process itself may permanently damage the tissues as a result of disruption of cell membranes by osmotic shock, which alters the original environment of the tissue lipids and brings them into contact with enzymes from which they are normally protected. Thawing of samples taken from frozen storage before extraction may enhance this deterioration. Therefore, tissue samples should be homogenized and extracted with solvents without being allowed to thaw. Lipolytic enzymes of animal and plant tissues are usually deactivated irreversibly by homogenization with polar solvents.

The efficiency with which lipids are extracted from a dried sample also depends on the size of the particles. Therefore, particle size reduction increases surface area and enhances lipid extraction (*e.g.*, grinding of oilseeds before lipid extraction). In some cases, instead of being subjected to these operations separately, the sample is homogenized together with the extracting solvent (or solvent system).

Neutral lipids are hydrophobically bound and can be extracted from tissues by non-polar sovents, whereas polar lipids, which are bound predominantly by electrosctatis forces and hydrogen bonding, require polar solvents capable of breaking such bonds. However, less polar neutral lipids such as TAGs and cholestrol esters may be extracted incompletely with non-polar solvents, probably because a significant part of these lipids is not accessible to the solvents. Lipids that are covalently bound to polypeptide and polysaccharide groups will not be extracted at all by organic solvents and will remain in the non-lipid residue. Therefore, acid hydrolysis may be required to release covalently bound lipids to render them fully extractable.

The type of solvent and the actual method of lipid extraction depend on both the chemical nature of the sample and the type of lipid extract (*e.g.*, total lipids, surface lipids of leaves) desired. The most important characteristic of the ideal solvent for lipid extraction is a high solubility of lipids coupled with low or no solubility of proteins, amino acids, and carbohydrates. The solvent should have a relatively low boiling point and should evaporate readily without leaving any residues when recovering lipids. Flammability and toxicity of the solvent are also important considerations to minimize potential hazards. The solvent should also readily penetrate sample particles. Cost and nonhygroscopicity are other important considerations. The ideal solvent or solvent mixture for extracting tissue lipids should be sufficiently polar to remove all such compounds from their association with cell membranes or lipoproteins but should not chemically react with them. At the same time, the solvent should not be so polar that TAGs and other nonpolar simple lipids fail to dissolve in it, hence are left adhering to the tissues. The extracting solvent may also prevent enzymatic hydrolysis of lipids, thus ensuring the absence of any side reactions. The solvents mostly used for isolation of lipids and ethers (diethyl ether, isopropyl ether, dioxane, tetrahydrofuran), hydrocarbons (hexane, benzene, cyclohexane, isooctane), halocarbons (chloroform, dichloromethane), alcohols (methanol, ethanol, isopropanol, *n*-butanol), acetone, and acetonitrile, or their mixtures. However, use of solvents such as benzene should be kept to a minimum because of their potential carcinogenicity to humans.

Solubility of lipids in organic solvents is dictated by the proportion of the nonpolar hydrocarbon chain of the fatty acids or other aliphatic moieties and polar functional group such as phosphate or sugar moieties in their molecules. Lipids containing no distinguishable polar groups (*e.g.*, TAGs or cholesterol esters) are highly soluble in hydrocarbon solvents such as hexane, benzene, or cyclohexane and in more polar solvent such as chloroform or diethyl ether, but remain insoluble in polar solvents such as methanol. The solubility of such lipids in alcoholic solvents increases with the chain length of the hydrocarbon moiety of the alcohol; therefore, they are more soluble in ethanol and completely soluble in *n*-butanol. Similarly, the shorter chain fatty acid residues in the lipids are better soluble in more polar solvents (*e.g.*, tributyrin is completely soluble in methanol while tripalmitin is insoluble). Polar lipids are only sparingly soluble in hydrocarbon solvents unless solubilized by association with other lipids; however, they dissolve readily in more polar solvents such as methanol, ethanol, or chloroform.

10.2 Separation of Lipids by TLC

In lipid analysis thin-layer chromatography is used as the main analytical tools.

TLC can be used for fractionation of complex lipid mixtures, assay of purity, and identification; it also offers a means of providing information on structure and for monitoring the extraction and separation of components via preparative column chromatography. The adsorbent generally used in TLC for lipid analysis, very fine grade silica gel, may contain calcium sulfate as a binder to ensure adhesion to the plate.

Samples of lipid extracts are applied as discrete spots or as narrow streaks, 1.5 to 2 cm from the bottom of the plate. The plate is then placed in a chamber containing the developing solvent or a solvent mixture. The solvent moves up the plate by capillary action, taking the various components with it at different rates, depending on their polarity and how tightly they are held by the adsorbent. The plate is removed from the developing chamber when the solvent is nearly at the top of it and dried in the air or under a flow of nitrogen.

The location of the corresponding lipid spots on a developed plate must be detected before the spots are isolated. This may be done by adding a reagent directly on the plate. The reagent may be specific to certain functional groups of the lipid molecules, or it may be a nonspecific reagent that renders all lipids visible. Nondestructive chemical reagents such as 2′,7′-dichlorouorescin in 95 per cent methanol (1 per cent, w/v) allow recovery of lipids after detection. Lipids exhibit a yellow color and, in the presence rhodamine 6G (0.01 per cent, w/v), produce pink spots under UV light. Water can also be used as a nondestructive spray for preparative TLC. Separated lipids may appear as white spots on a translucent background and as such are easily distinguished. Developed plates may also be subjected to saturated iodine vapor in a chamber, a technique that may produce brown spots when iodine reacts with unsaturated bonds of the lipid molecules; however, lipids so separated cannot be recovered and used for further analysis. Spraying of the plate with sulfuric acid (50 per cent v/v) and drying at 180°C for one hour will also make lipids visible as black deposits of carbon. Lipids detected by nondestructive methods can be recovered by scraping the bands of interest and dissolving them in an appropriate solvent. Complex lipid mixtures cannot be always separated by one-dimensional TLC, but two-dimensional TLC may resolve such mixtures.

The separation efficiency of TLC plates is affected by the degree of hydration of the adsorbent. Therefore, it is necessary to consider the activation of the adsorbent, which depends on the time and temperature and storage conditions of the plates and the relative humidity of the environment. Use of authentic lipid standards would allow direct comparison of the resolved lipids in an unknown mixture.

High performance TLC (HP-TLC) plates, which offer high resolving power and speed of separation, are available commercially as plates precoated with fine silica gel of uniform particle size. However, the amount of sample that can be applied to such HP-TLC plates is very small.

Silver nitrate impregnated (argentation) TLC may be used to separate triacylglycerols or fatty acid methyl esters according to the number of double bonds and also by virtue of the geometry (*e.g.*, *cis* or *trans*) and position of the double bonds in the alkyl chain. Structural isomers of TAGs (owing to their fatty acid constituents) may also be separated on this type of TLC plate. Impregnation of TLC plates with boric acid (3 per cent, w/v) prevents isomerization of mono- and diacylglycerols when neutral lipids are being separated. TLC is a preferred method of purification and separation of lipid classes before they are subjected to further separation to individual components.

Scanning densitometry has been one of the most popular methods of quantifying components separated by TLC. The spots detected by charring can be measured by means of scanning photodensitometry, and areas of peaks, which are proportional to the amount of original lipid present, are recorded. Scintillation counting is also possible after the introduction into the lipid of an appropriate scintillator [*e.g.*, a mixture of 2.5-diphenyloxazole (PPO) and 1,4,-bis(5-phenyl-2-oxazolyl)benzene (POPOP)]. Scanning fluorometry allows resolution of lipids on an adsorbent containing a fluorescent dye; alternatively, such reagents can be sprayed on. The measured fluorescence is directly related to the amount and the nature of the lipid present; thus, calibration curves prepared with standards are needed. Phospholipids separated on TLC plates can be accurately quantified by determining the phosphorus content of the resolved spots.

Table 10.1: Solvent Systems for Separation of Lipids by TLC

Lipid Component	*TLC Adsorbent*	*Solvent System*
Complex lipids (animal tissues)	Silica gel G	Chloroform-methanol-water, 25 : 10: 1, v/v/v
	Silica gel H	Chloroform-methanol-acetic acid-water, 25 : 15 : 4 : 2, v/v/v/v
	Silica gel H	First developing system, pyridine-hexane, 3: 1, v/v; second developing system, chloroform-methanol-pyridine-2 M ammonia, 35 : 12 : 65 : 1, v/v/v/v
Complex lipids (plant tissues)	Silica gel G	Acetone-acetic acid-water, 100 : 2 : 1, v/v/v
		Diisobutyl ketone-acetic acid, 40 : 25 : 3.7, v/v/v
Simple lipids	Silica gel G	Hexane-diethyl ether-formic acid, 80 : 20 : 2, v/v/v
	Silica gel G	Benzene-diethyl ether-ethyl acetate-acetic acid, 80 : 10 : 10: 0.2, v/v/v/v
Partial acyl-glycerol	Silica gel G containing 5% w/v boric acid	Chloroform-acetone, 96 : 4, v/v
Neutral plasmalo-gens	Silica gel G	Hexane-diethyl ether, 95 : 5, v/v, in first direction; hexane-diethyl ether, 80 : 20, v/v, in second direction

The latroscan (latron Laboratories, Japan) applies the same principles of TLC to separate lipid mixtures, which are then detected by means of a flame ionization detector (FID). The TLC medium is silica (7.5 µm thick) sintered onto quartz rods (0.9 mm i.d., 15 cm long, available commercially as Chromarods). Chromatographic resolution of lipid classes and composition of solvent systems employed in classical TLC are somewhat modified for use with Chromarods. Impregnation with copper sulfate, silver nitrate, boric acid, or oxalic acid has been reported to produce better resolution and increased responses in the determination. Parrish and Ackman have shown that stepwise scanning and developing in solvent systems of increasing polarity resolves individual lipid components in neutral lipids of marine origin. The FID gives a linear response that is proportional to the number of nonoxidized carbon atoms in the material entering the flame. The Chromarods are reusable up to 100 to 150 times but must be thoroughly cleaned after each use. The main advantage of the TLC-FID is the short analysis time, with the possibility of determining all components in a single analysis. The partial scanning and redevelopment possible with the TLC-FID system confers a unique advantage over any other lipid analysis technique. The TLC-FID system has been used to analyze different types of lipid.

10.3 Types of Chromatography

Let us begin with column chromatography.

Lipid extracts are usually fractionated by column chromatography on a preparative scale before they are subjected to detailed analysis. Solid-liquid (adsorption), liquid-liquid (partition), and ion exchange chromatography are among the widely used methods of lipid fractionation. In solid-liquid chromatography, separation is based on partioning and/or adsorption of the lipid components between solid and liquid (mobile) phases. Elution of the desired lipid class is achieved by varying the polarity and strength of the mobile phase. Common stationary phases for column chromatography are silica, alumina, and ion exchange resins.

Low pressure column chromatography using 50- to 500-mesh adsorbents is often chosen for separation of different lipid classes. The main parameters involved in column chromatography are the weight of the adsorbent, the conditioning of the adsorbent (moisture content), and the column size. It is generally accepted that long narrow columns give the best resolution, but large diameter columns increase sample capacity. Diameters over 5 cm and heights over 45 cm are not convenient for typical laboratory use. The preferred column material for lipid analysis is silicic acid or Florisil (magnesium silicate).

In adsorption chromatography, compounds are bound to the solid adsorbent by polar, ionic, and, to a lesser extent, nonpolar or van der Waals forces. Therefore, separation of lipid components takes place according to the relative polarities of the individual components, which are determined by the number and type of nonpolar hydrophobic groups. In general, elution of the column with solvents of increasing polarity separates the lipid mixture according to increasing polarity of its components in the following order: saturated hydrocarbons, unsaturated hydrocarbons, wax esters, steryl esters, long chain aldehydes, triacylglycerols, long chain alcohols, free fatty acids, quinones, sterols, diacylglycerols, monoacylglycerols, cerebrosides, glycosyl diacylglycerols, sulfolipids, acidic glycerophosphatides, phosphatidylethanolamine, tysophosphatidylethanolamine, phosphotidylcholine, sphingomyelin, and lysophosphatidylcholine. The procedure applicable to most lipid mixtures is eluting on a silicic acid column with the solvent sequence of chloroform, acetone, and methanol, to separate into neutral lipids, glycolipids, and phospholipids, respectively.

The ion exchange columns carry ionic groups that bind to the opposite charge of the ionic groups of lipids. Thus, a polymer with fixed cations will bind anionic lipids from a mixture, provided the pH of the solvent mixture allows ionization of the anionic groups. At the same time, the concentration of nonlipid anions in the solvent mixture should not compete for all the fixed ions. Diethylaminoethyl (DEAE) cellulose is used for separation of lipid classes, and triethylaminoethyl (TEAE) cellulose is useful for separation of lipids having only ionic carboxyl groups (*e.g.*, fatty acids, bile acids, gangliosides) or phosphotidylethanolamine from ceramide polyhexosides. In polar lipid analysis, DEAE cellulose in the acetate form is the most frequently used anion exchange material. It is most effective in the pH range of 3 to 6, and often separation of polar lipid is achieved by stepwise elution with ammonium acetate buffer in water-methanol. The cation exchanger carboxymethylcellulose (CMC) as its sodium salt has been used occasionally over the same pH range for separation of phospholipids.

The analysis of lipids can be done effectively by gas chromatography (GC).

The analysis of lipids by GC [or gas-liquid chromatography (GLC)] has been much studied in the literature. In this method, the components of the lipid mixture are partitioned in the vapor state between a mobile gas phase and a stationary nonvolatile liquid phase dispersed on an inert support.

Table 10.2: Ion Exchange Chromatography Materials Used in Lipid Analysis

Ionizing Group	*Commercial Classification*	*Analytical Use*
$-(CH_2)_2NH(C_2H_5)_2$ (diethylaminoethyl)	DEAE-(anionic exchanger)	Anionic lipids (phospholipids), sulfolipids, sialoglycolipids, fatty acids)
$-(CH_2)_2\overset{+}{N}(C_2H_5)_3$ (triethylaminoethyl)	TEAE-(anionic exchanger)	Anionic lipids
$-CH_2COO-$ (carboxymethyl)	CM-(cationic exchanger)	Phospholipid mixtures

Analysis of fatty acid composition by GC usually requires derivatization of fatty acids to increase their volatility. Fatty acid methyl esters (FAMEs) may be prepared by different transmethylation techniques and then separated on GC columns and detected by FID. The gas phase for GC is usually nitrogen or helium for packed columns and helium or hydrogen for capillary columns. The identification of chromatographic peaks is based on comparison of their retention times with authentic samples. GC analysis of, TAGs of food lipids may also provide Information about positional distribution of fatty acids in the molecules. Naturally occurring triacylglycerols, which are purified by TLC, can then be resolved without derivatization on the basis of their carbon number or molecular weight by means of capillary GC equipped with 8 to 15 m long columns coated with methylphenyl-, methyl-, or dimethylsilicones (nonpolar capillary). Use of helium or hydrogen as a carrier gas for separation of TAGs on such columns requires higher temperatures than are needed for separation of methyl esters. Mono- and diacylglycerols must be converted to trimethylsilyl (TMS) or *tert*-butyldimethylsilyl ethers (TBDMS) for complete resolution. A combined gas chromatographic-mass spectroscopic technique has been applied for determining molecular species on the glycerol esters.

High Performance Liquid Chromatography

HPLC is a highly efficient form of adsorption, partition, or ion exchange liquid chromatography which uses a very uniform, finely divided, microspherical (5–10 μm) support medium of controlled porosity and degree of hydration. The adsorbent is tightly packed into a stainless steel column (10–30 cm long, 2–4 mm diameter), and a high pressure pump ensures adequate and constant flow of solvent through the column. Elution of the column may be carried out either isocratically (with a solvent mixture of constant composition) or by gradient elution, in which the solvent composition is varied linearly or in a stepwise fashion with both binary and ternary solvent systems. The column eluates are continuously monitored by means of a flow-through detector, which should be insensitive to solvent flow rate, temperature, and composition.

Sample derivatization is employed to facilitate the separation and/or to enhance the limit of detection for, the HPLC analysis. Hydrolysis or saponification is done to cleave ester linkages and to obtain free fatty acids for their subsequent analysis. Although lipids do not possess specific UV absorption peaks, they can be detected in the region of 203 to 210 nm by monitoring for the presence of double bonds in the fatty acyl groups, or functionalities such as carbonyl, carboxyl, phosphate, amino, or quaternary ammonium groups. It is advisable to avoid chloroform-methanol mixtures, however, since they display a strong absorption below 245 nm. Frequently, diacylglycerols require preparation of UV absorbing derivatives *tert* benzoate, dinitrobenzoates, pentafluorobenzoates and lerl-butyldimethylsilyl ethers) for detection. Fatty acids can be analyzed by forming 9-anthryldiazomethane (ADAM) derivatives and employing a fluorescence detector. Fatty acids increase their hydrophobicity when ADAM is bound to the carboxyl group, thus the derivatives are retained longer on a reversed phase column than fatty acids, and good separation is obtained. Fluorescence detectors are specific for detection and quantification of tocopherols and fluorescent derivatives of fatty acids.

The evaporative light-scattering detector (ELSD) and the flame ionization detector have been in wide use for detection of all types of lipids following HPLC separation. These instruments are sometimes referred to as universal detectors. The principle of ELSD involves evaporation of the mobile phase of the separated lipid fraction by means of a nebulizer (spray the eluent stream with a large volume of nitrogen or air) to obtain droplets of solute (lipids). These solute droplets are directed through a light source (perhaps a laser light source); the degree of scattering of the light is proportional to the mass of the solute.

Normal phase HPLC also allows the separation of TAGs containing normal chain and hydroxy fatty acids. Normal phase HPLC on silver ion loaded anion exchange columns is currently used to resolve TAGs based on their degree of unsaturation. Reversed phase HPLC (using C-18 columns) is also widely applied for separation and quantification of tri-, di-, and monoacylglycerols.

Since glycerol possesses a prochiral carbon, asymmetric esterification of the primary positions leads to the formation of enantiomers. Although enantiomeric triacylglycerols cannot be resolved by normal HPLC, their diastereomeric naphthylethylurethane derivatives can be separated by HPLC on silica gel. Use of a stationary phase with chiral moieties to separate enantiomers of mono- and diacyl-*sn*-glycerols after derivatization with 3,5-dinitro-phenylurethane (DNPU) by HPLC separation of enantiomers has been reported.

Separation of phospholipids is very laborious in TLC analyses, however, HPLC provides a better means of separation and quantification. At present, gradient (binary or tertiary) elution on silica columns is frequently used for separation of different classes of phospholipids. Several polar solvent systems that are suitable for such separations are available. UV, flame ionization, or evaporative light-scattering, detectors are suitable for phospholipid identification and quantification. Separation of glycolipids can be achieved by means of a silica column with a binary gradient: hexane-IPA-2.8 mM ammonium acetate or reversed phase C-18 (ODS).

Isocratic, normal phase separation of cholesterol esters, free fatty acids, and free sterols is commonplace. Simultaneous analysis of nonpolar and polar lipids using HPLC silica gel column has also been reported. In addition to normal and reversed phase methods, size exclusion HPLC has been used to separate TAGs and other nonpolar lipids. This approach has been specifically employed for the analysis of polymerized lipids such as those generated during deep fat frying. Christopoulou and Perkins have described separation of monomers, dimers, and trimers of fatty acids in oxidized lipids by means of a size exclusion column with a refractive index detector, while Burkov and Henderson have reported the analysis of polymers in autoxidized marine oil by means of a similar column with ELSD.

Micro-HPLC columns, having a volume one-hundredth that of a conventional column, benefit from low consumption of materials used as stationary phase, sample, and mobile phase. This apparatus, which can be easily coupled with a mass spectrometer and a Fourier transform infrared (FTIR) detector, operates at a low flow rate and offers temperature programming.

Supercritical fluids may be used as the mobile phase in supercritical fluid chromatography. Let us learn more detail about SFL.

Carbon dioxide has been widely used as a mobile phase in SFC, since it has a low critical temperature (31.1°C) and pressure. Such temperatures are not detrimental to lipids, and also the operating pressure is much lower than HPLC. Since CO_2 is nonpolar, its supercritical fluid can dissolve less polar compounds and is suitable for the analysis of less polar species.

To analyze polar components, polar solvents such as methanol or ethanol may be added to the supercritical fluid to cover active sites (–Si–OH) on the surface of the supporting material and to increase the dissolving power of the mobile phase. Both packed and capillary columns are used for SFC. Packing materials developed for HPLC are suitable for SFC packed columns. Similarly, fused silica capillary tubes used for GC are suitable for SFC, and stationary phases may employ dimethylpolysiloxane, methylphenylpolysiloxane, diphenylpolysiloxane, and cyanopropylpolysiloxane.

Capillary SFC with FID or UV has been used for analysis of TAGs, free fatty acids, and their derivatives. Supercritical fluid argentation chromatography has been used for the separation of TAGs according to the number of double bonds, the chain length, and the nature of the double bonds. Detectors such as FID, UV, and FTIR developed for GC and HPLC are applicable to SFC.

Enzymatic methods of lipid analysis is also an effective method for determining triacylglycerol content of samples. The TAG is reacted with lipase to obtain glycerol and free fatty acids.

The glycerol so produced is then converted to α-glycerophosphate by means of glycerol kinase. The α-glycerophosphate dehydrogenase is then used to reduce nicotinamide adenine diphosphate (NAD) to NADH. The resultant NADH is then measured by a colorimetric reaction. Cholesterol oxidase is used for determining cholesterol concentration in blood plasma. Polyunsaturated fatfy acids (PUFAs) with *cis*-methylene groups between their double bonds (*e.g.*, linoleic, linolenic and arachidonic acids) can be quantitatively measured by reading the UV absorbance of conjugated diene hydroperoxides produced via lipoxygenase (lipoxidase), catalyzed oxidation. The extinction coefficient of diene hydroperoxides at 234 nm is the same for all PUFAs. Fatty acid esters must be saponified prior to such analysis. The phosphatidylcholine or lecithin content of samples (*e.g.*, as a measure of the egg content of foods) can be found by catalyzing the conversion of lecithin to phosphatidic acid and choline by means of lecithinase (phospholipase D).

The method of stereospecific analysis of TAGs described by Brockerhoff and Yurkowski uses pancreatic lipase that eventually removes fatty acids from the *sn*-1 and *sn*-2 positions of the TAG. This procedure was employed recently to determine the existing structural differences of fish oils and seal oil. Phospholipase A_2 is used to release fatty acids at the *sn*-2 position of the synthesized phosphotide during the analytical procedure.

Lands used a different approach to determine steric positions of the fatty acids. TAG is hydrolyzed with lipase and the products are separated by TLC. The *sn*-3 hydroxyl of glycerol is then phosphorylated with diacylglycerol kinase to produce 3-phosphoryl monoacylglycerol. In the following analytic step, phospholipase A_2 is used to remove the fatty acids only from the *sn*-2 position. The fatty acids in the *sn*-1 position can be released by saponifying the resultant 1-acyl lysophosphatidic acid. The fatty acids in positions 1 and 2 can be identified by GC analysis. The composition of fatty acids in position 3 can be calculated by comparing these results with those from the fatty acid composition determined for the TAG.

Chapter 11
Crystallization of Fats

11.1 Introduction

Crystallization can be considered a subset of overall solidification. Crystallization requires a solute concentration greater than the concentration of the saturated solution.

The crystallization process from solution first requires supersaturation; supercooling is a prerequisite for crystallizing from a melt. These phenomena lead to nucleation and crystal growth.

Once formed, crystals can have different shapes, which are called "habits" or morphologies. Stable crystals modify their habit, whereas metastable ones undergo phase transitions. Both these processes result in polymorphic behavior, a behavior common to fats and other lipids. Further, most crystals "ripen" and disappear, as a result of changes in the degree of supersaturation. Supersaturation "evolves" as crystal growth proceeds, and the liquid phase becomes less supersaturated. This reduced supersaturation results in a stability requirement for larger crystals, since crystals below a critical size will return to solution or the melt.

While studying about crystallization it is very much needed to learn about supercooling of materials.

Only when sufficient supercooling has been achieved can crystallization centers or nuclei form. A crystal nucleus is the smallest crystal that can exist in a solution with defined temperature and concentration. Crystal nuclei were further distinguished from crystal "embryos" as molecular aggregates that continue to grown rather than redissolve.

Generally, it is accepted that both homogenous and heterogeneous forms of nucleation occur. The former occurs in the absence of any foreign particle surfaces. The latter type occurs in practical systems when nuclei develop on the surface of solid impurities that are present. Chemical nucleation has also been referred to as a third, distinct type of nucleation. Unlike heterogeneous nucleation, chemical nucleation results when organic agents that were added to a system dissolve and chemically react, activating the polymer. The ionic chain ends, which develop from polymer chain scission, aggregate and form nuclei. Therefore, it appears that chemical nucleation simply refers to chemical induction of heterogeneous nuclei rather than the more direct physical addition of solid heterogeneous nuclei.

Both primary and secondary nucleation have been described by many authors. Secondary nucleation is relatively simple to understand. It results from fracture of growing crystals into smaller stable crystal nuclei. Nuclei stability is a function of nuclei solubility. This solubility, in turn, depends on the nuclei size at a given temperature. Nucleation theory addresses the concept of critical size regarding the development of stable primary nuclei.

An elegant discussion of these concepts was presented by Timms. Two opposing actions exist when molecules attempt to aggregate. The first, energy evolution due to the heat of crystallization, favors crystallization as energy is released from the aggregating embryo. The second action, that of molecular surface enlargement, requires energy input to overcome, surface tension or pressure. A stable nuclei will form only when the heat of crystallization is greater than the energy required to overcome surface energy.

$$\Delta G_{embryo} = 4\pi r^2\lambda - \frac{4\pi r^3 \Delta G_v}{3V_m}$$

where, the first term on the right represents the surface energy contributions and the second depicts the volume contributions of the heat of fusion. Defining terms, ΔG_{embryo} is the Gibbs free energy of the embryo, *r* is the radius of the nucleus, λ is the surface free energy per unit surface area. ΔG_v is the molar free energy change resulting from the melt-solid phase change, and V_m is the molar volume. The overall free energy will reach a maximum at some critical embryo size. Free energy will then tend toward a minimum in all embryos. This occurs through melting of smaller embryos or continued growth of embryos greater than the critical size.

During crystallization process crystal growth takes place. Now let us study about crystal growth and crystal geometry.

Crystal growth continues as the properly configured crystallizing molecule diffuses to the proper place on the growing crystal surface. The rate of growth is directly proportional to supercooling and varies inversely with viscosity, since molecular diffusion is reduced as melt viscosity increases. These crystals will be relatively perfect, since the crystal attachment surface can become more precisely configured with the slower crystallization rate. Greater supercooling, by contrast, affords faster crystallization (provided the viscosity remains adequate for molecular diffusion) but also results in more crystal faults. Faults can arise as molecules from the melt attach to the crystal surface and have insufficient time to become optimally arranged before new attachments are made.

Boistelle described one effect of the variable consequences resulting from different amounts of supercooling–the formation of "metastable phases." These are simply thermodynamically unstable phases formed as a consequence of kinetic preference. In the case of crystallizing materials, the stable phase rarely forms first; rather, the free energy of the first crystal is closest to the free energy of the original melt. The kinetic preference for the unstable, phase is a consequence of a lower difference in surface free energy between the melt and the crystal surface compared to the difference in the stable phase versus the melt. As with nucleation, the rate of growth is proportional to the degree of supersaturation. Since the solubility of a stable phase is less than that of a metastable phase, the nucleation and growth rates of the stable phase are expected to be faster. However, as discussed earlier, solubility curves are developed from thermodynamic data, whereas the initial nucleation and growth are governed by kinetics.

As crystallization continues, the degree of supersaturation in the system decreases. This then causes the critical crystal size to be become greater. Therefore, smaller crystals will dissolve and only larger crystals will grow. Eventually, only one crystal size will be stable. This process, called Ostwald ripening, occurs over a very long time possibly for years.

Sintering has been described as the formation of crystal bridges between preformed crystals in a semisolid dispersion of crystals and liquid. In fats, the bridges are fat crystals having different thermodynamic and kinetic crystallization parameters relative to the precrystallized material. The bridge material can result from fractional crystallization and can consist of triacylglycerols or minor lipid compounds that fail to crystallize with the predominant lipids. Sintering has been measured in flocculation studies that assumed the adhesion between performed crystals (sintering) resulted in greater flocculate volumes. Sintering is important especially relative to the final texture and consistency of fat-based foods.

The process of sintering, like Ostwald ripening, occurs over a long time. As such it and Ostwald ripening are often called postcrystallization processes. Since, however, each describes crystallization phenomena, each is more precisely a crystal maturation process.

Crystals are solids with atoms arranged in a periodic three-dimensional pattern. Representation of the arrangement as a "point lattice" depicts each point having identical surroundings. All cells of a particular lattice are identical; therefore anyone cell's dimensions and angles describe the lattice constants or lattice parameters of the unit cell. The unit cell is the repeating unit of the whole structure. This compares to the concept of the subcell. As the name implies, a subcell is a smaller periodic structure within the real unit cell which defines the repeating unit of the whole structure. This has particular relevance to long chain hydrocarbons, where the real unit cell is large and the subcell geometry is measured with X-ray diffraction. Only seven different cells are necessary to include all the possible point lattices.

Table 11.1: The Seven Crystal Systems

System	*Angles and Axial Lengths*
Triclinic	All axes unequal and none at right angles $a \neq b \neq c$ and $\alpha \neq \beta \neq \gamma$ and $\neq 90^\circ$
Orthorhombic	All axes unequal and all at right angles $a \neq b \neq c$ and $\alpha = \beta = \gamma$ and $= 90^\circ$
Hexagonal	Two axes = 120° and the third at 90° relative to them $a = b \neq c$ and $\alpha = \beta = 90^\circ$ and $\gamma = 120^\circ$
Cubic	All axes equal and all at right angles $a = b = c$ and $\alpha = \beta = \gamma$ and $= 90^\circ$
Tetragonal	Two of three axes equal and all at right angles $a = b \neq c$ and $\alpha = \beta = \gamma$ and $= 90^\circ$
Rhombohedral	All axes equal and none at right angles $a = b = c$ and $\alpha = \beta = \gamma$ and $\neq 90^\circ$
Monoclinic	Three unequal axes having one pair not = 90° $a \neq b \neq c$ and $\alpha = \gamma = 90^\circ \neq \beta$

Solids of the same composition that can exist in more than one form are referred to as polymorphic. Crystal habit has simply been defined as the overall shape of the crystal. From a crystallographic perspective, habit reflects the directional growth within a crystal, whereas morphology describes the set of faces determined by means of the symmetry elements of the crystal. This subtle distinction allows crystals of the same morphology to nevertheless occupy different habits.

Polymorphism is defined in terms of an ability to reveal different unit cell structures resulting from varied molecular packing. Polytypism refers to altered stacking direction of the crystal lamellae. Each lamella is configured as identical polymorphs, but the direction of tilt of the hydrocarbon axis from the methyl end-group plane varies.

11.2 Fat Crystallization and Crystal Packing Structure

All fats are lipids, but the converse is not true. Therefore, a distinction must be drawn to provide an accurate perspective when considering fat crystallization as discussed earlier. Fats are defined

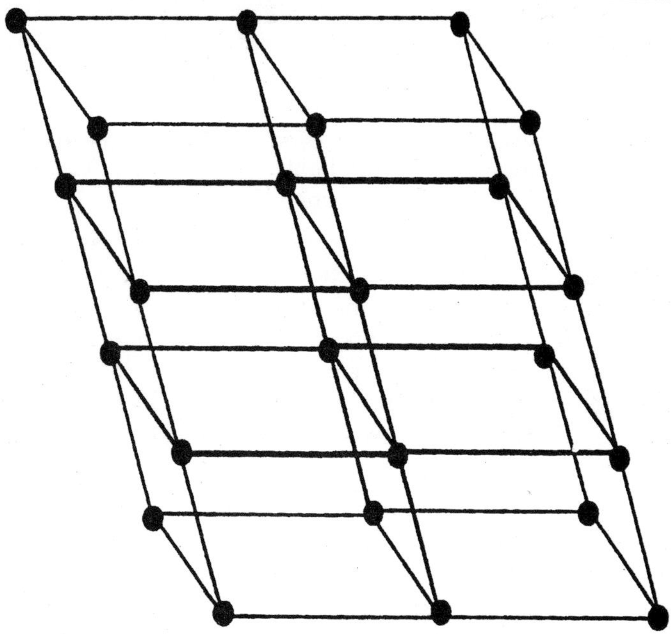

Figure 11.1: A Point Lattice

as follows: "A glyceryl ester of higher fatty acids. . . . .Such esters are solids at room temperature and exhibit crystalline structure. . . . .The term 'fat' usually refers to triacylglycerols specifically, whereas 'lipid' is all-inclusive." Included in the more general term "lipid" are hydrocarbons, steroids, soaps, detergents, all acylglycerols, phospholipids, gangliosides, and lipopolysaccharides. Since all these lipids crystallize with many degrees of complexity, this chapter primarily addresses the crystallization of triacylglycerols, making only limited reference to other lipids.

Lipids have been classified based on their interaction with water. Triacylglycerols belong to class I polar lipids. These are insoluble, nonswelling amphiphiles. Triacylglycerols will spread at the aqueous interface and form a stable monolayer. They have a low affinity for water compared to other class I polar lipids such as diacylglycerol and cholesterol. Class II polar lipids include many of the phospholipids, as well as glycolipids and monoacylgycerol. These, too, are insoluble; however they swell because water is soluble in their polar moieties. A result of this interaction with water is the ability of class II polar lipids to undergo lyotropic mesomorphism and develop into liquid crystals. It has been suggested that mesomorphism of phospholipids influences the nucleation and solidification behavior of cocoa butter.

Both the technical and biological functions of lipids are better understood with knowledge of, their structural composition. X-ray analysis provides much of the structural information known regarding all lipids. Shipley presented a brief history of the X-ray analysis of triacylglycerol single crystals. One of the first direct studies performed, the examination of the triclinic, β-form of trilaurin, set the groundwork for determining the conformational nature of other, β-form triacylglycerols.

Of the seven crystal systems referred to in Section I.D, three predominate in the crystalline triacylglycerols. Usually, the most stable form of triacylglycerols has a triclinic subcell with parallel hydrocarbon-chain planes (T_1). A second common subcell is orthorhombic with perpendicular chain phases (O). The third common subcell type is hexagonal (H) with no specific chain plane conformation. Therefore the hexagonal form exhibits the lowest stability and has the highest Gibbs free energy, closest to the original melt.

Interpretation of X-ray crystallography data from trilaurin and tricaprin resulted in representation of triacylglycerols in a tuning fork conformation when crystalline. The fatty acid esterified at the *sn*-1 and *sn*-2 positions of glycerol are extended and almost straight. The *sn*-3 ester projects 90° from *sn*-1 and *sn*-2, folds over at the carboxyl carbon, and aligns parallel to the *sn*-1 acyl ester. Molecules are packed in pairs, in a single layer arrangement, with the methyl groups and glycerol backbones in separate regions.

The polymorphic structures described for these simple triacylglycerols are valid for natural fats that contain complex triacylglycerols, given their common X-ray short spacings (axes normal to the chain direction). However, this bilayer structure does not exist in all triacylglycerols. Indeed, a trilayer structure has been demonstrated. This trilayer structure occurs when the *sn*-2 position of the triacylglycerol contains a fatty acid ester that is either

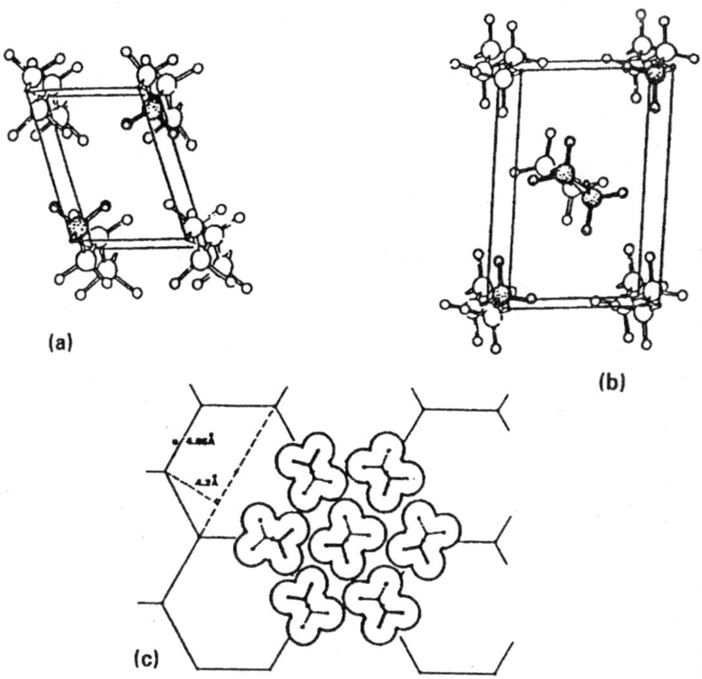

Figure 11.2: Schematic Representations of (a) Triclinic Parallel; (b) Orthorhombic Perpendicular, and (c) Hexagonal Subcells

cis-unsaturated or of a chain length different by four or more carbons from those on the *sn*-1 and *sn*-2 positions.

When unsaturation results in a trans configuration around the carbon-carbon double bond, the crystal structure exhibits the normal bilayer appearance. The trans carbon-carbon double bond results in a linear chain configuration, unlike the bent chain configuration observed in the cis-unsaturated molecule. In fact, trielaidin (*trans*-C18 : 1) has the same polymorphic configuration as tristearin (C18 : 0); although the phase transition temperatures of trielaidin are 30°C below that of tristearin. Synthesized, stereospecific 1,2-dioleoyl-3-acyl-sn-glycerides with even carbon-saturated fatty acyl chains of 14 to 24 carbons have a trilayer, packing structure.

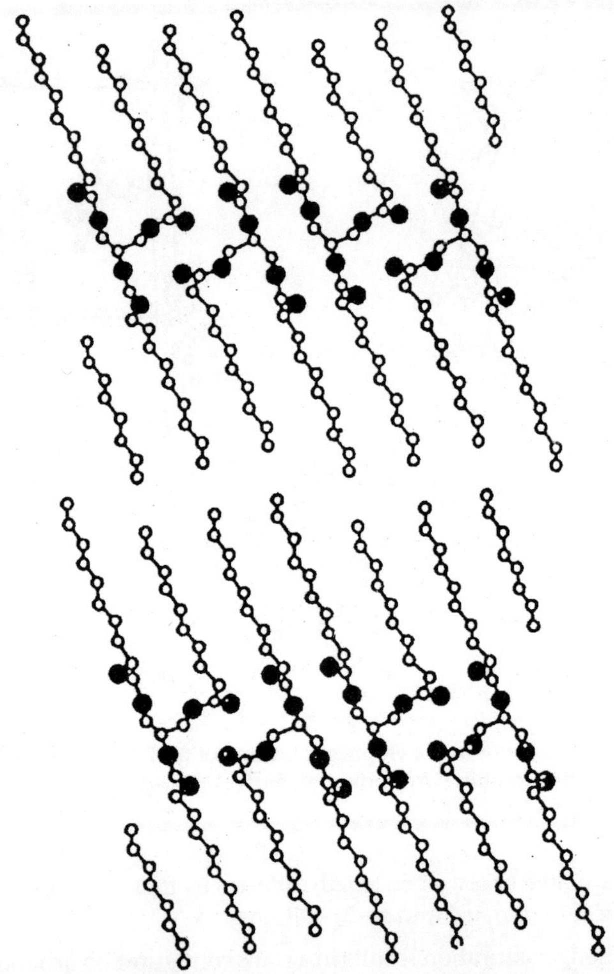

Figure 11.3: Schematic Representation of the Crystal Structure of the Triclinic Form of Tricaprin Projected on the *bc* Plane

11.3 Milk Fat and Phase Behavior of Fats

In milk fat at least 168 different triacylglycerol species have been identified. This complexity results in three distinct endotherms on heating of samples held at 26°C.

As with many natural fats, the temperature at which crystallization occurs influences milk fat firmness, crystalline conformation, and percentage of solid fat. Shear effects on milk fat crystallization have also been investigated. Shear rate was directly related to crystal growth rate at crystallization temperatures of 15 and 20°C. At 30°C, shear was negatively related to growth rate, indicating the possibility of different growth mechanisms. However, the lower amount of supercooling at 30°C necessitates a larger crystal diameter for stability. Therefore, shear may simply interrupt the crystal growth preventing development of the critical crystal size.

Fats may be thermally treated with a process referred to as tempering, to produce specific attributes in a finished product. Tempering normally results in the formation of stable crystals having the proper size and in the proper amount. Hardness variability in milk fat results from different thermal treatments. Less solid fat results when dairy butter is cooled slowly. This is better understood considering the presence of three milk fat fractions with largely independent solid solutions. A DSC thermogram depicts these independent fractions. The fractions are neatly defined as high-, middle-, and low-melting fractions (HMF, MMF, LMF, respectively). LMF is liquid at ambient temperature. Stable polymorphs of MMF

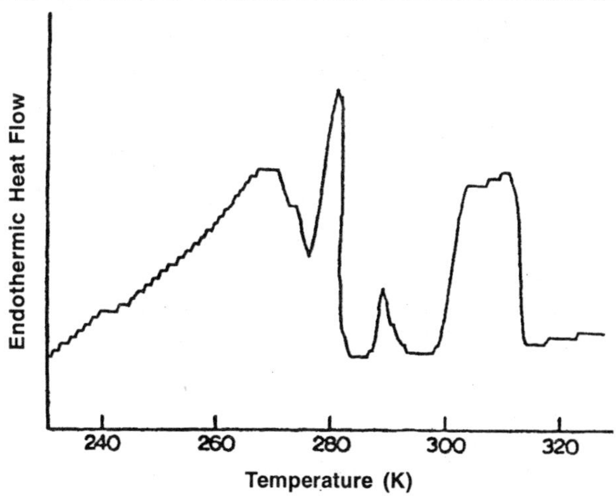

Figure 11.4: DSC Melting Curve of Milk Fat Showing Independent Solid Solutions

and HMF were found to be a mixture of β′–2 + β′–3, and β–2, respectively. The role of LMF in the polymorphic transformations in milk fat fractions was investigated. LMF was found to facilitate the transformation of HMF to the β–2 form but has no effect on MMF, which stayed in the β′ form.

The various milk fat fractions and their different stable forms provide opportunities regarding applications to food and non-food systems. Anhydrous milk fat fractions were evaluated regarding their effect on chocolate tempering. Chocolate with HMF addition required normal chocolate tempering temperatures (< 30°C) to produce adequate temper for molding, even though a high melting endotherm existed after tempering at 31.1°C. This high-melting endotherm was attributed to the high-melting acylglycerols of HMF, which crystallized separately from cocoa butter.

Now let us try to learn about phase behaviour of fats, in brief.

Timms provided a complete review of the phase behavior of fats. A phase is a physical state of matter that is homogeneous and separated from other phases by a definite boundary. Therefore, if only the solid and liquid states of matter are considered, solid-solid, liquid-solid, and liquid-liquid phases can exist. Triacylglycerol mixtures are ideally miscible when liquid and therefore, show no heat or volume changes when mixed. As a consequence, distinct liquid-liquid phases are not apparent. Many natural fats clearly exhibit several distinct solid phases with partial miscibility, which leads to compound crystals and solid solutions. Identification of these solid solutions with X-ray powder diffraction is effective, particularly with fats in equilibrium or relatively stable metastable forms. Thermal analysis using differential scanning calorimetry (DSC) provides for observation of the effect of changing temperature.

An idealized set of DSC thermograms is depicted in Figure 11.5. Melting profiles can be correlated to definitive X-ray determination, which allows subsequent estimation of polymorphic form using DSC alone. However, polymorph analysis using DSC alone depends on thermal history, and other techniques may be necessary to sort out the complexity.

11.4 The Phase Behaviour of Cocoa Butter

Two reviews of the composition of cocoa butter provide a good understanding of the heterogeneity of this natural fat. More than 98

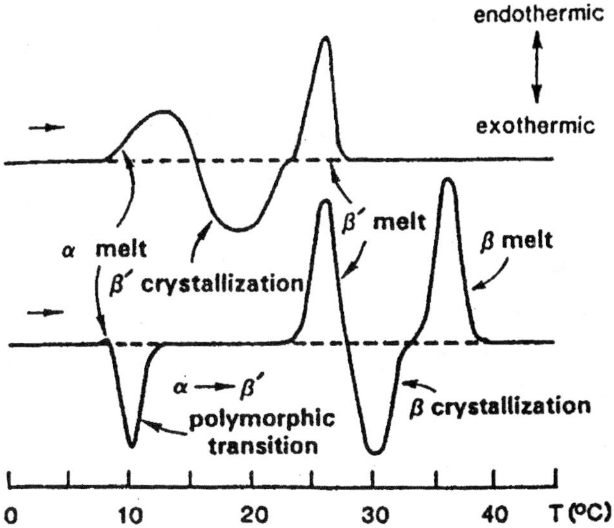

Figure 11.5: Schematic Representation of Polymorph Identification by Means of Differential Scanning Calorimetry (DSC)

per cent of cocoa butter is simple lipid. More than 95 per cent of this is triacylglycerol. Three triacylglycerols predominate: 1-palmitoyl-2-oleoyl-3-stearoyl-*sn*-glycerol (POS) composes about 40 per cent, 1,3-distearoyl-2-oleoyl-*sn*-glycerol (SOS) makes up 27.5 per cent, and 1,3-dipalmitoyl-2-oleoyl-*sn*-glycerol (POP) makes up about 15 per cent. At least 15 minor triacylglycerols have also been identified. Free fatty acid values are reported to be 1.5 per cent, and the concentration of mono- and diacylglycerol is about 2 per cent of the simple lipid fraction. The concentration of phospholipid and glycolipids in cocoa butter varies by climate and method of analysis. These complex lipids constitute about 1 per cent of original cocoa butter.

More is known about the phase behavior of cocoa butter than that of any other fat. X-ray diffraction techniques have been used to define six polymorphs, each having a distinctive melting point. Interestingly, refined cocoa butter was used in this work, and the possible effects of the compounds removed by refining appear to have been discounted. The existence of six polymorphs in a study with 12 different cocoa butters was confirmed. Thermal analysis alone was used to evaluate the polymorphism of cocoa butter and mixtures of cocoa butter with other fats. That study led to some of the confusion with terminology regarding cocoa butter polymorphs in that the previously defined designations (from lowest, I, to highest, VI, melting point) were reversed regarding melting temperature. Also, six cocoa butter polymorphs were found in work combining microscopy and thermal analyses. Table 11.2 lists the nomenclature and melting points of the cocoa butter polymorphs as determined by several groups. The definitions of cocoa butter polymorphs follow several conventions. Consistency throughout this chapter requires the Wille and Lutton "I through VI" system if thermal data are referenced, and the Hernqvist system if X-ray data are referenced. When appropriate, only the Wille and Lutton system is used.

Table 11.2: Nomenclature and Melting Point (ºC) of Cocoa Butter Polymorphs

Wille and Lutton		*Lovegren et al.*		*Hicklin et al.*		*Davis & Dimick*		*Hernqvist*	*Larson*
Form	*mp*	*Form*	*mp*	*Form*	*mp*	*Form*	*mp*		
I	17.3	VI	13.0	I	17.9	I	13.1	sub-α	β'_2 or γ
II	23.3	V	20.0	II	24.4	II	17.1	α	α
III	25.5	IV	23.0	III	27.7	III	22.4	β'_2	$\alpha + \beta$
IV	27.3	III	25.0	IV	28.4	IV	26.4	β'_1	α'
V	33.8	II	30.0	V	33.0	V	30.7	β	β_2
VI	36.3	I	33.5	VI	34.6	VI	33.8	β	β

A good description of the procedures necessary to form the six cocoa butter polymorphs is available. The microscopic analyses used in this work also provide insight into the crystal habit of the forms. Each crystal formation procedure required a different method of tempering. Other than rapid cooling of the melt to 0°C or lower, no reference was made by any of the authors regarding the rate of cooling. Neglect of this crystallization parameter or the assumption that the crystallization temperature was most important is one explanation for the lack of reference to cooling rate. In fact, van Malssen *et al.* found that crystallization temperature, not rate, determined which form would crystallize from the melt. However, their data indicate a possible discrepancy. They stated at 0.25°C/min cooling, β′ crystals formed between 22 and 26°C. At a cooling rate of 1°C/min, the α form crystallized at less than 23°C. Further, at a cooling rate of 6°C/s, both α and γ (sub-α) forms crystallized at less than 3°C. These authors' observation of no β crystallization after 10 days at 28°C is supported by recent real-time X-ray data, which showed no direct crystallization of β (form VI) from cocoa butter melts. As discussed above, cocoa butter transforms monotropically, from the least stable to the more stable polymorphs.

One unique consequence of the polymorphism of cocoa butter is called fat "bloom." Investigations of bloom formation were published as early as 1937. Bloom occurs in chocolate, where it appears as a thin coating or scattered white patches on the surface. Neville *et al.* attempted to describe the mechanism of bloom formation in terms of physical expansion of melting fat pushing higher melting solid fat to the surface. Full presented a good review of the current understanding of bloom. Three possible mechanisms of bloom formation have been described. Schlichter-Aronhime and Garti argued that because bloom results from a form IV to V conversion and from a V to VI conversion, the condition is caused by liquefaction of fat and does not depend strictly on the appearance of a specific polymorphic form. Regardless of the actual mechanism or mechanisms of bloom formation, the polymorphism and phase behavior of cocoa butter underlie the phenomenon.

Schlichter-Aronhime and Garti compared the addition of low percentages of surfactants to cocoa butter to samples of cocoa butter blended with other fats. They noted that fats can change the melting ranges and the number and type of polymorphs, dependent on system compatibility, whereas surfactants may affect the rate of transformation without sensibly altering the crystal lattice. These authors' generalization of work regarding surfactants and polymorphism stated that surfactants stabilize metastable polymorphs, thereby delaying transformation to the most stable form.

Chocolate bloom was found to be inhibited by sorbitan monostearate. Also, sorbitan tristearate was shown to inhibit the transformation of form V cocoa butter to form VI. However, in the same series of experiments, these authors found that sorbitan tristearate hastened the transformation of the metastable polymorphs to form V. They reported that this occurs as a result of the liquefaction effect of the emulsifier and the liquid-mediated transformation process to form V. They reasoned that form V is stabilized because it transforms to form VI via the solid state, simply by expulsion of trapped liquid. Therefore, the emulsifier promotes only liquid-mediated transformations.

The effectiveness of the emulsifier relative to its state appears conflicting. Schlichter-Aronhime and Garti stated that solid emulsifiers efficiently delay transformations, whereas liquid ones have no effect.

The crystal morphology of cocoa butter during isothermal crystallization has been detailed. A higher concentration of StOSt was found in crystals that developed early versus the original StOSt concentration in the melt. It was hypothesized that subsequent crystallization occurred on these crystal "seeds." Such fractional crystallization may produce crystal seeds, rich in StOSt, which have been shown to accelerate cocoa butter crystallization.

Compounds other than triacylglycerols, found in cocoa butter, have also been investigated for their potential to act as crystal seeds. High-melting crystals isolated from crystallizing cocoa butter melts during the very early stages of crystallization were found to contain very high phospholipid and glycolipid concentrations relative to the original cocoa butter. Dimick proposed that these amphiphilic compounds may associate with the small amount of water in cocoa butter and serve as the nuclei for crystallization. The phospholipid species found in both the crystal seed and the original butter were identified. The data indicated that faster crystallizing cocoa butters contained a relatively high percentage of phosphatidylcholine and phosphatidylglycerol, whereas the slower crystallizing samples had more phosphatidylinositol and significantly less phosphatidylcholine.

The effects on cocoa butter crystallization from simple degumming and those due to added phospholipid seed material were subsequently investigated in the same laboratory. Degummed Bahlan and Côte d'Ivoire cocoa butters both had significantly slower crystallization rates relative to the untreated buffers. Addition of pure *sn*-1,2-distearoyl-phosphatidylcholine completely inhibited the crystallization of the original butters. However, addition of 0.1 per cent of this phospholipid to the degummed butters increased their crystallization rates to match the untreated samples. After tempering, the original Côte d'Ivoire butter containing pure *sn*-1,2-distearoyl-phosphatidylcholine had greater solids at 30°C (via NMR) than the untreated butter. Pure *sn*-1,2-dioleoyl-phosphatidylcholine increased the crystallization rate of the Côte d'Ivoire butter but inhibited the crystallization of the Bahian butter.

Chapter 12

Emulsion and Emulsifiers

12.1 Introduction

An emulsion is a dispersion of droplets of one liquid in another liquid with which it is incompletely miscible. Now let us try to understand more information about emulsion.

Emulsions are thermodynamically unstable systems because of the positive free energy required to increase the surface area between the oil and water phases. The origin of this energy is the unfavorable interaction between oil and water, which exists because water molecules are capable of forming strong hydrogen bonds with other water molecules, but not with oil molecules. Thus emulsions tend to reduce the surface area between the two immiscible liquids by separating into a system that consists of a layer of oil (lower density) on top of a layer of water (higher density). This is clearly seen if one tries to *homogenize* pure oil and pure water together: initially an emulsion is formed, but after a few minutes phase separation occurs.

Emulsion instability can manifest itself through a variety of physico-chemical mechanisms, including creaming, flocculation, coalescence, and phase inversion. To form emulsions that are kinetically stable for a reasonable period of time (a few weeks, months, or even years), chemical substances known as *emulsifiers* must be added prior to homogenization. Emulsifiers are surface-active molecules that adsorb to the surface of freshly formed droplets during homogenization, forming a protective membrane that prevents the droplets from coming close enough together to aggregate. Most food emulsifiers are *amphiphilic* molecules; that is, they have both polar and nonpolar regions on the same molecule. The most common types used in the food industry are lipid-based emulsifiers (small molecule surfactants and phospholipids) and amphiphilic biopolymers (proteins and polysaccharides).

Most food emulsions are more complex than the simple three-component (oil, water, and emulsifier) system described above. The aqueous phase may contain water-soluble ingredients of many different kinds, including sugars, salts, acids, bases, surfactants, proteins, and polysaccharides. The oil phase may contain a variety or lipid-soluble components such as triacylglycerols, diacylglycerols, monoacylglycerols, fatty acids, vitamins, and cholesterol. The interfacial membrane may be comprised of surface-active components of a variety of types, including small molecule surfactants, phospholipids, polysaccharides, and proteins. Some of the ingredients in food emulsions are not located exclusively in one phase, but are distributed between the oil, water, and interfacial

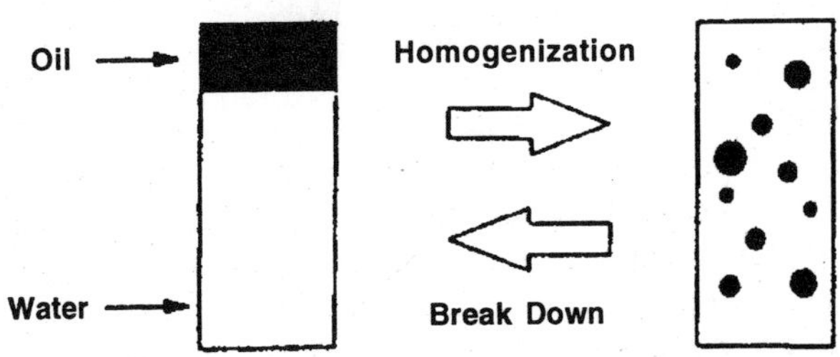

Figure 12.1: Emulsions are Thermodynamically Unstable Systems that Tend to Revert Back to the Individual Oil and Water Phases with Time.
To produce an emulsion, energy must be supplied.

phases according to their partition coefficients. Despite having low concentrations, many of the minor components present in an emulsion can have a pronounced influence on its bulk physico-chemical properties. Food emulsions may consist of oil droplets dispersed in an aqueous phase (*e.g.*, mayonnaise, milk, cream, soups), or water droplets dispersed in an oil phase (*e.g.*, margarine, butter, spreads). The droplets and/or the continuous phase may be fluid, gelled, crystalline, or glassy. The size of the droplets may vary from less than a micrometer to a few hundred micrometers, and the droplets themselves may be more or less polydisperse.

To complicate matters further, the properties of food emulsions are constantly changing with time because of the action of various chemical, physical, and biological processes. In addition, during their processing, storage, transport, and handling, food emulsions are subjected to variations in their temperature (*e.g.*, via sterilization, cooking, chilling, freezing) and various mechanical forces (*e.g.*, stirring, mixing, whipping, flow through pipes, centrifugation) that alter their physico-chemical properties. Despite the compositional, structural, and dynamic complexity of food emulsions, considerable progress has been made in understanding the major factors that determine their bulk physico-chemical properties.

12.2 The Unfolding of Biopolymers

The conformation and unfolding of biopolymers at oil-water interfaces depends on their molecular structure.

Biopolymers that have a high proportion of nonpolar groups tend to be surface-active, that is, they can accumulate at oil-water interfaces. The major driving force for adsorption is the hydrophobic effect. When the biopolymer is dispersed in an aqueous phase, some of the nonpolar groups are in contact with water, which is a thermodynamically unfavorable condition. By adsorbing to an interface, the biopolymer can adopt a conformation of nonpolar groups in contact with the oil phase (away from the water) and hydrophilic groups located in the aqueous phase (in contact with the water). In addition, adsorption reduces the number of contacts between the oil and water molecules at the interface, thereby reducing the interfacial tension. The conformation a biopolymer adopts at an oil-water interface, and the physico-chemical properties of the membrane formed, depend on its molecular structure. Flexible random coil biopolymers adopt an arrangement in which the predominantly nonpolar segments protrude into the oil phase, the predominantly polar segments protrude into the aqueous phase, and the neutral regions lie flat against the interface. The membranes formed by molecules of these types tend to have relatively open structures, to be relatively thick, and to have low viscoelasticities. Globular biopolymers (usually proteins) adsorb to an interface so that the predominantly nonpolar regions on their surface face the oil phase; thus they tend to have a definite orientation at an interface. Once they have adsorbed to an interface, biopolymers often undergo structural rearrangements that permit them to maximize the number of contacts between nonpolar groups and oil.

Random coil biopolymers have flexible conformations and therefore rearrange their structures rapidly, whereas globular biopolymers are more rigid and therefore, unfold more slowly. The unfolding of a globular protein at an interface often exposes amino acids that were originally located in the hydrophobic interior of the molecule, which can lead to enhanced interactions with neighboring protein molecules through hydrophobic attraction or disulfide bond formation. Consequently, globular proteins tend to form relatively thin and compact membranes, high in viscoelesticity. Thus, membranes formed from globular proteins tend to be more resistant to rupture than those formed from random coil proteins.

To be effective emulsifiers, biopolymers must rapidly adsorb to the surface of the emulsion droplets formed during homogenization and provide a membrane that prevents the droplets from

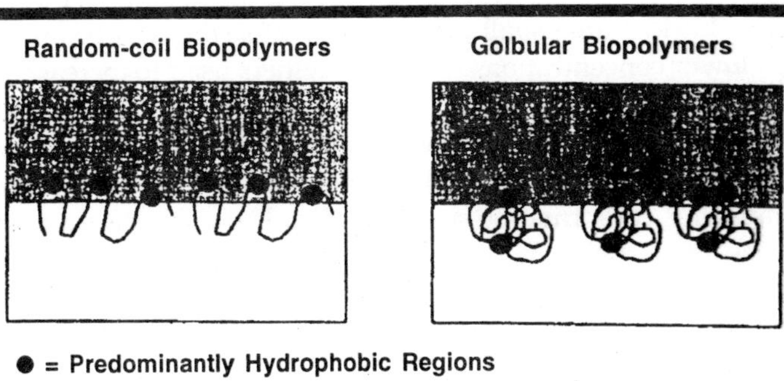

Figure 12.2: The Conformation and Unfolding of Biopolymers

aggregating. Biopolymer membranes can stabilize emulsion droplets against aggregation by a number of different physical mechanisms. All biopolymers are capable of providing short-range steric repulsive forces that are usually strong enough to prevent droplets from getting sufficiently close together to coalesce. If the membrane is sufficiently thick, it can also prevent droplets from flocculating. Otherwise, it must be electrically charged so that it can prevent flocculation by electrostatic repulsion. The properties of emulsions stabilized by charged biopolymers are particularly sensitive to the pH and ionic strength of aqueous solutions. At pH values near the isoelectric point of proteins, or at high ionic strengths, the electrostatic repulsion between droplets may not be large enough to prevent the droplets from aggregating.

Proteins are commonly used as emulsifiers in foods because many of them naturally have a high proportion of nonpolar groups. Most polysaccharides are so hydrophilic that they are not surface-active. A small number of naturally occurring polysaccharides, however, have some hydrophobic character (*e.g.*, gum arabic) or have been chemically modified to introduce nonpolar groups (*e.g.*, some modified starches), and these biopolymers can be used as emulsifiers.

12.3 The Role of Biopolymers in Food Emulsions

The major role of biopolymers in food emulsions is to increase the viscosity of the aqueous phase. This modifies the texture and mouth feel of the food product ("thickening"), as well as reducing the rate at which particles sediment or cream ("stabilization"). Both proteins and polysaccharides can be used as thickening agents, but polysaccharides are usually preferred because they can be used at much lower concentrations. The biopolymers used to increase the viscosity of aqueous solutions are usually highly hydrated and extended molecules or molecular aggregates. Their ability to increase the viscosity depends principally on their molecular weight, degree of branching, conformation, and flexibility. The viscosity of a dilute solution of particles increases as the concentration of particles increases.

$$\eta = \eta_0(1 + 2.5\phi)$$

where η is the viscosity of the solution, η_0 is the viscosity of the pure solvent, and ϕ is the volume fraction of particles in solution.

Biopolymers are able to enhance the viscosity of aqueous solutions at low concentrations because they have an *effective* volume fraction that is much greater than their actual volume fraction. A biopolymer rapidly rotates in solution because of its thermal energy, and so it sweeps out a spherical volume of water that has a diameter approximately equal to the end-to-end length of the molecule. The volume of the biopolymer molecule is only a small fraction of the total volume of the sphere swept out, and so the effective volume fraction of a biopolymer is much greater than its actual volume fraction. Consequently, small concentrations of biopolymer can dramatically increase the viscosity of a solution. The effectiveness of a biopolymer at increasing the viscosity increases as the volume fraction it occupies within the sphere it sweeps out decreases. Thus large, highly extended linear biopolymers increase the viscosity more effectively than small compact or branched biopolymers.

In a dilute biopolymer solution the individual molecules (or aggregates) do not interact with each other. When the concentration of biopolymer increases above some critical value; the viscosity increases rapidly because the spheres swept out by the biopolymers overlap with each another. This type of solution is known as a *semidilute* solution, because even though the molecules are interacting with one another, each individual biopolymer is still

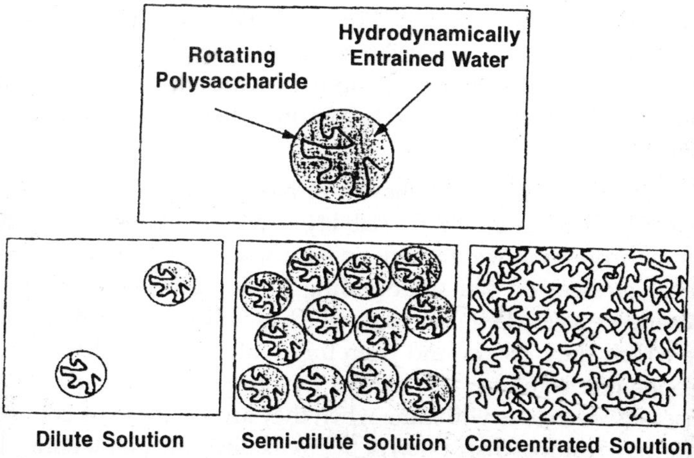

Figure 12.3: Extended Biopolymers in Aqueous Solutions

largely surrounded by solvent molecules. At still higher polymer concentrations, the molecules pack so close together that they become entangled, and the system has more gel-like characteristics. Biopolymers that are used to thicken the aqueous phase of emulsions are often used in the semidilute concentration range.

Solutions containing extended biopolymers often exhibit strong shear-thinning behavior; that is, their apparent viscosity decreases with increasing shear stress. Some biopolymer solutions even have a characteristic yield stress. When a stress is applied below the yield stress, the solution acts like an elastic solid, but when it exceeds the yield stress the solution acts like a liquid. Shear thinning tends to occur because the biopolymer molecules become aligned with the shear field, or because the weak physical interactions responsible for biopolymer-biopolymer-interactions are disrupted. The characteristic rheological behavior of biopolymer solutions plays an important role in determining their functional properties in food emulsions. For example, a salad dressing must be able to flow when it is poured from a container, but must maintain its shape under its own weight after it has been poured onto a salad. The amount and type of biopolymer used must therefore be carefully selected to provide a low viscosity when the salad dressing is poured (high applied stress), but a high viscosity when the salad dressing is allowed to sit under its own weight (low applied stress).

The viscosity of biopolymer solutions is also related to the mouth-feel of a food product. Liquids that do not exhibit extensive shear-thinning behavior at the shear stresses experienced in the mouth are perceived as being "slimy." On the other hand, a certain amount of viscosity is needed to contribute to the "creaminess" of a product.

The shear-thinning behavior of biopolymer solutions is also important for determining the stability of food emulsions to creaming. As oil droplets move through an emulsion, they exert very small shear stresses on the surrounding liquid. Consequently, they experience a very large viscosity, which greatly slows down the rate at which they cream and therefore, enhances stability. Many biopolymer solutions also exhibit thixotropic behavior (*i.e.*, their viscosity decreases with time when they are sheared at a constant rate) as a result of disruption of the weak physical interactions that cause biopolymer molecules to aggregate. A food manufacturer must therefore, select an appropriate biopolymer or combination of biopolymers to produce a final product that has a desirable mouth-feel and texture.

Biolpolymers are used as functional ingredients in man food emulsions (*e.g.*, yogurts, cheeses, desserts, egg and meat products) because of their ability to cause the aqueous phase to gel. Gel formation imparts desirable textural and sensory attributes, as well as preventing the droplets from creaming. A biopolymer gel consists of a three-dimensional network of aggregated or entangled biopolymers that entraps a large volume of water, giving the whole structure some "solidlike" characteristics. The appearance, texture, water-holding capacity, reversibility, and gelation temperature of biopolymer gels depend on the type, structure, and interactions of the molecules they contain.

Gels may be transparent or opaque, hard or soft, brittle or rubbery, homogeneous or heterogeneous; they may exhibit syneresis or have good water-holding capacity. Gelation may be induced by a variety of different methods, including altering the temperature, pH, ionic strength or solvent quality; adding enzymes; and increasing the biopolymer concentration. Biopolymers may be cross-linked by covalent and/or non-covalent bonds.

It is convenient to distinguish between two types of gel: particulate and filamentous. Particulate gels consist of biopolymer aggregates (particles or clumps) that are assembled together to form a three-dimensional network. This type of gel tends to be formed when there are strong attractive forces over the whole surface of the individual biopolymer molecules. Particulate gels are optically opaque because the particles scatter light, and they are prone to syneresis because the large interparticle pore sizes mean that the water is not held tightly in the gel network by capillary forces. Filamentous gels consist of filaments of individual or aggregated biopolymer molecules that are relatively thin and tend to be formed by biopolymers that can form junction zones only at a limited number of sites on the surface of a molecule, or when the attractive forces between the molecules are so strong that they stick firmly together and do not undergo subsequent rearrangement. Filamentous gels tend to be optically transparent because the filaments are so thin that they do not scatter light significantly, and they tend to have good water-holding capacity because the small pore size of the gel network means that the water molecules are held tightly by capillary forces.

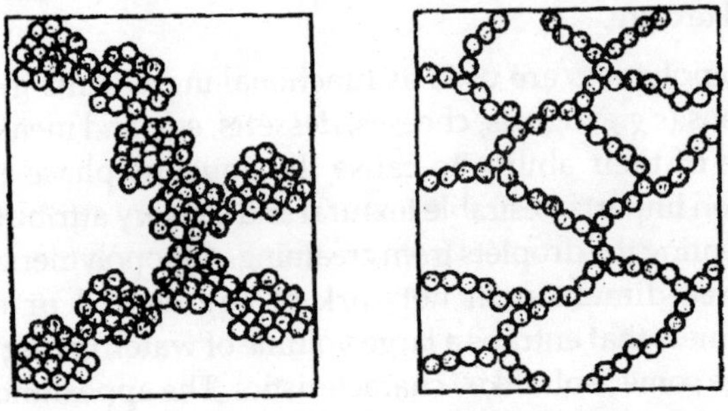

Figure 12.4: Biopolymer Molecules

In some foods a gel is formed upon heating (*heat-setting* gels), while in others it is formed upon cooling (*cold-setting* gels). Gels may also be either *thermoreversible* or *thermoirreversible,* depending on whether gelation is or is not reversible. Gelatin is an example of a cold-setting thermoreversible gel: when a solution of gelatin molecules is cooled below a certain temperature, a gel is formed, but when it is reheated the gel melts. Egg white is an example of heat-setting thermoirreversible gel: when an egg is heated above a temperature at which gelation occurs, a characteristic white gel is formed; when the egg is cooled back to room temperature, however the gel remains white (*i.e.,* it does not revert back to its earlier liquid form). Whether a gel is reversible or irreversible depends on the changes in the molecular structure and organization of the molecules during gelation. Biopolymer gels that are stabilized by non covalent interactions and do not involve large changes in the structure of the individual molecules prior to gelation tend to be reversible. On the other hand, gels that are held together by covalent bonds or involve large changes in the structure of the individual molecules prior to gelation tend to form irreversible gels.

The type of force holding the molecules together in gels varies from biopolymer to biopolymer. Some proteins and polysaccharides (*e.g.,* gelatin, starch) form helical junction zones through extensive hydrogen bond formation. This type of junction zone tends to form when a gel is cooled, becoming disrupted when it is heated, and thus, it is responsible for cold-setting gels. Below the gelation temperature, the attractive hydrogen bonds favor junction zone formation, but above this temperature the configurational entropy favors a random coil type of structure. Biopolymers with extensive nonpolar groups (*e.g.*, caseins, denatured whey proteins) tend to associate via hydrophobic interactions. Electrostatic interactions play an important role in determining the gelation behavior of many biopolymers, and so gelation is particularly sensitive to the pH and ionic strength of the solution containing the biopolymers. For example, at pH values sufficiently far from their isoelectric point, proteins may be prevented from gelling because of the electrostatic repulsion between the molecules. However, if the pH of the same solution is adjusted near to the isoelectric point, or salt is added, the proteins gel.

The addition of multivalent ions, such as Ca^{2+}, can promote gelation of charged biopolymer molecules by forming salt bridges between the molecules. Proteins with thiol groups are capable of forming covalent linkages through thiol-disulfide interchanges, which help to strengthen and enhance the stability of gels. The tendency for a biopolymer to form a gel under certain conditions, and the physical properties of the gel formed, depend on a delicate balance of biopolymer-biopolymer, biopolymer-solvent, and solvent-solvent interactions of various kinds.

12.4 The Formation of an Emulsion

The formation of an emulsion may involve a single step or a number of consecutive steps, depending on the nature of the starting material.

Oil-soluble ingredients, such as certain vitamins, coloring agents, antioxidants, and surfactants, are mixed with the oil, while water-soluble ingredients, such as proteins, polysaccharides, sugars, salts, and some vitamins, coloring agents, antioxidants and surfactants are mixed with the water. The intensity and duration of the mixing process depends on the time required to solvate and uniformly distribute the ingredients. Adequate solvation is important for the functionality of a number of food components, If the lipid phase contains many crystalline material, it is usually necessary to warm it before homogenization to a temperature at which all the fat melts; otherwise it is difficult, if not impossible, to efficiently create a stable emulsion.

The process of converting two immiscible liquids into an emulsion is known as *homogenization.* and a mechanical device designed to carry out this process is called a *homogenizer.* To distinguish the nature of the starting material, it is convenient to divide homogenization into two categories, The creation of an emulsion directly from two separate liquids will be referred to as *primary* homogenization, whereas the reduction in size of droplets in an existing emulsion will be referred to as *secondary* homogenization. The creation of a food emulsion may involve the use of one or the other form of homogenization, or a combination of both. For example, salad dressing is formed by direct homogenization of the aqueous and oil phases and is therefore, an example of primary homogenization, whereas homogenized milk is manufactured by reducing the size of the fat globules in natural milk, hence is an example of secondary homogenization.

In many foods processing operations and laboratory studies, it is more efficient to prepare an emulsion using two steps. The separate oil and water phases are converted to a coarse emulsion, with fairly large droplets, using one type of homogenizer (*e.g.*, high speed blender). Then the droplet size is reduced by means of another type of homogenizer (*e.g.*, colloid mill, high pressure valve homogenizer). In reality, many of the same physical processes that occur during primary homogenization also occur during secondary homogenization, and there is no clear distinction between them. Emulsions that have undergone secondary homogenization usually contain smaller droplets than those that have undergone primary homogenization, although this is not always the case. Some homogenizers (*e.g.*, ultrasound, microfluidizers, membrane homogenizers) are capable of producing emulsions with small droplet sizes directly from separate oil and water phases.

To highlight the important physical mechanisms that occur during homogenization, it is useful to consider the formation of an emulsion from pure oil and pure water. When the two liquids are placed in a container, they tend to adopt their thermodynamically most stable state, which consists of a layer of oil on top of the water. This arrangement is adopted because it minimizes the contact area between the two immiscible liquids and because the oil has a lower density than the water. To create an emulsion, it is necessary to mechanically agitate the system, to disrupt and intermingle the oil and water phases. The type of emulsion formed in the absence of an emulsifier depends primarily on the initial concentration of the two liquids. At high oil concentrations, a water-in-oil emulsion tends to form, but at low oil concentrations, an oil-in-water emulsion tends to form. In this example, it is assumed that the oil concentration is so low that an oil-in-water emulsion is formed. Mechanical agitation can be applied in a variety of ways, the simplest being to vigorously shake the oil and water together in a sealed container. An emulsion is formed immediately after shaking, and it appears optically opaque (because light is scattered from the emulsion droplets). With time, the system rapidly reverts back to its initial state–a layer of oil sitting on top of the water. This is because the droplets formed during the application of the mechanical agitation are constantly moving around and frequently collide and coalesce with neighboring droplets.

To form a stable emulsion, one must prevent the droplets from merging after they have been formed. This is achieved by having a sufficiently high concentration of a surface-active substance, known as an *emulsifier,* present during the homogenization process. The emulsifier rapidly adsorbs to the droplet surfaces during homogenization, forming a protective membrane that prevents the droplets from coming close enough together to coalesce. One of the major objectives of homogenization is to produce droplets as small as possible, because this usually increases the shelf life of the final product. It is therefore, important for the food scientist to understand the factors that determine the size of the droplets produced during homogenization. It should be noted that homogenization is only one step in the formation of a food emulsion, and many of the other

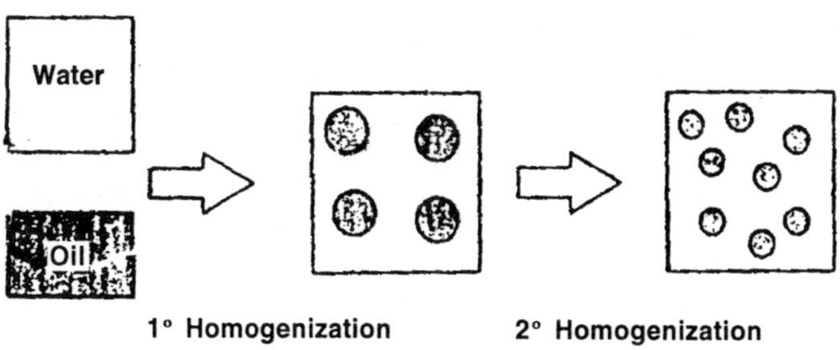

Figure 12.5: The Homogenization Process

unit operations (*e.g.*, pasteurization, cooking, drying, freezing, whipping) also affect the final quality of the product.

12.5 The Effectiveness of Emulsifiers

The effectiveness of emulsifiers at creating emulsions containing small droplets depends on a number of factors.

The concentration of emulsifier present relative to the dispersed phase, the time required for the emulsifier to move from the bulk phase to the droplet surface, the probability that an emulsifier molecule will be adsorbed to the surface of at droplet during a droplet-emulsifier encounter (*i.e.*, the adsorption efficiency), the amount by which the emulsifier reduces the interfacial tension, and the effectiveness of the emulsifier membrane at protecting the droplets against coalescence.

It is often assumed that small emulsifier molecules adsorb to the surface of emulsion droplets during homogenization more rapidly than larger ones. This assumption is based on the observation that small molecules diffuse to the interface more rapidly than larger ones under quiescent conditions. It has been demonstrated that under turbulent conditions large surface active molecules tend to accumulate at the droplet surface during homogenization preferentially to smaller ones.

There are a wide variety of food emulsions, and each one is created from different ingredients and must have different final characteristic properties. Consequently, a number of homogenization devices have been developed for the commercial production of food emulsions, each with its own particular advantages and disadvantages, and each having a range of foods to which it is most suitably applied. The choice of a particular homogenizer depends on many factors, including the equipment available, the site of the process (*i.e.*, a factory or a laboratory), the physico-chemical properties of the starting materials and final product, the volume of material to be homogenized, the throughput, the desired droplet size of the final product, and the cost of purchasing and running the equipment.

12.6 Types of Homogenizers

Many types of homogenizers are used in food industries. They can be explained as follows:

High Speed Blenders

High speed blenders are the most commonly used means of directly homogenizing bulk oil and aqueous phases. The oil and aqueous phase are placed in a suitable container, which may contain as little as a few milliliters or as much as several liters of liquid, and agitated by a stirrer that rotates at high speeds. The rapid rotation of the blade generates intense velocity gradients that cause disruption of the interface between the oil and water, intermingling of the two immiscible liquids, and breakdown of larger droplets to smaller ones. Baffles are often fixed to the inside of the container to increase the efficiency of the blending process by disrupting the flow profile. High speed blenders are particularly useful for preparing emulsions with low or intermediate viscosities. Typically they produce droplets that are between 1 and 10 μm in diameter.

Colloid Mills

The separate oil and water phases are usually blended together to form a coarse emulsion premix prior to their introduction into a colloid mill, because this increases the efficiency of the homogenization process. The premix is fed into the homogenizer, where it passes between two disks separated by a narrow gap. One of the disks is usually stationary, while the other rotates at a high speed, thus generating intense shear stresses in the premix. These shear stresses are large enough to cause the droplets in the coarse emulsion to be broken down. The efficiency of the homogenization process can be improved by increasing the rotation speed, decreasing the flow rate, decreasing the size of the gap between the disks, and increasing the surface roughness of the disks. Colloid mills are more suitable than most other types of homogenizer for homogenizing intermediate or high viscosity fluids (*e.g.*, peanut butter, fish, or meat pastes), and they typically produce emulsions with droplet diameters between 1 and 5 μm.

High Pressure Valve Homogenizers

Like colloid mills, high pressure valve homogenizers are more efficient at reducing the size of the droplets in a coarse emulsion premix than at directly homogenizing two separate phases. The coarse emulsion premix is forced through a narrow orifice under high pressure, which causes the droplets to be broken down because of the intense disruptive stresses (*e.g.*, it impact forces, shear forces, cavitation, turbulence) generated inside the homogenizer. Decreasing the size of the orifice increases the pressure the emulsion experiences which causes a greater degree of droplet disruption and therefore, the production of smaller droplets. Nevertheless, the throughput is reduced and more energy must be expended. A food manufacture must therefore select the most appropriate homogenization conditions for each particular application, depending on the compromise between droplet size, throughput, and energy expenditure. High pressure valve homogenizers can be used to homogenize a wide variety of food products, ranging from low viscosity liquids to viscoelastic pastes, and can produce emulsions with droplet sizes as small as 0.1 μm.

Ultrasonic Homogenizer

A fourth type of homogenizer utilizes high intensity ultrasonic waves that generate intense shear and pressure gradients. When applied to a simple containing oil and water, these waves cause the two liquids to intermingle and the large droplets formed to be broken down to smaller ones. There are two types of ultrasonic homogenizer commonly used in the food industry: piezoelectric transducers and liquid jet generators. Piezoelectric transducers are most commonly found in the small bench top ultrasonic homogenizers used in many laboratories. They are ideal for preparing small volumes of emulsion (a few milliliters to a few hundred milliliters) a property that is often important in fundamental research when expensive components are used. The ultrasonic transducer consists of a piezoelectric crystal contained in some form of protective metal casing, which is tapered at the end. A high intensity electrical wave is applied to the transducer, which causes the piezoelectric crystal inside to oscillate and generate an ultrasonic wave. The ultrasonic wave is directed toward the tip of the transducer, where it radiates into the surrounding liquids, generating intense pressure and shear gradients (mainly due to cavitational affects) that cause the liquids to be broken up into smaller fragments and intermingled with one another. It is usually necessary to irradiate a sample with ultrasound for a few seconds to a few minutes to create a stable emulsion. Continuous application of ultrasound to a sample can cause appreciable heating, and so it is often advantageous to apply the ultrasound in a number of short bursts.

Ultrasonic jet homogenizers are used mainly for industrial applications. A stream of fluid is made to impinge on a sharp-edged blade, which causes the blade to rapidly vibrate, thus generating an intense ultrasonic field that breaks up any droplets in its immediate vicinity though a combination of cavitation, shear, and turbulence. This device has three major advantages: it can be used for continuous production of emulsions; it can generate very small droplets; and it is more energy efficient than high pressure valve homogenizers (since less energy is needed to form droplets of the same size).

Membrane Homogenizers

Membrane homogenizers form emulsions by forcing one immiscible liquid into another through a glass membrane that is uniform in pore size. The size of the droplets formed depends on the diameter of the pores in the membrane and on the interfacial tension between the oil and water phases. Membranes can be manufactured with different pore diameters, with the result that emulsions with different droplet sizes can be produced. The membrane technique can be used either as a batch or a continuous process depending on the design of the homogenizer. Increasing numbers of applications for membrane homogenizers are being identified and the technique can now be purchased for preparing emulsions in the laboratory or commercially. These instruments can be used to produce oil in water, water in oil, and multiple emulsions. Membrane homogenizers have the ability to produce emulsions with very narrow droplet size distributions, and they are highly energy efficient, since there is much less energy loss due to viscous dissipation.

The choice of a homogenizer for a given application depends on a number of factors such as, volume of sample, energy requirement etc.

Even after the most suitable homogenization technique has been chosen, the operator must select the optimum processing conditions, such as temperature, time, flow rate, pressure, valve gaps, rotation rates, and sample composition. If an application does not require that the droplets in an emulsion be particularly small, it is usually easiest to use a high speed blender. High speed blenders are also used frequently to produce the coarse emulsion premix that is fed into other devices.

To create an emulsion that contains small droplets (<1 μm), either industrially or in the laboratory, it is necessary to use one of the other methods. Colloid mills are the most efficient type of homogenizer for high viscosity fluids, whereas high pressure valve, ultrasonic, or microfluidization homogenizers are more efficient for liquids that are low or intermediate in viscosity. In fundamental studies one often uses small volumes of sample, and therefore a number of laboratory homogenizers have been developed that are either scaled-down versions of industrial equipment or instruments specifically designed for use in the laboratory. For studies involving ingredients that are limited in availability or expensive, an ultrasonic piezo-electric transducer can be used because it requires only small sample volumes. When it is important to have monodisperse emulsions, the use of a membrane homogenizer would be advantageous.

The efficiency of the homogenization process can be calculated by comparing the energy required to increase the surface area between the oil and water phases with the actual amount of energy resulted to create an emulsion. The difference in free energy between the two separate immiscible liquids and an emulsion can be estimated by calculating the amount of energy needed to increase the interfacial area between the oil and aqueous phases.

12.7 Emulsion Properties

Food manufacturers want to produce a high quality product at the lowest possible cost. To achieve this goal they must have a good appreciation of the factors that determine the properties of the final product. This knowledge in turn, is used to formulate and manufacture a product with the desired characteristics (*e.g.*, appearance, texture, mouth feel, taste, shelf life). These bulk physico-chemical and sensory properties are determined by such molecular and colloidal properties of emulsions as dispersed volume fraction, droplet size distribution, droplet-droplet interactions, and interfacial properties. Consequently, a wide variety of experimental techniques have been developed to characterize the molecular, colloidal, microscopic, and macroscopic properties of food emulsions. Analytical techniques are needed to characterize the properties of food emulsions in the laboratory, where they are used to improve our understanding of the factors that determine emulsion properties, and in the factory, where they are used to monitor the properties of foods during processing to ensure that the manufacturing process is operating in an appropriate manner. The subsections that follow highlight some of the most important properties of food emulsions, and outline experimental techniques for their measurement.

The structural oganization and interactions of the droplets in an emulsion often play an important role in determining the properties of a food. For example, two emulsions may have the same droplet concentration and size distribution, but very different properties, because of differences in the degree of droplet flocculation. Various forms of microscopy are available for providing information about the microstructure of food emulsions. The unaided human eye can resolve objects that are farther apart than about 0.1 mm (100 μm). Most of the structural components in food emulsions (*e.g.*, emulsion droplets, surfactant micelles, fat crystals, ice crystals, small air cells, protein aggregates) are much smaller than this lower limit and cannot therefore be observed directly by the eye.

Optical microscopy can be used to study components that have sizes between about 0.5 and 100 μm. The characteristics of specific components can be highlighted by selectively staining certain ingredients or by using special lenses. Electron microscopy can be used to study components that have sizes down to about 0.5 nm. Atomic force microscopy can be used to provide information about the arrangements and interactions of single atoms or molecules. All these techniques are burdened by sample preparation steps that often are laborious and time-consuming, and subject to alter the properties of the material being examined. Nevertheless, when carried out correctly the advanced microscopic techniques provide extremely valuable information about the arrangement and interactions of emulsion droplets with each other and with the other structural entities found in food emulsions.

The size of the droplets in an emulsion influences many of their sensory and bulk physico-chemical properties, including rheology, appearance, mouth feel, and stability. It is therefore, important for food manufacturers to carefully control the size of the droplets in a food product and to have analytical techniques to measure droplet size. Typically, the droplets in a food emulsion are somewhere in the size of range of 0.1 to 50 μm in diameter.

Food emulsions always contain droplets that have a range of sizes, and so it is usually important to characterize both the average size and the size distribution of the droplets. The droplet size distribution is usually represented by a plot of droplet frequency (number or volume) versus droplet size (radius or diameter). Some of the most important experimental techniques for measuring droplet size distributions are included in Table 12.1.

Table 12.1: Experimental Techniques for Characterizing the Physico-chemical Properties of Food Emulsions

Disersed phase volume fraction	Proximate analysis, density, electrical conductivity light scattering, NMR, ultrasound
Droplet size distribution	Light scattering (static and dynamic), electrical conductivity, optical microscopy, electron microscopy, ultrasound, NMR
Microstructure	Optical microscopy, electron microscopy, atomic force microscopy
Creaming and sedimentation	Light scattering, ultrasound, NMR, visual observation
Droplet charge	Electrokinetic techniques, electro-acoustic techniques
Droplet crystallization	Density, NMR, ultrasound, differential scanning calorimetry, polarized optical microscopy
Emulsion rheology	Viscometers, dynamic shear rheometers
Interfacial tension	Interfacial tensiometers (static and dynamic)
Interfacial thickness	Ellipsometry, neutron reflection, neutron scattering, light scattering, surface force apparatus

Light-scattering and electrical conductivity techniques are capable of providing a full particle size distribution of a sample in a few minutes. Since, however, these techniques usually require that the droplet concentration be very low ($\phi < 0.001$), samples must be diluted considerably before analysis. Optical and electron microscopy techniques, which provide the most direct measurement of droplet size distribution, are often time-consuming and laborious to operate, and sample preparation can cause considerable artifacts in the results. In contrast, recently developed techniques based on NMR and ultrasonic spectroscopy can be used to rapidly and non-destructively measure the droplet size distribution of concentrated and optically opaque emulsions. These techniques are particularly useful for on-line characterization of emulsion properties.

The physical state of the components in a food emulsion often has a pronounced influence on its overall properties. For example, oil in water emulsions are particularly prone to partial coalescence when the droplets contain a certain percentage of crystalline fat. Partial coalescence leads to extensive droplet aggregation, which decreases the stablity of emulsions to creaming and greatly increases their viscosity. In water-in-oil emulsions, such as margarine or butter, the formation of a network of aggregated fat crystals provides the characteristic rheological properties. The most important data for food scientists are the temperature at which melting or crystallization begins, the temperature range over which the phase transition occurs, and the value of the solid fat content at any particular temperature. Phase transitions can be monitored by measuring changes in any property (*e.g.*, density. compressibility, heat capacity, absorption or scattering of radiation) that is altered upon conversion of an ingredient from a solid to a liquid. The density of a component often changes when it undergoes a phase transition, and so melting or crystallization can be monitored by measuring changes in the density of a sample with temperature or time.

Phase transitions can also be monitored by measuring the amount of heat released or absorbed when a solid melts or a liquid crystallizes, respectively. This type of measurement can be carried out by means of differential thermal analysis or differential scanning calorimetry. These techniques also provide valuable information about the polymorphic form of the fat crystals in an emulsion. More recently, rapid instrumental methods based on nuclear magnetic resonance and ultrasound have been developed to measure solid fat contents. These instruments are capable of nondestructively determining the solid fat content of a sample in a few seconds and are extremely valuable analytical tools for rapid quality control and online procedures. Phase transitions can be observed in a more direct manner by means of polarized optical microscopy.

Chapter 13

Cholesterol and Human Health

13.1 Introduction

Cholesterol has assumed a central role in experimental and human atherosclerosis and people are being exhorted to know their "Cholesterol numbers".

However, cholesterol levels tend to vary diurnally and with season, and single measurements may not be accurate indicators of risk. This is especially true if the single determined value is near one of the accepted cut points. Low levels of cholesterol may lead to increased risk of death due to non-cardiovascular causes. Low cholesterol may become a problem at levels below 160 mg/dL.

Since cholesterol feeding has been shown to be atherogenic in some animal species, since elevated levels of cholesterol are a risk factor, and because the substance is relatively easy to measure, cholesterol has borne the brunt of the attack on CHD. The effect of dietary cholesterol on levels of blood cholesterol appears to be small. In 1950 Gertler *et al.* isolated from a large study of coronary disease four groups of 10 men each. They were the men with the lowest or highest serum cholesterol and those who ingested the most or the least dietary cholesterol. These groups were compared with similar groups selected from the control subjects. In every group the men with coronary disease exhibited significantly higher cholesterol levels than the controls, but in no case was any relation to cholesterol intake seen. Early in the Framingham study it was found that plasma cholesterol levels were unrelated to diet, a finding also reported from the Tecumseh study. Several groups have reported that addition of eggs to the diet of free-living subjects did not affect their cholesterol level. Gordon *et al.* analyzed and compared the diets of men who did or did not have coronary disease in three large prospective coronary disease studies–Framingham, Puerto Rico, and Honolulu. Men who had coronary disease ingested fewer total calories, less carbohydrate, and less alcohol. Intakes of cholesterol and ratios of polyunsaturated to saturated dietary fat were similar for men who did or did not have coronary disease. McNamara reviewed data from 68 clinical studies relating to effect of dietary cholesterol on plasma cholesterol. He concluded that there was a mean rise of 2.3 ± 0.2 mg/dL of plasma cholesterol for every 100 mg of ingested cholesterol. Hopkins reviewed that complexity of the association between cholesterol intake and plasma cholesterol. He found that the magnitude of the change in plasma cholesterol as a function of dietary cholesterol is influenced by baseline cholesterol intake.

In contrast to dietary cholesterol, there is a little question that the saturation of dietary fat exerts a profound influence on blood cholesterol levels. Ahrens *et al.* fed a number of subjects a liquid formula diet containing 45 per cent of energy as fat. In general, plasma cholesterol rose as the saturation of the fat rose. Keys *et al.* and Hegsted *et al.* studied the effects of changes in dietary fat on change in blood cholesterol levels in man and offered formulas to predict changes in cholesterolemia based on chances in dietary fat. Both groups found fat saturation to have the greatest effect. Stearic acid did not appear to fit the formula, and direct experiments in human subjects have shown this to be true. Hayes and Khosla have hypothesized that the two most important fatty acids related to cholesterol levels are myristic acid (which raises cholesterol levels at every concentration) and linoleic acid (which exerts an increasing hypocholesterolemic effect until it reaches a dietary levels of 6–7 per cent of energy). McNamara *et al.* fed normal subjects diets high or low in cholesterol and containing saturated or unsaturated fat. The major factor determining cholesterolemia was the saturation of the fat, which was about four times greater than that of dietary cholesterol.

In addition to fatty acid saturation, the position of a specific fatty acid in the triglyceride molecule is important. In an effort to test cholesterolemic effects of specific fatty acids, McGandy *et al.* fed human subjects diets that contained fats into which high levels of specific saturated fatty acids (lauric, myristic, palmitic, or stearic) had been incorporated by interesterification with dietary fats. They found no differences. A similar study carried out earlier with rabbits showed no effects on cholesterolemia or atherosclerosis.

Table 13.1: Plasma Cholesterol Levels in Subjects Fed High or Low Levels of Cholesterol with Saturated or Unsaturated Fat

Fat in Diet		*Cholesterol*	
	Fat (P/S)	*Dietary (mg)*	*Plasma (mg/dL)*
Low cholesterol			
Saturated	0.31 ± 0.18	288 ± 64	243 ± 50
Unsaturated	1.90 ± 0.90	192 ± 60	218 ± 46
High cholesterol			
Saturated	0.27 ± 0.15	863 ± 161	248 ± 51
Unsaturated	1.45 ± 0.50	820 ± 102	224 ± 46

During the randomization process, the structure of fats is changed so that each component fatty acid is present in each position of the triglyoceride at one-third of its total concentration. This change appears to influence the atherogenicity of a fat. Both tallow and lard contain about 24 per cent of total fatty acids as palmitic acid. In lard about 90 per cent of the component palmitic acid is at the 2-position of the triglyceride, whereas in tallow no more than 15 per cent. Lard is significantly more atherogenic for rabbits than tallow. When the fats are randomized, each contains 33 per cent of its palmitic acid at the 2-position and they are of equivalent atherogenicity.

In every decade, concern surfaces about the atherogenic effects of fats containing trans-unsaturated double bonds (trans fat). Generally, trans fats are metabolized in a manner similar to that of their cis counterparts. They accumulate in tissues upon feeding and disappear relatively soon after cessation of feeding. In rabbits they lead to increased cholesterolemia but not to more severe atherosclerosis. This is true in rabbits fed cholesterol or a cholesterol-free atherogenic diet.

Table 13.2: Influence of Native and Randomized Tallow and Lard on Atherosclerosis in Rabbits a Fed Semipurified Diet Containing 0.4 per cent Cholesterol for 60 days

	Groups			
	Tallow	*Randomized Tallow*	*Lard*	*Randomized Lard*
Number	7/8	7/8	8/8	8/8
% 16 : 0 at sn-2	*3.7*	*8.5*	*20.8*	*7.6*
Plasma				
Cholesterol (mg/dL)	1177 ± 156	1189 ± 166	926 ± 184	834 ± 153
Aorta				
Arch	1.29 ± 0.24	1.50 ± 0.53	2.69 ± 0.28	1.50 ± 0.28
Thoracic	0.79 ± 0.26	0.79 ± 0.28	1.75 ± 0.28	0.69 ± 0.19

In men fed diets high in elaidic acid, Lp(a) levels are increased, but this effect is not seen when the dietary trans fat is low in elaidic acid. These findings suggests specific metabolic roles for specific trans fatty acids. In rats, trans fats exert no deleterious effects if the diet contains sufficient levels of linoleic acid. In man, the cholesterolemic effects of trans fats seem to be related to the ratio of dietary trans fat to linoleic acid, the cholesterol levels rising as the ratio falls. Several studies have shown that the tissue levels of trans fatty acids are no higher in subjects with coronary disease than in controls. Houtsmuller suggested in 1978 that trans fat be regarded as a quasi-saturated fat.

In 1985 the Life Sciences Research Office of the Federation of American Societies of Experimental Biology published a report that concluded that there was little reason for health concerns at the reported intake level of trans fat (8 g/person/day). Two years later the British Nutrition Foundation published a report with the same conclusion. A decade later reports sponsored by the International Life Sciences Institute and the British Nutrition Foundation indicate that these organizations found no reasons to alter their earlier conclusions. All the reports contained the safe suggestion that more research was needed.

Several new players have appeared on the heart disease stage, and they may ultimately affect our views of fat and cholesterol as major factors in CHD. Almost 30 years McCully suggested that homocysteinemia could be a major risk for coronary disease, and his work is being recognized today. Studies comparing lipid levels in European countries show little relation of these levels to CHD mortality. Ischemic heart disease mortality is four times higher in Belfast, Ireland, than in Toulouse, France, despite general similarities in subjects, diets (the French ingest significantly more cholesterol and alcohol) and in their risk factor profiles. Similarly, mortality from heart disease in men 50 to 54 years old is four times higher in Vilnius, Lithuania, than it is in Linkoping, Sweden, despite only small differences in traditional risk factors. In the years 1985 to 1987, age-specific mortality from ischemic heart disease for men aged 45 to 54 years was 237/100,000 in Belfast and 56/100,000 in Toulouse. In men aged 55 to 64, the rates per 100,000 were 761 in Belfast and 175 in Toulouse. In the younger age group, total cholesterol levels were significantly lower in the French (230 ± 41 mg/dL vs. 240 ± 41 mg/dL) and HDL cholesterol levels were higher (54 ± 15 mg/dL vs. 47 ± 12 mg/dL). There were no differences in the older men. Energy intake was virtually the same in the two groups. The French ingested significantly more protein and cholesterol and significantly less carbohydrate, while fat intake was similar. In 1977 the CHD mortality per 100,000 men aged 50 to 54 years was 300 in Lithuania and 220 in Sweden. In 1994 mortality was 445 in Lithuania (102 percent increase) and 110 in Sweden (a 50 per cent decrease). Data from the Russian Lipid Research Clinics show an appreciable number of deaths in men with low levels of LDL and high levels of HDL.

Furthermore, the isolation of *Chlamydia pneumoniae* from atherosclerotic, but not normal arteries and the finding of cytomegalovirus in diseased arteries may shed light on mechanisms underlying the onset of the disease.

Polyunsaturated fatty acids also affect blood pressure and platelet function, both of which play a role in CHD.

13.2 The Effects of Fat in Cancer

The effect of fat in cancer need to be stratified by fat type, fat quantity and total caloric intake. There is too frequently a rush to judgement concerning effects of specific nutrients.

Much of the recent interest in diet and cancer can be traced to a major paper in which Doll and Peto suggested that as much as 35 per cent of deaths due to cancer might be associated with diet. They then made the following comment: "It must be emphasized that the figure chosen (of 35 per cent of cancers related to diet) is highly speculative and chiefly refers to dietary factors which are not yet reliably identified." They also stated that "there is no evidence of any generalized increase (in deaths due to cancer) other than that due to tobacco."

A fatty acid that has been known for decades has recently emerged as a potent inhibitor of carcinogenesis and atherosclerosis. It is octadeca-dienoic acid (conjugated linoleic acid: CLA). CLA has been shown to inhibit chemically induced forestomach tumors in mice when given intragastrically and mammary tumors when included in the diet of rats. It inhibits the growth and metastasis of human tumors when injected into immune-deficient mice and inhibits growth of tumor cells *in vitro*. CLA has also been shown to reduce the severity of atherosclerosis in cholesterol-fed rabbits. CLA is a component of animal fats, especially cheeses.

Table 13.3: Effects of Calories and Fat on Incidence of Methylcholanthrene-Induced Skin Tumors in Mice

Level of		*Tumor Incidence (%)*
Calories	*Fat*	
Low	Low	0
High	Low	54
Low	High	28
High	High	66

More recently, the association between dietary fat and risk of breast cancer appears to be weakening. Goodwin and Boyd reviewed a large number of studies and reported that 7 of 13 international comparisons found a correlation between fat intake and breast cancer risk but only one of 14 case control studies did. Hirohata *et al.* found no association between dietary fat intake and breast cancer in Japan or Hawaii. The NHANES I reported on 99 cases of breast cancer, as opposed to 5386 non-cases, and found no differences in fat or fatty acid intake. Willett *et al.* studied a cohort of over 89,000 American women whose fat intake ranged from 32 to 44 per cent of calories and found a slight decrease in relative risk with increasing fat intake.

As in the case of breast cancer, local (case control) studies of fat intake and colon cancer risk show minimal correlations with fat intake, whereas international studies find strong correlation. Rogers and Longnecker reviewed diet and cancer and in a summary of 24 cases of colon cancer found a small, but inconsistent, association between fat intake and risk of colon cancer. Stocks and Karn in an early (1933) study of colon cancer and diet in England found dairy foods to be negatively correlated with risk. Jensen *et al.* in a study comparing colon cancer and diet in Finland and Denmark found saturated fat to be inversely correlated with risk. Stemmermann *et al.* made a similar observation in Hawaii. Others reported that risk increased with increasing intake of saturated fat. Tuyns in a Belgian study suggested that the dietary factor leading to increased risk was oligosaccharides rather than fat. It should be evident from the foregoing that there is no consensus with regard to the effects of dietary fat on colon cancer risk. Is it fat or the calories from fat that affects cancer risk? In 1927 Hoffman suggested that the increase in cancer incidence seen then was due to overnutrition. About 50 years later Berg made a similar suggestion, namely, that increasing risk of hormone-related tumors was due to increased caloric intake.

Moreschi showed in 1909 that transplanted tumors did not grow as well in underfed mice as they did in freely fed ones. A few years later Rous demonstrated that neither spontaneous nor transplanted tumors showed optimal growth in rodents whose food intake was restricted. Lavik and Baumann studied dietary effects on chemically induced skin tumors in mice. When the diet was low in both calories and fat, tumor incidence was nil. Tumor incidence in mice fed diets low in fat but high in calories was 48 per cent lower than in those whose diet was low in calories but high in fat and only 18 per cent lower than in mice fed a high-fat diet, high in calories. Our work showed that caloric restriction by 40 per cent led to significantly reduced incidence of chemically induced breast or colon tumors in rats, even when the restricted diet contained twice as much fat. At 10 per cent caloric restriction, incidence of induced breast tumors in rats was unchanged, but tumors per tumor-bearing rat were reduced by 36 per cent and total weight of tumors by 47 per cent. Rats fed a diet containing 26.7 per cent corn oil but total fat restricted by 25 per cent exhibited lower tumor incidence, fewer tumors, and smaller tumors than rats freely fed a diet containing only 5 per cent fat. In man both colon and gastric cancers have been correlated positively with caloric intake. Overweight in humans is clearly correlated with increasing cancer risk.

13.3 Risk Factors for Heart Disease

Among the major risk factors for coronary heart disease (CHD), also termed cardiovascular disease (CVD), are cigarette smoking, elevated cholesterol level, elevated blood pressure, obesity, and maleness. The first of these is correctable and the last is an unalterable fact of life; but the others may, to some extent, be amenable to nutritional intervention. Hopkins and Williams have described over 200 risk factors, indicating the variety of the factors that have been linked to the development of CHD by at least one investigator.

Although an experimental relationship between dietary cholesterol and atherosclerosis had been adduced in 1913 when Anitschkow was able to produce aortic lesions in rabbits interest in fat in the diet and its relation to this disease did not begin to blossom for several decades. In 1950 Gofman and his colleagues developed a method for the ultracentrifugal separation of plasma lipoproteins, showed how these fractionated particles could be related to heart disease, and implicated diet as a factor. In 1953 Keys described what was the beginning of his Seven Countries Study and named atherosclerosis as a new public health problem.

Table 13.4: Classification of LDL Particles

Class	*Subfraction Density (g/mL)*	*Particle Diameter (nm)*
LDL I	1.025–1.035	26–27
LDL II	1.032–1.038	25.5–26
LDL IIIA	1.038–1.050	24.7–25.6
LDL IIIB		24.2–24.6
LDL IVA	1.048–1.065	23.3–24.2
LDL IVB		21.8–23.2

The lipoproteins are lipid-protein agglomerates rather than real compounds. They are described by their hydrated densities (a physical property) but may differ in size and composition. Thus, although the chemical analyses of lipoproteins of various densities are often published, they represent average values and are less precise indicators of identity than melting points or spectral data. A research continues, we continue to discover sub-fractions of lipoproteins that affect CHD risk. Lipoprotein (a) [Lp(a)], which was described first by Berg is a low density lipoprotein (LDL) in which apoprotein B is linked to an apoprotein unit [apoprotein (a)] via a disulfide bridge. Elevated Lp(a) levels have been associated with a high risk of CHD Lp(a) interferes with fibrinolysis and its levels in the blood appear not to be affected by diet or drugs.

Chapter 14

Immune System and Dietary Fats

14.1 The Immune System

The immune system is the most dispersed network of cells in the body that are not in physical contact yet cooperating functionally.

The immune system is comprised of both fixed and mobile cells that protect against invading organisms and the development of abnormal (malignant) cells. If, however, it over reacts to self, the result is autoimmunity; if it overreacts to non-self, there is an allergic response. Some of these immune cells have very short lives but others, which form the basis of immunologic memory, exist for many years and perhaps for the life of the host. Fixed cells of the reticuloendothelial system (RES) are found primarily in the liver, spleen, bone marrow, thymus, lymph nodes, lungs and intestines. Mobile cells are found throughout the circulation in high numbers, which increase quickly in response to the presence of foreign antigens. It is estimated that 80 per cent of mobile leukocytes are normally sequestered in organs of the RES, while the remainder float in the bloodstream.

Functionally, the components of the immune system are often divided into response elements. That is, humoral (or circulating) and cell-mediated immunity are the most commonly used concepts for describing immune functions, although there is communication between these facets of immune cells via a variety of soluble products called cytokines. In addition to the distinctions of humoral and cell-mediated immunity, a number of other components of the immune system play a role in protecting against pathogenic organisms. These include the polymorphonuclear leukocytes, which are divided into neutrophils, basophils, and eosinophils based on their staining characteristics in a smear of blood treated with Giemsa or Wright's

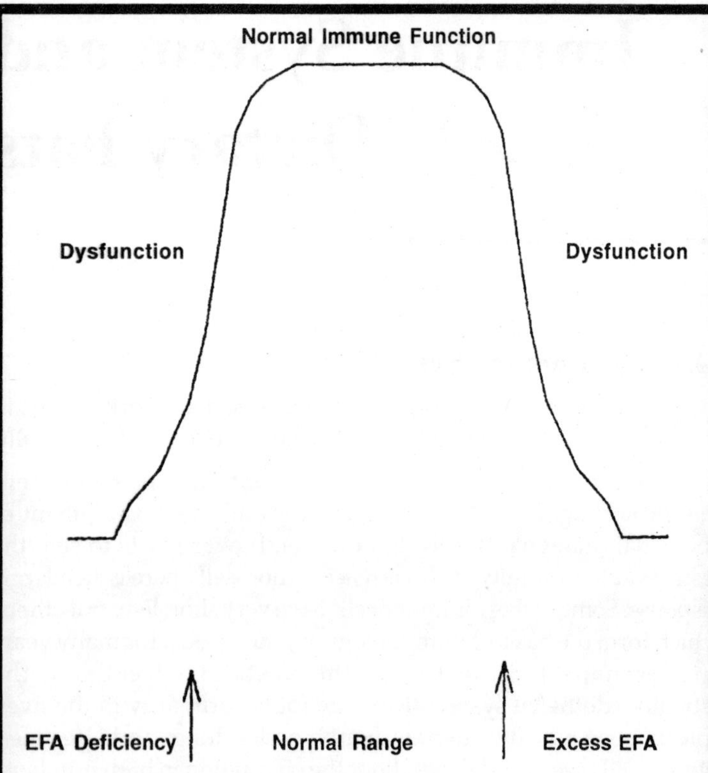

Figure 14.1: Immune Responses as a Function of Dietary Essential Fatty Acid

stain. Neutrophils participate in acute inflammatory reactions against bacteria and other foreign bodies. Basophils and eosinophils are components of allergic reactions. Eosinophils are elevated also in response to parasitic infections.

There is a wide range of normal immune functions and a wide range of normal numbers of circulating immune cells. For example, the normal range of leukocytes in the blood of humans is 5000 to 10,000/mm^3. Within this range there are no discernible differences in susceptibility to infection. In fact, there is little increased risk of infection unless total white blood cells drop below 1000/mm^3.

Functional terms used to describe the immune system include natural (innate) and specific (acquired). Natural immunity is a result of molecules such as cytokines or complement, or a result of the function of natural killer (NK) cells or phagocytes, or barriers such as mucosal surfaces. The natural immune system is able to attack a substance to which it has had no previous exposure; subsequent exposure to that stimulus does not increase the response. On the other hand, the specific immune response recognizes a specific feature (antigen) of the foreign substance or cell and generates an immunologic memory, which amplifies responses during subsequent exposures. The specific immune system can recruit portions of the innate system to function simultaneously. Lymphocytes are the primary effectors of the specific immune response. These cells "direct" much of the immunologic activity by the production of cytokines of many types.

Cytokines are soluble protein molecules that function as messengers to other leukocytes. They are produced and secreted primarily by monocytes, and their main targets are lymphocytes and other monocytes. Some of the important cytokines include interleukins, interferons, and tumor necrosis factors (TNFs). There are at least 15 distinct forms of interleukins, 3 interferons, and 2 tumor necrosis factors.

14.2 Types of Immunity

Eicosanoids and Immunity

Since eicosanoids are soluble mediators of the inflammatory response, their role is critical in many types of immune reaction. One of the most widely used drugs in the world, aspirin, is a cyclooxygenase inhibitor that reduces prostaglandin synthesis. The anti-inflammatory effect of aspirin and other nonsteroidal anti-inflammatory drugs (NSAIDs) is due to blocking of the metabolism of arachidonic acid by cyclooxygenase. Some animal studies suggest that high doses of other prostaglandin inhibitors, like indomethacin, may enhance some mononuclear cell dependent responses. There are limited reports that humans with deficient immune responses respond to indomethacin treatment whereas some normal individuals do not benefit from treatment with this drug and others show an increase in certain antibody responses. The variation in response might be a function of which eicosanoid is involved in the altered immune responses. Burn injury leads to elevations of prostaglandin E and immune hyporesponsiveness. Administration of prostaglandin inhibitors lowers the PGE levels and partially corrects the immune response. NSAID administration appears to prevent colon cancer in both humans and experimental animals; in addition, prostaglandin synthesis is decreased by these drugs. It is unknown how these compounds inhibit the growth of colon cancer. While there may be direct effects of these prostaglandin inhibitors on several aspects of colon carcinogenesis, including changes in cell metabolism, the cell cycle, and expression of tumor suppressor proteins, one cannot rule out alterations in immune response.

Cell-Mediated Immunity

Cell-mediated immunity is based on the interactions of T lymphocytes and monocytic cells. T cells have been divided into many subclasses based primarily on cell surface receptors, which vary in relation to the functions these cells perform. The generally accepted classification of T cells is the CD (cluster designation or cluster of differentiation) scheme, which depends diagnostically on antibodies to distinguish these cellular sets of lymphocytes. Therefore, T cells can be classified as CD^{4+} and CD^{8+} but these cells are often referred to-by their effect or functions as helper and suppressor cells, respectively. In addition, there are NK cells, LAK (lymphokine-activated killer) cells, and lymphocytes responsible for delayed-type hypersensitivity.

The monocytes and their derived macrophages are cells that protect against certain bacteria and tumor cells. While monocytes circulate in the blood, once they have migrated into tissues they usually become fixed in place and are called macrophages. There are macrophages normally lining venules in the liver (Kupffer cells), spleen, and lymph nodes; these permanently fixed cells are termed histiocytes. The relative proportions of leukocytes in the blood differ across species. While the predominant circulating cells in rodents are lymphocytes, granulocytes are the most common in humans. This difference is relevant in comparing the experimental work done predominantly in rodent species with studies in humans. Macrophages are one of the most important sources of eicosanoids in the body, playing a major role in a variety of inflammatory and immunologic responses. Membranes of these cells are particularly enriched with arachidonic acid. Different inflammatory stimuli have the ability to induce secretion of eicosanoids of varied forms. Lymphocytes do not seem able to produce eicosanoids but can release arachidonic acid and certainly respond to the effects of these mediators. Many research studies describe use of mononuclear cells-these are mixed populations of lymphocytes and monocytes, usually obtained by differential centrifugation of blood or other biological fluids.

Humoral Immunity

Humoral immunity is a result of the circulation in the plasma of antibodies and the activity of B lymphocytes. The B stands for bursa of Fabrius in birds, which is the immunologic equivalent of the bone marrow in mammals. Humoral immunity characterizes the production of antibodies by the interaction of B lymphocytes with plasma cells. Circulating antibodies are the products of these cells, but they cannot be produced without interaction with T, or thymus-derived, cells. In addition to antibody production, the complement system is part of the immune response. Complement is a series of proteins in the plasma that participate in antigen-antibody reactions, displaying a wide range of actions ranging from aiding in phagocytosis by leukocytes to killing of tumor cells. Complement serves to stimulate various functions of the leukocytes.

14.3 The Study of Dietary Cholestrol and Dietary Fats

Dietary cholesterol has been studied extensively for its effects on immune functions, primarily in animals and *in vitro*. A fair number of studies in humans have been conducted, however. The rationale for this work is that cholesterol and fatty acids in the cell membrane are the primary determinants of membrane microviscosity, which in turn controls a number of events at the cell surface, including enzymatic activities such as those of phospholipase and cyclooxygenase. There is considerable variation in serum cholesterol concentrations among humans, but two factors have contributed to the relatively small number of studies showing a positive correlation between dietary cholesterol and changes in the immune response. First, dietary cholesterol is relatively weak in its ability to increase serum cholesterol. Second, the populations studied have not exhibited a broad enough range of serum cholesterol concentrations to permit observation of large effects on immunologic functions *in vivo* (although some studies have noted differential responses *in vitro*). Therefore, experimental manipulations in animals or *in vitro* that entailed extreme differences in exposure to cholesterol have resulted in significant effects on immunologic responses.

It is impossible to singularly characterize the effect of cholesterol on immunologic responses. This is because cholesterol is delivered to cells lipoproteins, complex aggregates of cholesterol, triglyceride, phospholipid, and one or more apoproteins to a variety of immunologic cells. It has been shown that isolated apolipoprotein E (apoE) inhibits lymphocyte proliferation, but its effectiveness depends on interaction with cholesterol and phospholipid. Recent interest in the role of oxidized lipoproteins in the development of atherosclerosis suggests that some of the immunologic effects attributed to cholesterol may have been due to oxidized cholesterol molecules, In fact, there is considerable evidence that oxidized low density lipoprotein (LDL) or cholesterol reduces immune responses by lymphocytes.

The change in immunologic response elicited by elevated cholesterol concentrations depends both on the concentration of cholesterol to which cells are exposed and the type of immunologic response being measured. There is no single direction. Most studies in this area have looked at lymphocyte functions, and it can be concluded that results are primarily dependent on lipoprotein concentration.

Studies done in hypercholesterolemic rabbits and monkeys suggest that both B- and T-cell-dependent functions are elevated. However, hypercholesterolemic guinea pig showed increased immune function with a doubling of baseline serum cholesterol but a significant reduction when concentrations of cholesterol were elevated fourfold. This suggests that there may be species specificity as well as specific ranges of serum cholesterol levels that modulate immune responses. Most human studies of lymphocyte function *in vivo* or *in vivo*, have shown decreased functions in the presence of high total or LDL cholesterol. However, the total number of circulating T-lymphocyte subsets $CD3^+$, $CD4^+$, and $CD8^+$ in hypercholesterolemic children correlates with LDL-cholesterol concentrations over a threefold range. Since T lymphocytes are found in significant numbers in atherosclerotic plaques, it is presumed that the elevated LDL levels are associated with both development of arterial lesions and changes seen with chronic inflammation.

Monocyte and macrophage functions are reduced in most studies that have examined the effect of excess cholesterol. Similarly, functions of the polymorphonuclear leukocytes have also been reduced in response to a surplus of cholesterol. The functions are, to a great extent, dependent on plasma membrane microviscosity; excess cholesterol incorporation will stiffen the membranes, thereby decreasing the ability of these cells to engulf microbes. A variety of other immune functions have been examined under the influence of different concentrations of cholesterol, and most studies show generally reduced immune cell responses; however, there is considerab;e disagreement in the research literature on the overall effect of cholesterol on immune reactions. Reports generally agree that there is reduced resistance to bacterial or viral infections, but some specific components of the immune system are suppressed while others are enhanced in the presence of hypercholesterolemia.

With the view that atherosclerosis shares many traits of a chronic inflammatory condition gaining widespread acceptance only in the late 1980s, the role of cholesterol in the human immune response as it relates to the development of arterial lesions has been of considerable interest. Since a large percentage of the fat-filled foam cells in atheroma are derived from macrophages, it is logical to assume that cytokines and classical inflammatory repair mechanisms are at work as part of the atherogenic process.

Similar to cholesterol, fatty acids also play an important role in determining the physical state of the cell membrane.

The more polyunsaturated a fatty acid molecule is, the more fluidity it imparts to the cell membrane. Also, the more polyunsaturated fat in the diet, the lower the immune response, within certain limits. That is, the immune response does not go down to zero in the presence of very high intake of polyunsaturated fat; it is, however, lower relative to that seen with ingestion of saturated fats. In both humans and animals, cell membrane phospholipids reflect dietary intake within limits of the membrane to accommodate a certain range of fatty acids. Dietary fat primarily alters storage of fatty acids in adipose tissue, but smaller changes in the membranes of many other cell types, including those of the RES, have been reported. A variety of experimental studies have demonstrated that both the amount and type of fatty acids consumed have roles to play in altering immune responses. Although it is easier to discuss specific fatty acids in the diet, it should be remembered that there are very few free fatty acids consumed and most are in triacylglycerol molecules, which are usually a mixture of two or three different fatty acids attached to a glycerol molecule. The majority of dietary fats usually supply four to six different fatty acids.

There is considerable evidence that both ω6 and ω3 polyunsaturated fatty acids (PUFAs) can modulate a variety of immunologic activities. There was considerable interest in the clinical use of ω6-enriched vegetable oils during the 1970s for patients who had received kidney transplants, as well as those with multiple sclerosis and autoimmune disorders. Development of better immunosuppressive agents and other drugs for these conditions led investigators away from this area of research. However, there remains considerable interest in the effects of dietary fats on immune response, particularly in relation to development of cancer but also for cardiovascular disease and general health. It is fair to state that beyond the generally accepted immunosuppressive effects of polyunsaturated fats, there is considerable disagreement on whether the specific fatty acid, ratios between different fatty acids, or the quantity of dietary fat has the greatest impact on immunologic responses. In general, when the long chain fatty acids increase in a cell membrane, there is a concomitant decrease in arachidonic acid concentration.

Sources of ω3 fatty acids used most commonly in studies of immune function include fish oils, evening primrose oil, and flaxseed oil. Plant-derived oils generally supply 18 : 3, while the fish oils have the longer chain fatty acids EPA and DHA. There is substantial debate concerning the optimal range of the ω6/ω3 ratio in the diet with many researchers favoring a range of 1.0–7.0 based on studies in rats. The 10th edition of the U.S. Recommended Dietary Allowances does not make specific recommendations for any essential fatty acids in humans but indicates the desirability of establishing these in the near future.

One study of the immune response in rats fed blends of different proportions of sun- flower oil (rich in linoleic acid) and flaxseed oil (rich in linolenic acid) found that the higher the ω3/ω6 ratio of fatty acids in the plasma, the greater the reduction of lymphocyte-dependent immune responses, including T-cell blastogenesis and natural killer cell activity. These authors concluded that α-linolenic acid was as potent as fish oil for suppressing immune responses.

While most animal studies simply use feeding of a single source of fat, or a blend, throughout an experiment, at least one study has tried to mimic the normal human intake pattern of ω3 PUFA. Mice were fed a diet containing safflower oil and switched to a sardine/olive oil mixture for periods of 1 to 7 days per week. Peritoneal macrophages had phospholipid compositions that reflected the diet. Synthesis of leukotrienes E_4 and C_4 decreased, while E_4 and C_5 levels increased with more frequent consumption of fish oil. Prostaglandin $F_{1\alpha}$ also decreased with increasing consumption of fish oil. The summary finding of this study-that fish oil must be consumed at least twice a week to produce significant changes from the control diet touches on the current debate about how frequently a person has to consume fish to derive a health benefit. As with most questions on diet and risk of chronic diseases, there is no definitive answer, but some studies have suggested maximal benefit with consumption of at least two servings of fatty fish weekly.

14.4 The Effect of Dietary Fat on Energy Balance

Dietary fat can have a profound effect on energy balance and ultimately body weight. However, reports on its effects on food intake have been variable and appear to be influenced by how fat is presented as well as other parameters, such as metabolic state of the experimental subject. Studies designed to investigate the effect of dietary fat on food intake generally employ one of two approaches: short-term studies examining the influence of dietary fat on meal size or frequency, and longer term investigations of the effects of dietary fat on energy intake over days or weeks. Short-term effects are typically examined in one of two ways:

1. Fat is given as a "preload" and subsequent feelings of hunger or food intake are reported, or
2. Fat is given as part of a mixed diet and concurrent food intake is measured. In the first instance, "satiety" (*i.e.*, the ability of a substance to suppress further eating) is measured by the time elapsed, or amount of food eaten at thc next meal.

In the second case, "satiation" is assessed and is defined as the size of the current meal.

Many studies have been carried out examining the effect of a "preload," often in the form of a liquid, on subsequent intake, typically over a short time period. Others have investigated the effects of high- or low-fat meals or snacks, or meals supplemented with fat or carbohydrate, on subsequent feelings of fullness and food intake. The results of these studies have been quite variable. Many reports indicate that with a preload, individuals do not fully compensate for the calories ingested; that is, they do not decrease their intake in accurate proportion to the calories previously consumed. This seems to be truer for fat versus carbohydrate preloads, though some studies have indicated that fat has satiety value equal to that of carbohydrate or even greater. In addition, compensation seems to occur initially, then decreases over time.

It is important that the volume and sensory characteristics of the preloads be similar when the satiety effect of fat is compared to that of carbohydrate or protein. A number of studies have not controlled for both these factors. However, when preloads are similar, investigators have found that individuals vary substantially in their response to pre loads and that body weight may play a role in these responses. For example, some have reported that males of normal weight who are not concerned about their body weight or food intake ("unrestrained eaters") appear to compensate adequately for the caloric content of a preload. Others have shown that in obese individuals, fat exerts only a weak action on satiety. Data such as these have led to the speculation that obesity may be the result of an insensitivity to satiety signals generated by ingestion of fat.

In studies of satiation, dietary fat is an integral variable of the test diet(s). Studies show that caloric intake is greater with dietary fat than with carbohydrate. When subjects are fed diets in which fat, fiber, and simple sugars are manipulated to obtain low-energy versus high-energy diets, energy intake is greater on the high-energy diet. Caputo and Mattes reported that individuals consuming high-fat meals consume more calories than those consuming meals high or low in carbohydrate or low in fat. Others have shown that individuals consuming high-fat or high-carbohydrate diets for 1 week consume more calories on the former regime. This hyperphagic effect of dietary fat has been observed by others.

Relatively few long-term studies on humans have been conducted because of difficulties such as inability of the investigator to control for the subjects' current or past food intake, activity level, or genetic background. However, Kendall *et al.* reported that women .consuming either a low- or high-fat diet for two separate 11-week periods consume more calories on the high-fat diet. Lissner *et al.* also showed that subjects eating a diet either .low, medium, or high in fat for 2 weeks consume the most calories on the high-fat diet.

In addition, animal studies have shown that the type of fat may influence the satiation effects of dietary fat. In both chickens and rats medium chain triglycerides have been shown to have a greater satiating effect than long chain triglycerides. Also, the physical form as well as type of fat contributes to its satiation effects. Overeating occurs with diets containing saturated fats as well as mixed fats.

Some have suggested that the hyperphagic effect of high-fat diets is not due solely to the fat but is influenced by the presence of carbohydrate and overall caloric density. Ramirez and Friedman fed rats diets varying in carbohydrate, fat, cellulose, or caloric density and found that energy intake varies directly as a function of caloric density regardless of the fat or cellulose content of the diets. They concluded that high levels of fat, carbohydrate, and energy interact to produce overeating in animals fed high fat diets. In support of this hypothesis, Emmett and Heaton who examined food records from 160 subjects who had weighed their food for 4 days, reported that an increase in refined sugar intake is associated with a linear increase in the intake of fat combined with carbohydrate. This result suggests that refined sugar may act as a vehicle for fat intake by increasing fat palatability.

It is generally that a high level of fat in the diet may induce adipose cell hypertrophy and/or hyperplasia. The influence of dietary fat type on adipose tissue cellularity and the development of obesity is however less definitive, particularly with respect to diet-induced alterations in fat cell hyperplasia. Several studies reported a greater effect of unsaturated fat versus saturated fat diet on increasing fat cell number. Specifically, an increase in the DNA content of the epididymal fat pad (indicative of an increased cell number) is observed in growing rats fed diets containing 20 per cent sunflower oil, while an increase in fat cell size is observed in those fed diets containing 20 per cent lard. It is suggested that the multiplication rate of adipose tissue cells may be increased when the degree of unsaturation of the dietary lipids is greater. However, this suggestion is contradicted by other investigations that report a greater degree of fat cell hyperplasia with saturated compared to unsaturated fat diets. Lemmonier *et al.* observed an increase in fat cell number in the perirenal depots of rats fed high levels of saturated fat and an increase in fat cell size in those rats fed an unsaturated fat diet. In the epididymal fat pad, a similar degree of hypertrophy (and no cell hyperplasia) is observed in rats fed either of the high-fat diets compared to the low-fat-fed controls. Bourgeois *et al.* also reported responses specific to adipose tissue site and to sex for various types of high-fat diet, with the greatest increases in fat cell number observed in animals fed the saturated fats.

In contrast, several studies suggested that the alterations in fat cell size and number associated with high-fat feeding are not influenced by type of dietary fat. Thus, Kirtland *et al.* observed that long-term feeding of high-fat versus low-fat diets (20 and 3 per cent w/w, respectively) in guinea pigs results in an increase in fat pad weight and fat cell size. However, quite similar results are observed with inclusion of either beef tallow or corn oil in the high-fat diet formulation. Likewise, non-obese mice respond to high-fat diets containing either corn oil or tallow with similar increases in body and fat pad weight and adipocyte volume over the low-fat-fed controls.

The discrepancies observed with regard to the influence of type of dietary fat on adipose tissue cellularity may be related to differences in the level of dietary fat, in the species, strain, sex, or age of the animals, or in the length of time of the various studies. Differences in the level of total energy or the metabolic utilization of fuels may also determine the impact of dietary fat type on adipose tissue cellularity. Kirtland and Gurr found that the number of cells in developing adipose tissue is not influenced by the nature of dietary fat, but rather by the energy intake early in life. Likewise, changes in adipose tissue cellularity associated with fish oil feeding are believed to be due to differences in the oxidation, storage, and/or mobilization of this metabolic fuel.

It is unknown whether differences in total energy intake or metabolic fuel utilization would explain the discrepancies observed in the various investigations of type of dietary fat on adipose tissue cellularity, since many of the studies fail to include such measurements. However, several studies report effects of various types of dietary fat on fat cell size and number at similar levels of total energy intake. Likewise, depot-dependent differences in response to various types of dietary fat are sometimes observed in the same animal. Thus, the effects of dietary fat type on adipose tissue cellularity must be associated with factors other than or in addition to alterations in total energy intake or metabolic utilization.

Chapter 15
Use of Microorganisms for the Production of Lipids

15.1 Introduction

Some microorganisms such as eukaryotic yeasts, molds and algae are well known to produce triacylglycerols and some lipids.

Biomasses of oleaginous yeasts, contain high levels of lipids, predominantly triacylglycerols, having fatty acid composition resembling several edible fats and oils, Oleaginous yeasts can grow well on a wide variety of substrates, such as pure sugars (*e.g.*, glucose, sucrose, fructose), mixed sugars contained in molasses, lactose contained in whey, and ethanol. Whey, a soluble liquor byproduct of the cheese-making industry, contains up to 5 per cent w/v lactose in addition to non-coagulated protein. The protein can be separated from lactose by ultrafiltration, and the resulting permeate containing lactose serves as an excellent substrate for the oleaginous yeast, *Candida curvata* D (= *Apiotrichum curvatum* ATCC 20509). The fat in the resulting biomass resembles the high-priced natural cocoa butter (in fatty acid composition and triglyceride structure. Use of *Candida curvata* D for the production of a cocoa butter substitute from whey permeate is being considered as a commercial process.

Some oleaginous molds make high proportions of lipids, predominantly triacylglycerols; the synthesis, which occurs in the biomass, uses glucose as carbon source. The fatty acid composition of the lipids of different molds varies widely. Whereas some molds (*e.g., Aspergillus terreus, Tolyposporium ehrenbergii*) produce single-cell oils similar to edible plant oils in fatty acid composition, others produce substantial proportions of less common fatty acids. For example, *Claviceps purpurea* produces large proportions of ricinoleic acid, and some *Mucor* species produce substantial proportions of α-linolenic acid.

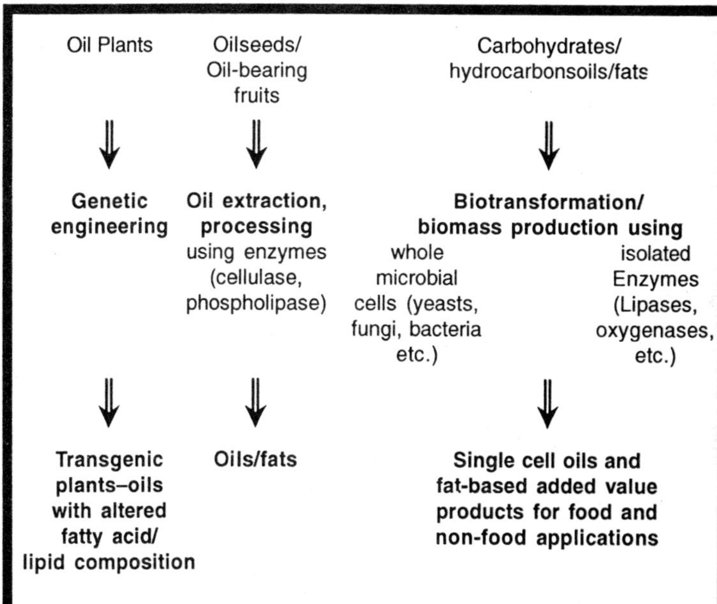

Figure 15.1: Major Areas of Lipid Biotechnology

Table 15.1: Lipid Content and Levels of Major Constituent Fatty Acids in Biomass of Lipid-Accumulating Yeasts

Organism	*Carbon Source*	*Lipid Content (% w/w of biomass)*	*Fatty Acid (% w/w of Total)*[a]				
			16:0	*16:1*	*18:0*	*18:1*	*18:2*
Candida curvata D (*Apiotrichum curvatum*)	Whey (lactose)	58	32		15	44	8
Cryptococcus albidus	Glucose, ethanol	65	16	1	3	56	
Lipomyces lipofer	Ethanol, glucose	64	37	4	7	48	3
Lipomyces starkeyi	Ethanol, glucose	63	34	6	5	51	3
Rhodosporidium toruloides	Glucose, mixed sugars	66	18	3	3	66	0
Trichosporon pullulans	Ethanol, glucose	65	15		2	57	24
Rhodotorula glutinis	Glucose, mixed sugars	72	37	1	3	47	8
Yarrowia lipolytica (*Candida lipolytica*)	Glucose	32–36	11	6	1	28	51

[a] Fatty acids are designated by number of C atoms (left of colon) and number of *cis* double bonds (right of colon)

Table 15.2: Lipid Content and Levels of Major Constituent Fatty Acids in Biomass of Lipid-Accumulating Molds Grown of Glucose as Carbon Source

Organism	*Lipid Content (% w/w of biomass)*	*Fatty Acid (% w/w of Total)*[a]					
		16 : 0	*16 : 1*	*18 : 0*	*18 : 1*	*18 : 2*	*18 : 3*
Aspergillus terreius	57	23			14	40	21
Claviceps purpurea[b]	31–60	23	6	2	19	8	
Mucor ramannianus	56	19	3	4	28	14	31[c]
Tolyposporiumm ehrenbergii	41	7	1	5	81	2	

[a] Fatty acids are designated by number of C atoms (left of colon) and number of *cis* double bonds (right of colon)

[b] Contains 42 per cent 12-hydroxy-*cis*-9-octadecenoic (ricinoleic) acid

[c] 18 : 3ω6 (γ-Linolenic acid)

Fat + Microbial cells ⇒ Incubation

⇒ Separation of products from culture broth and cells (Downstream processing)

⇒ Modified fat-based product

Fat + Isolated enzymes (soluble or immobilized) ⇒ Incubation

⇒ Separation of products from enzyme and unreacted substrate (Downstream processing)

⇒ Modified fat-based product

Figure 15.2: General Principles of Modification of Fats and other Lipids by Biotransformation using Whole Microbial Cells or Enzymes Isolated from Various Organisms

Bacteria that have been reported to produce substantial proportions of triacylglycerols in their biomass include Actinomycetales, especially *Mycobacteria*, *Corynebacteria*, and *Nocardia*. The biomass of an *Arthrobacter* species contain as much as 80 per cent lipids, of which about 90 per cent are triacylglycerols. However, cell yields of such organisms are rather low, and the presence of potentially toxic lipids in their biomass restricts the usefulness of such microorganisms as producer to triglyceride-rich, single-cell oils.

15.2 Triacylglycerol Lipases

Triacylglycerol lipases catalyze the hydrolysis, and they also catalyze the reverse reaction of lipolysis.

In media with low water content, moreover, lipases catalyze a wide variety of interesterification and transesterification reactions: for example, between triacylglycerols or between a triacylglycerol and a fatty acid (acidolysis) or an alcohol (alcoholysis) or glycerol (glycerolysis). The ability of lipases to catalyze the hydrolysis, esterification, and transesterification of lipids has been utilized in commercial processes for the modification of fats and other lipids.

The substrate specificities and regioselectivities in the hydrolysis of triacylglycerols, catalyzed by some common and commercially available lipases. It is evident that the triacylglycerol lipases have widely varying substrate specificities, preferring substrates with long and medium chain fatty acids over the short chain ones, and vice versa. Moreover, specificity of lipases for the fatty acids esterified at the *sn*-1, *sn*-2, and *sn*-3 positions of the glycerol backbone varies widely, ranging from nonspecificity for either of the three *sn*-1, *sn*-2, and *sn*-3 positions to strong *sn*-1,3 or *sn*-3 specificity.

Table 15.3: Specificity of Triacylglycerol Lipases from Different Sources

Source of Lipase	*Fatty Acid Specificity*[a]	*Positional Specificity*
Micoorganisms		
Aspergillus niger	S, M, L	*sn*-1,3 » *sn*-2
Candida antarctica	S > M, L	*sn*-3
Candida rugosa (syn. *C. cylindracea*)	S, L > M	*sn*-1,2,3
Chromobacterium viscosum	S, M, L	*sn*-1,2,3
Rhizomucor miehei	S > M, L	*sn*-1,3 » *sn*-2
Penicillium roquefortii	S, M » L	*sn*-1,3
Pseudomonas aeruginosa	S, M, L	*sn*-1
Pseudomonas fluorescens	S, L > M	*sn*-1,2,3
Rhizopus delemar	S, M, L	*sn*-1,2,3
Rhizopus oryzae	M, L > S	*sn*-1,3 » *sn*-2
Plants		
Rapeseed (*Brassica napus*)	S > M, L	*sn*-1,3 > *sn*-2
Papaya (*Carica papaya*) latex		*sn*-3
Animal tissues		
Porcine pancreatic	S > M, L	*sn*-1,3
Rabbit gastric[b]	S, M, L	*sn*-3

[a] S: short chain; M: medium chain; L: long chain.

In general, the substrate specificity and the positional specificity of triacylglycerol lipases observed in the hydrolysis reactions are also maintained in the reverse reaction of hydrolysis or in interesterification and transesterification reactions. The foregoing properties of triacylglycerol lipases permit their use as biocatalysts for the preparation of specific lipid products of definite composition and structure that often cannot be obtained by reactions carried out using chemical catalysts. This section outlines some current commercial applications and potentially interesting uses of lipase-catalyzed reactions for the production of specialty products from oils and fats.

Some typical applications of lipase-catalyzed interesterification reactions include the preparation, from inexpensive starting materials, of products resembling cocoa butter in their triacylglycerol structure and physical properties. Commercial processes for the preparation of cocoa butter substitutes involve interesterification of palm oil midfraction with stearic acid or ethyl stearate using *sn*-1,3-specific lipases.

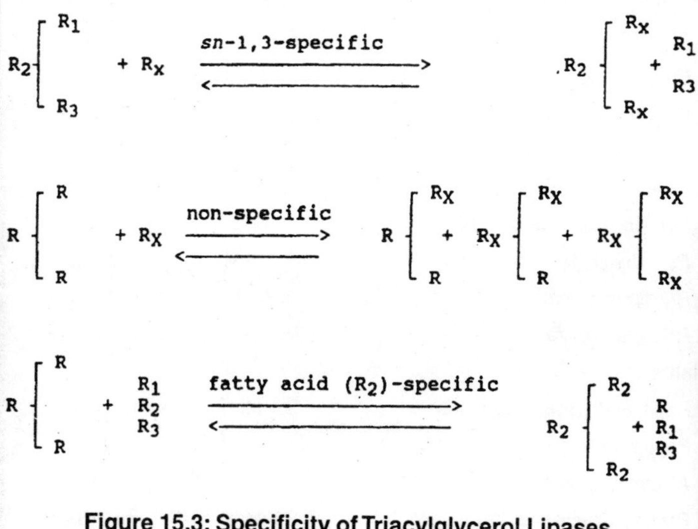

Figure 15.3: Specificity of Triacylglycerol Lipases in Interesterification and Transesterification: R_1, R_2, R_3, Fatty Acids/acyl Moieties

The triacylglycerols of human milk contain the palmitic acid esterified predominantly at the *sn*-2 position. Structured triacylglycerols resembling triacylglycerols of human milk are produced by transesterification of tripalmitin, derived from palm oil, with oleic acid or poly-unsaturated fatty acids, obtained from plant oils, using *sn*-1,3-specific lipases as biocatalyst. Such triacylglycerols are used in infant food formulations.

Possible applications of interesterification reactions catalyzed by *sn*-1,3-specific lipases include the preparation of structured triacylglycerols for use in specific dietetic products (nutraceuticals). Thus, interesterification of a common plant oil, such as sunflower oil, with a medium chain fatty acid by means of an *sn*-1,3-specific lipase would yield triacylglycerols containing medium chain acyl moieties at the *sn*-1,3 positions and long chain acyl moieties at the *sn*-2 position.

Evidence has accumulated lately that nutritional properties of triacylglycerols can be altered beneficially by structuring such lipids (*e.g.*, by inserting certain fatty acyl moieties at specific positions of the glycerol backbone to yield structured triacylglycerols. In particular, structured triacylglycerols prepared by lipase-catalyzed transesterification in which the physiologically active ω3 and ω6 polyunsaturated fatty acids, such as docosahexaenoic /(DHA) and

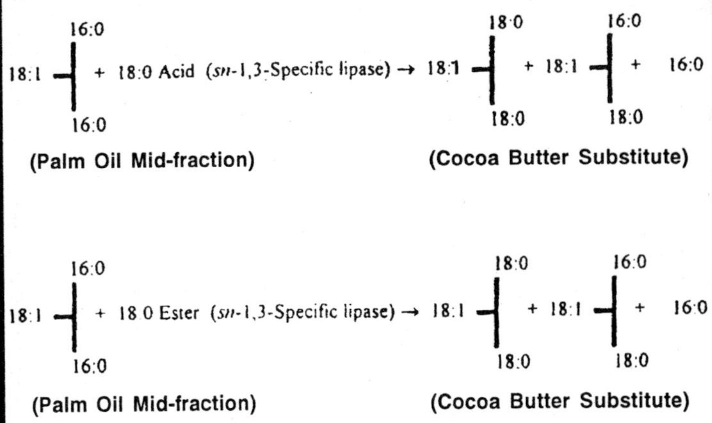

Figure 15.4: Lipase-catalyzed Transesterification for the Production of Cocoa Butter Substitutes

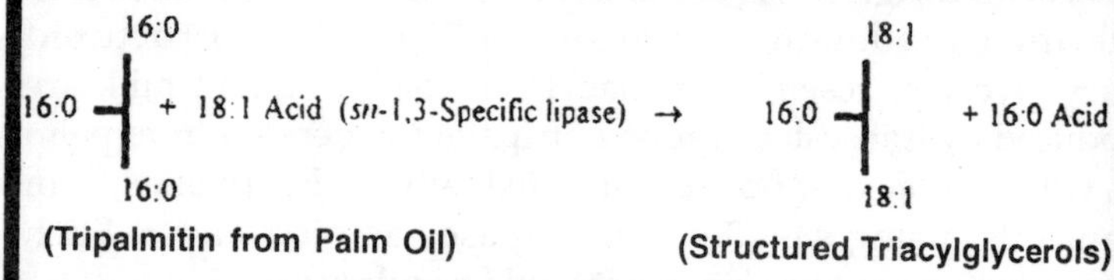

Figure 15.5: Preparation of Structured Triacylglycerols for Use as Human Milk Fat Replacers by Lipase-catalyzed Transesterification

γ-linolenic (GLA) acids (eicosanoid precursors), are esterified at specific positions of glycerol backbone, are envisaged to exhibit interesting biological properties that might enable their use in specific nutraceutical products and infant formulas.

Structured triacylglycerols containing two molecules of caprylic acid and one molecule of erucic acid have been prepared by lipase-catalyzed esterification of caprylic acid to monoerucin; upon hydrogenation of the erucoyl moieties to behenoyl moieties, the resulting triacylglycerols yield products resembling Caprenin, a commercially available low-calorie triglyceride. Structured triacylglycerols of the monoglyceride diacetate and diglyceride monoacetate types are prepared using *sn*-1,3-specific lipase by transesterification of plant oils with triacetin or an alkyl acetate.

Lipase-catalyzed esterification has been carried out for the synthesis of fatty acyl esters of carbohydrates that can be used as emulsifiers. Esters of monosaccharides and disaccharides as well as those of sugar alcohols and other polyols have been prepared in

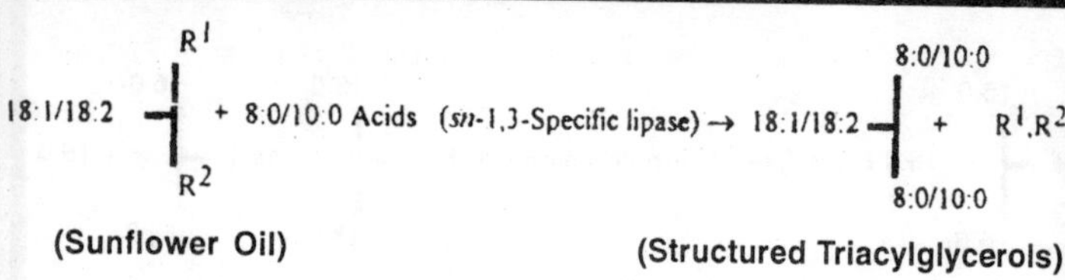

Figure 15.6: Preparation of Structured Triacylglycerols for use as Nutraceuticals by Lipase-catalyzed Transesterification

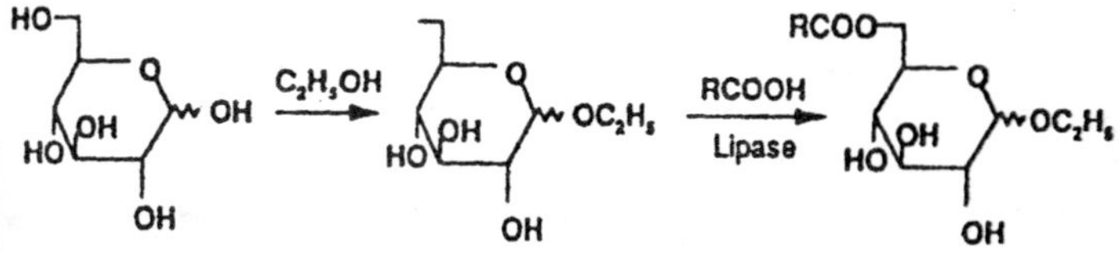

Figure 15.7: Preparation of Esters of Alkylglycosides by Lipase-catalyzed Esterification

good yields and excellent regioselectivity by esterification catalyzed by microbial lipases. Transesterification of sugars with short chain alkyl esters of fatty acids also provide sugar esters in good yields.

Triacylglycerols, contained in common fats and oils, as well as wax esters of jojoba oil, have been transesterified with various sugar alcohols in pyridine using lipases; the products were primary monoesters of sugar alcohols having excellent surfactant properties. Transesterification and esterification reactions, catalyzed by lipases, have also been applied for the preparation of fatty acid esters of alkyl glycosides.

Fatty acid esters of polyols are useful as surfactants. Polyglycerol-fatty acid esters have been prepared in good yields by transesterification of polyglycerol, adsorbed on silica gel, with methyl esters of fatty acids.

Propylene glycol monoesters, suitable as emulsifiers, have been prepared in good yields by reacting a fatty acid anhydride with 1,2-propanediol in the presence of lipase from a *Pseudomonas* species. Esterification of eicosapentaenoic and docosahexaenoic; acids with 1,2-propanediol in the presence of lipase from *Rhizomucor miehei* provides propylene glycol monoester emulsifiers that are potentially beneficial to health.

Polyethylene glycol esters of fatty acids, widely used as nonionic surfactants, have been prepared in essentially quantitative yields by esterification of oleic acid with polyethylene glycol, catalyzed by lipase from *Rhizomucor miehei*.

Reaction of a triacylglycerol mixture, such as soybean oil, with lysine, catalyzed by lipase from *Rhizomucor miehei* yields acyl amides (*i.e.*, N-ε-acyllysines). Reaction of ethyl octanoate with ammonia (ammonolysis), catalyzed by lipase from *Candida antarctica*, provides octanamide in near quantitative yields. There is a report of one-pot enzymatic synthesis of octanamide in high yield via esterification of octanoic acid with ethanol, followed by ammonolysis of the resulting ethyl octanoate; both reactions were conducted using the lipase from *Candida antarctica*.

Lipase-catalyzed hydrolysis can be applied for the production of fatty acids from fats using, for example, a nonspecific lipase preparation from *Candida cylindracea*. *Pseudomonas* sp. *Aspergillus* sp. *Thermomyces lanuginosus*, or *Chromobacterium viscosum* as alternative mild processes compared to drastic "steam splitting". Specifically, fatty acids have been obtained by lipase-catalyzed hydrolysis of technically important fats, such as animal fats castor oil, palm stearin, and the thermally labile perilla oil, which is high in α-linolenic acid. Now enzymatic "presplitting" of oils prior to steam splitting is being considered as an economically feasible alternative to complete enzymatic hydrolysis or steam splitting along.

Table 15.4: Specificity of Triacylglycerol Lipases from Different Sources Toward Various Fatty Acids/Acyl Moieties

Source of Lipase	*Discrimination Against*[a]
Microorganisms	
Candida rugosa (syn. *C. cylindracea*) *Penicillium cyclopium* *Penicillium* sp. (lipase G) *Rhizomucor miehei* *Rhizopus arrhizus*	all-*cis*-4,7,10,13,16,19-DHA, petroselinic (*cis*-6-octadecenoic) acid, GLA (all-*cis*-6,9,12-octadecadienoic) acid, stearidonic (all-*cis*-6,9, 12, 15-octadeca-tetraenoic) acid, dihomo-γ-linolenic (all-*cis*-,6,9,12-eicosatrienoic) acid
Plants	
Rape (*Brassica napus*) seedlings	DHA, petroselinic, GLA, stearidonic, and dihomo-γ-linolenic acid
Papaya (*Carica papaya*) latex	
Animal tissues	
Porcine pancreas	DHA, petroselinic acid, GLA, and stearidonic acid

[a] DHA: Docosahexaenoic acid; GLA: γ-linolenic acid.

Triacylglycerol lipases from various organisms have one common feature in their selectivity toward groups of fatty acids/acyl moieties having olefinic bonds at definite positions. Thus, several lipases from microorganisms, plants, and animal tissues discriminate against fatty acids/acyl moieties having a *cis*-4, *cis*-6, or *cis*-8 double bond as substrates in hydrolysis, esterification, and interestetification reactions.

A + B + C + Water (Lipase discriminating against A) ⟹
(Acyl moieties of triacylglycerols)

A	**+**	**B + C**
(Enrichment of A in acylglycerols)		**(Enrichment of B and C in fatty acids)**

A + B + C + Water (Lipase with preference for A) ⟹
(Acyl moieties of triacylglycerols)

A	**+**	**B + C**
(Enrichment of A in fatty acids)		**(Enrichment of B and C in acylglycerols)**

Figure 15.8: Principle of Kinetic Resolution via Lipase-catalyzed Hydrolysis

15.3 Bioconversion of Fats

Bioconversion of fats using whole microbial cells is a promising approach for large-scale economical production of fat-based products. Cells of the alga *Euglena gracilis* and other algae as well as microorganisms such as coryne bacterium can be used to catalyze the esterification of fatty acids.

Numerous examples of microbial oxidation of unsaturated fatty acids for the preparation of hydroxy fatty acids have become known. The microbial enzymes involved in oxidation of unsaturated fatty acids to hydroxy acids via hydration, termed monoxygenases, catalyze the incorporation of one atom of molecular oxygen into a substrate while the other atom is reduced to water. Dioxygenases

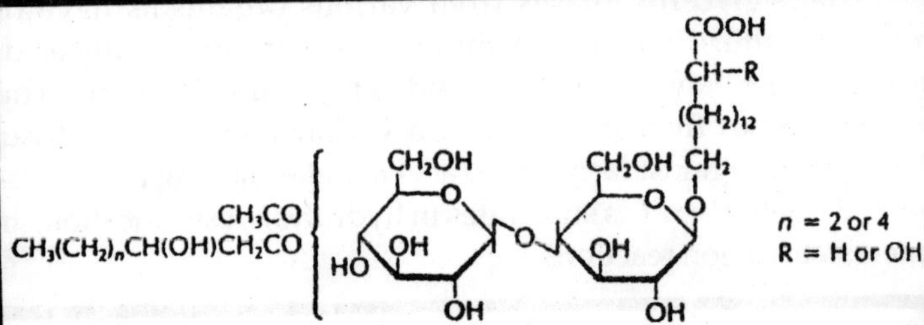

Figure 15.9: Common Forms of Microbial Cellobiose Lipids

catalyze the incorporation of both atoms of molecular oxygen into the substrate.

Since the initial reports by Wallen *et al.* on the hydration of the olefinic bond of oleic acid to 10-hydroxystearic acid by *Pseudomonas* sp., there have been reports of the hydration of oleic acid catalyzed by species of *Nocardia*, *Corynebacterium*, and *Micrococcus.*

Anaerobic microbial hydration of *cis*-9-olefinic bond has been reported to yield 10-hydroxy-12-cis-octadecenoic acid from linoleic acid, 10-hydroxy-12-*cis*,15-*cis*-octadecadienoic acid from α-linolenic acid, and 10,12-dihydroxystearic acid from ricinoleic acid. Recently, hydration catalyzed by *Nocardia cholesterolicum* has been shown to produce 10-hydroxy-12-cis-octadecenoic acid from linoleic acid in

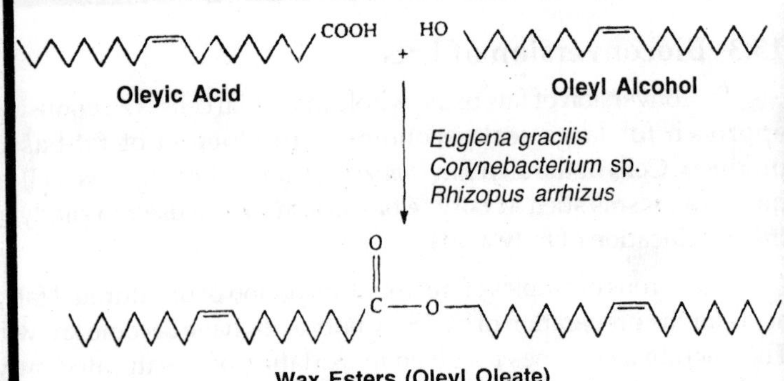

Figure 15.10: Microbial Production of Wax Esters

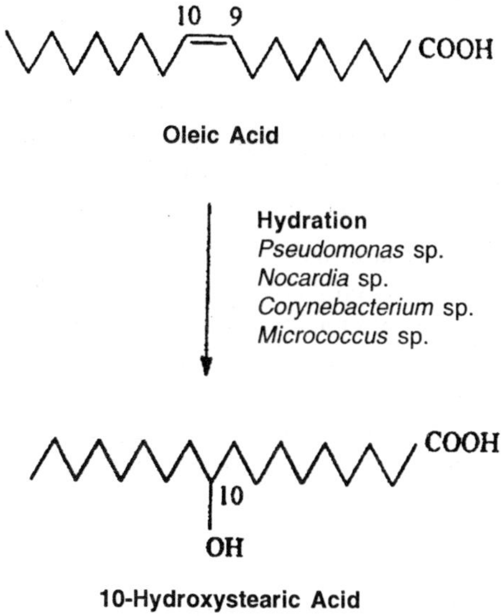

Figure 15.11: Microbial Production of 10-hydroxy Stearic Acid by Hydration of Oleic Acid

71 per cent yield and 10-hydroxy-12-*cis*,15-*cis*-octadecadienoic acid from α-linolenic acid in 77 per cent yield.

A cyanobacterium, *Phormidium tenue*, converts linoleic acid via 9-hydroperoxy and 13-hydroperoxyoctadecadienoic acids to the corresponding hydroxyoctadecadienoic acids. Moderate proportions of 15-hydroxy-, 16-hydroxy-, and 17-hydroxy-*cis*-9-octadecenoic acids have been prepared from oleic acid by *Bacillus megaterium* and *Bacillus pumilus*.

15.4 The Processing Technology of Oils and Fats

Enzymes gaining importance as processing aids in the technology of oil and fats. An enzymatic process developed very recently for the degumming of oils and fats in large-scale commercial operation.

Conventional techniques for the recovery of oils from seeds and fruits involve grinding and conditioning by heat and moisture to disintegrate the oil-bearing cells, followed by mechanical pressing in hydraulic presses or expellers or extraction by organic solvents, such as hexane. Fats from animal tissues are frequently recovered by rendering (*i.e.*, heat treatment with live steam). Lately, aqueous extraction processes have become known in which the seeds or fruits are first ground with water to disrupt the oil-bearing cells, then centrifuged to separate the oil from the solids and the aqueous phase.

Extensive mechanical rupturing of the cells of oil-bearing seeds is the prerequisite for efficient oil extraction. Several recent reports suggest the use of enzymes for the rupture of the plant cell walls to release the oil contained in the cell prior to recovery of the oil by mechanical pressing, solvent extraction, or aqueous extraction. Plant cell walls are generally composed of unlignified cellulose fibers to which strand of hemicellulose are attached; the cellulose fibers are often embedded in a matrix of pectic substances linked to structural protein. Since substantial differences are observed in the polysaccharide composition of the cell walls of different plant species, different combinations of cell wall degrading enzymes (carbohydrases and proteases) have to be used for individual seeds or fruits. Enzymes used in cocktails for cell wall degradation include amylase, cellulase, polygalacturonase, pectinase, hemicellulase, galactomanase, and proteases.

Enzyme pretreatment followed by mechanical expelling for improved oil recovery has been used for rapeseed and soybean. Typically, treatment of flaked rapeseed with commercial enzyme preparations [SP 249, Novo Nordisk Bioindustrial Inc., Danbury, CT; Olease, Biocon (US) Inc., Lexington, KY) at 30 per cent moisture and 50°C for 6 hours followed by drying and expelling gave 90 to 93 per cent recovery of oil, versus 72 per cent recovery of the controls not treated with the enzymes.

Enzyme-assisted expelling is now a commercial process for partial oil recovery from rapeseed. In the commercial process, the oil release is substantially enhanced and the resulting rapeseed cake with a superior nutritional value is preferred in a number of animal feeds.

Enzyme pretreatment followed by solvent extraction for enhanced oil recovery has been used for melon seeds and rapeseed. Incubation of autoclaved and moistened rapeseed flakes (30 per cent moisture) with carbohydrases for 12 hours, followed by drying to 4 per cent moisture and extraction with hexane, resulted in 4.0 to 4.7 per cent enhancement of oil extraction with the different enzyme preparations in the following order: mixed activity enzyme > β-glucanase > pectinase > hemicellulase > cellulase. A disadvantage of the process is the rather long incubation time.

Table 15.5: Approximate Composition (%) of Cell Wall Polysaccharides of Some Oil-bearing Materials

Polysaccharide	*Rapeseed*	*Coconut*	*Corn Germ*
Pectic substances	39	–	< 1
Mannans	–	61	–
Galactomannans	–	26	–
Arabinogalactans	8	some	–
Celluloses	22	13	39
Hemicelluloses			
Xyloglucans	29	–	–
Arabinoxylans	–	–	40
Others	2	–	10

Since the earlier publication of Lanzani *et al.* on rapeseed, sesame seed, sunflower seed, soybean, and peanut, several reports have appeared on enzyme pretreatment followed by aqueous extraction for enhanced oil recovery from coconut, corn germ, avocado, olives, and cocoa beans. Enzyme-assisted aqueous extraction has been extensively studied on rapeseed using the mixed enzyme preparation SP-311 (Novo); the product, an oil with low phospholipid content (phosphorus < 3 ppm), can be subjected to physical refining without complex degumming procedures. The protein recovered by aqueous enzymatic extraction has higher nutritive/market value than the rapeseed obtained by conventional pressing followed by solvent extraction; however, the oil yield in aqueous enzymatic processing is somewhat lower (90–92 per cent) than that afforded by the conventional process (about 97.5 per cent).

Addition to macerated fish of plant proteases, such as papain or bromelain, followed by incubation at 65°C for 30 minutes and at 50 to 55°C for 1 to 2 hours, respectively, has been shown to improve the recovery of fish oils. Thermostable microbial proteases operating at higher temperatures (70–75°C) appear to be more suitable for this purpose, since they reduce the risk of microbial growth and contamination.

15.5 The Principle of Enzymatic Degumming

Degumming is the an important processing step during the refining of vegetable oils, such as soybean and rapeseed (canola). Untreated vegetable oils contain varying proportions of phospholipids (phosphatides) that impair the quality and stability of the oils if not removed. These natural contaminants include phosphatidylcholines (PC), -inositols (PI), and -ethanolamines (PE) and phosphatidic acids (PA). Especially for vegetable oils that are subjected to physical refining, it is very important to eliminate the phospholipids prior to treatment of the oil with steam under vacuum for the removal of free (unesterified) fatty acids.

In the current industrial practice of degumming, the raw oils are treated with water or aqueous solutions of phosphoric acid, citric acid, or an alkali to hydrate the phospholipids that subsequently flocculate and thus, can be removed by centrifugation. Among the different classes of phospholipids, PC and PI are readily hydratable, whereas both PE and PA are hydratable when combined with potassium and unhydratable when complexed with divalent metals; moreover, PA occurs as a partially dissociated acid and is nonhydratable. Phospholipase D plays an important role in the formation of nonhydratable PA. Especially during the processing of soybeans, phospholipase D can remain fairly active even at 65°C in the presence of water-saturated hexane, thereby catalyzing the hydrolysis of hydratable phospholipids (*e.g.*, PC to nonhydratable PA); in aqueous media, however, this enzyme is readily deactivated by heat.

Recently, Lurgi AG (Frankfurt, Germany) developed an enzymatic process (EnzyMax) for degumming of oils in which the raw oil is treated with phospholipase A_2 that cleaves the fatty acids esterified at the *sn*-2 position of the diacylglycerophospholipids yielding *sn*-1-acyl-lysoglycerophospholipids, including *sn*-1-acyl-lyso PE and *sn*-1-acyl-lyso-PA, which can be readily hydrated and removed.

The enzyme used is a commercially available phospholipase A_2 (lecitase activity 10,000 IU/mL), isolated from porcine pancreas. Recently, a microbial phospholipase A_2 that is more economical than the porcine enzyme was used successfully for degumming. The process consists of three stages. In the first stage, solutions of citric acid (50 per cent w/v) and sodium hydroxide (3 per cent w/v) are dispersed with the crude oil or water-degummed oil and the pH is adjusted to approximately 5.0. The resulting mixture, maintained at 60°C, is intimately mixed with the enzyme solution (enzyme concentrate diluted with water to 0.2 per cent v/v) and passed through two or more enzyme reactors. The residence time of the mixture in the reactors varies from 1 to 6 hours, depending on the phosphatide content of the starting material and the quality requirement imposed on the degummed oil. The effluent from the enzyme reactors is finally admitted to a separator to obtain the streams of degummed oil and aqueous phase containing the phosphatide sludge. Some active enzyme remains in the sludge and is recycled into the process, generally several times.

Operational data from commercial plants for degumming of 540 tons/day rapeseed oil and 400 tons/day soybean oil show a likely average consumption of 10 g of enzyme concentrate per ton of oil and residual phosphorus content of the degummed oil of less than 10 ppm. The enzymatically degummed oil can be subjected to physical refining after bleaching. Phosphorus content is low and soaps are absent, so enzymatically degummed oils, unlike those obtained by other degumming processes, can be bleached with standard bleaching earths instead of high activity acidic bleaching

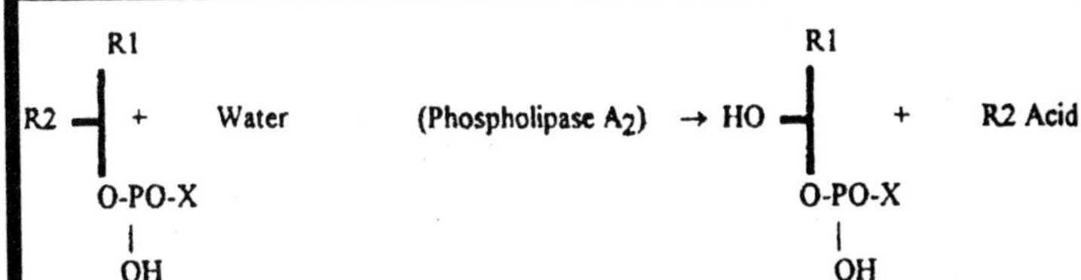

Diacylglycerophospholipid, X = Choline, ethanolamine etc.) **1-Acyl-lysoglycerophospholipid (hydratable)**

Figure 15.12: Principle of Enzymatic Degumming of Oils and Fats using Phospholipase A_2 for converting Diacylglycerophospholipids to easily Hydratable Lysophospholipids

earths. Further economic advantages of enzymatic degumming versus chemical neutralization and soapstock splitting are realized as a result of lower oil loss, less consumption of water, and very sparse effluent wastes of environmental concern.

Chapter 16

Trends in Production and Demand of Fats and Oils

As in past years, global demand for oils/fats, both for food and non-food purposes, is anticipated to expand significantly: during 2006-07, world consumption was expected to increase by almost 6 million tonnes or 4 per cent. An increasingly important demand factor is the fast growing use of oils/fats as fuels and as feedstock for biodiesel production (Agri Export Advantage, Export-Import Bank of India, Mumbai). Such utilization is expected to expand further in the European Union and the United States, while production is starting in various other countries, including Argentina, Brazil, Canada, Indonesia, Malaysia and the Philippines. The key oils concerned are soy and rapeseed oil, but palm and coconut oil as well as animal fats are also being used. Private sector investment into the development of biodiesel industries continues to be strong, irrespective of the uncertain development of mineral oil prices and notwithstanding the possibility that plants may not be running at full capacity. Government incentives and other public support measures, together with existing or prospected mandatory blending requirements explain this trend. According to private sources, global utilization of oils/fats as biofuels should have exceeded 10 per cent of total consumption in 2006-07.

16.1 Global Scenario Regarding Production and Demand of Major Oilseeds and Oils

Looking at last five years' major World oilseed production, we find that while there has been significant growth between 2002-03 and 2004-05 from 330.26 million tonnes to 381.29 million tonnes, the next two years have not seen such a growth. For 2006-07, provisional figures released by USDA were 395.50 million tonnes. Soybean and Cottonseed were the major contributors to the growth between 2002-03 and 2004-05 as per World Oilseeds Production given below in Table 16.1 and World's country wise major oilseeds and oils production are given in Table 16.2. Source of this information is USDA.

Table 16.1: World Major Oilseeds Production in Last Five Years (in million tonnes)

Oilseeds	*2002-03*	*2003-04*	*2004-05*	*2005-06*	*2006-07(p)*
Soybean	197.03	186.77	215.95	218.04	226.78
Rapeseed	32.91	39.43	46.14	48.55	46.78
Groundnut	30.83	32.78	33.39	33.71	31.59
Sunflower	23.93	26.76	25.30	29.77	30.54
Palm Kernel	7.76	8.43	9.51	9.98	10.85
Cottonseed	32.67	35.60	45.40	42.51	43.77
Copra	5.12	5.38	5.59	5.79	5.19
Total	330.26	335.16	381.29	388.33	395.50

Rapeseed/mustard plays a significant role in total vegetable oil pool. The World production of rapeseed / mustard oilseed and oil is likely to be as shown in Table 16.3.

Table 16.2: World Major Vegetable Oilseeds and Oils Production (Country-wise)

Name of the Country	*Oilseed 2006-07 (MT)*	*Oil 2006-07 (MT)*
United States	97.3	10.34
Brazil	58.36	6.05
China	56.55	15.21
Argentina	46.79	8.09
India	30.55	6.88
Indonesia	N.A.	20.09
Malaysia	N.A.	18.58
Eu-25	N.A.	13.21
Others	106.23	25.27
Total	395.50	123.71

1. Major Oilseeds includes Copra, Cottonseed, Palm Kernel, Peanut, Rapeseed, Soybean and Sunflowerseed.
2. Major Veg. Oils includes Coconut, Cottonseed, Olive, Palm, Palm Kernel, Peanut, Rapeseed, Soybean and Sunflowerseed Oil.

Table 16.3: World Mustard/Rapeseed/Canola Oil Production (Estimated in Million Tonnes)

Name of the Country	*Oilseed 2006–07*	*Oil 2006–07*
EU-25	15.59	6.69
China	12.50	4.55
India	6.5	2.29
Canada	9.1	1.45
Others	3.08	2.89
Total	46.78	17.87

Coming back to Indian scenario of mustard, the area under mustard this year is down from last year's 72.08 lakh hectares to 65.82 lakh hectares and consequently mustard production is likely to be proportionately down. There are concerns about rapeseed/mustard productivity this year because of weather and a clearer picture will emerge only after some time. However, given the current position, one may assume that production this year may be somewhere between 60 to 62 lakh tonnes. With a stock of about 10 lakh tonnes which will be available with NAFED in March this year, rapeseed available for crushing is likely to be between 70 to 72 lakh tonnes.

Mustard and olive oils are good vegetable sources of omega 3 fatty acids, so they must be included in the diet. A global body representing olive producers based in Madrid said recently that it planned to launch marketing campaigns in China and India to promote the consumption of olive oil in the two emerging markets. The International Olive Council will launch the promotional campaign in India in September and in China next year, its president Habib Essid told a news conference in the Spanish capital (Agri Export Advantage, Export-Import Bank of India, Mumbai July 18, 2007).

In India, olive oil is used mostly for manufacturing medicines and cosmetics and consumption remains very small. Palm oil is preferred for cooking. India, a nation of over one billion people, currently consumes just 2,500 tonnes of olive oil per year, but Essid said the potential existed to greatly increase exports of the product to the increasingly affluent country. "There are 200 million Indians who are potential buyers of olive oil," he said. Consumption of olive oil is on the rise around the world because of its reputation as a key part of a healthy Mediterranean diet, he said. Some three million tonnes of olive oil were produced during the 2006-07 season, according to the figures from the International Olive Council.

Spain is the world's biggest producer of olive oil, with production of 1.1 million tonnes in the 2006-07 season. Italy is the world's second-biggest producer with 650,000 tonnes, followed by Greece with 450,000 tonnes. Syria, Tunisia and Turkey each produce about 150,000 tonnes of olive oil per season. Production of olive oil is expected to reach 3.3 million tonnes in the 2007-08 season, thanks to the abundant rain which fell in Spain, Italy and Greece during the first half of this year, said Essid. The United States, Japan, Brazil and Australia are the biggest net importers of olive oil while Italy and Spain are the world's biggest consumers (Agri Export Advantage, Export-Import Bank of India, Mumbai July 18, 2007).

Growth in World Oilseed Production to Slow Down

Global 2006-07 oilseed production is currently forecast to increase by less than one per cent, a considerable slow-down compared with past two seasons. The slowdown comes from an only modest increase in soybean oil production and falling rapeseed, groundnut and sunflower oil output. By contrast, global palm oil production is expected to again rise markedly in 2007. World soybean output is anticipated to expand by 2-3 per cent, reaching a new record, but this is expected to be offset by declining world production of rape, groundnut and sunflower seed.

Regarding soybeans, a record crop has been harvested in the United States, thanks to a rise in both area and yield. In South America, where soybean plantings are still underway, production is tentatively forecast to grow at a below average rate of 2 per cent. Brazil's output is forecast to fall by about 2 per cent, owing to a sizeable reduction in plantings, for the second consecutive year, caused by high production costs and lack of funding. Brazil's shortfall should be offset by Argentina, where soybean plantings are expected to exceed the 2005-06 record level, leading to an output projection of 42 million tonnes, exactly double the amount produced in 2000. In China, the world's fourth largest producer of soybeans, output is reported to have fallen for the second consecutive year. World rapeseed output is forecast to decline sizeably after three years of record breaking crops, mainly because unfavourable weather conditions have lead to a drop in yields. Lower production is expected in four out of the five major producers, namely Australia, Canada, China and India. In India, production fell because farmers preferred to shift some land to grains and pulses, which were more attractively priced. European Union rapeseed output is estimated to have remained unchanged, whereas production has expanded in Ukraine. Also world groundnut production has fallen after three years of good performance. Lower output has been reported from India and the United States. With regard to sunflower seed, the drop in production in the United States has only partly been offset by rises elsewhere.

Compared with overall demand, global supplies of oils/fats continue to be ample, thanks to the availability of large stocks. However, 2006-07 production *per se* would not be sufficient to meet demand, therefore necessitating a reduction in global inventories of about 3 per cent. This development reverses the trend observed in the last two seasons, when global oils / fats production was in excess of demand, causing oils/fats stocks to surge to record levels. The projected reduction in inventories should mainly concern rape and sun oil, and be concentrated in China, India, the European Union and North America. Current forecasts for 2006-07 imply a reduction in the global stock-to-utilization ratio by one full percentage point, thus explaining why oil / fat prices are anticipated to remain high or even strengthen further during the course of this season. Early projections for next season, which point towards a further tightening of the supply and demand situation in 2008, could lend additional support to prices.

Global Demand for Oilseeds and Oils

It is of course evident that the demand for vegetable oils continues to grow and the major contribution to this growth is from developing countries, China and India playing a major role, both because of increasing population, per capita consumption and rising incomes.

Now another major factor which is going to have an impact on vegetable oils is the use of these for bio-fuels. Use of vegetable oils for bio-fuels has been talked about for several years, as an alternative source for fossil fuels. Last two years have seen an increasing diversion of vegetable oils for bio-diesel. Europe uses large quantities of diesel and some countries, for instance Germany have even made blending of bio-diesel mandatory. Almost entire rapeseed crop in Europe is now going for making bio-diesel. In the US, while usage of bio-diesel is less than in Europe because they use more gasoline blended with ethanol, the use of soybean for bio-diesel is still increasing thereto and the trend is catching up elsewhere also. Palm oil being the cheapest oil available, its use for bio-diesel is increasing at a rapid rate.

As a result, manufacturing plants have come up in significant numbers and from a production capacity of just about 6 million tonnes at the beginning of 2006, it was estimated to go upto over 20 million tonnes by the end of 2006. Of course, use of vegetable oils for bio-fuels and the quantities will depend to a large extent on crude prices. But it is clear that large chunk of vegetable oils is being taken out of the global food use and this is going to have a profound impact on the vegetable oil availability and prices.

Prices to Remain Firm in the Oilseed Complex

During the second half of the 2005-06 season (October-September), prices in the oilseed complex moved upward in response to forecasts of only modest growth in 2006-07 oilcrop production combined with a rise in demand to new record levels. The market reacted to the prospect of less ample supplies and reduced inventories compared with the three preceding years, when supplies were abundant relative to demand. In the last quarter of 2005-06, the FAO price indices for oils/fats and meals/cakes were respectively 10 and 30 per cent above the levels recorded during the same period of the previous season. The rise was more pronounced for meal/cake prices which have also come under the influence of strongly rising world prices of wheat and feed grains.

The latest forecasts for 2006-07 confirm that production of oilseeds may not be sufficient to satisfy global demand for oils/fats, thus necessitating a sizeable reduction in stocks. Such an outlook suggests that prices for oils/fats may continue rising during the current season. By contrast, the increase in prices for meals/cakes could come to a halt as global output of meals is currently anticipated to exceed demand, causing inventories to rise further. However, the price depressing effect of large stocks could be offset by continued strength in feed grain prices, which eventually should stimulate oilmeal demand. The futures market tends to point into this direction: by late November 2006, the CBOT March contract for soybeans was about US$50 per tonne (or 23 per cent) higher than the corresponding value of 2005 and, since September 2006, the development of soybean futures prices has been strongly influenced by maize futures.

Over the next few months, prices in the oilseed complex will be affected by the progress of the southern hemisphere crop, which will be harvested early in 2007. Later in the season, prices will increasingly reflect 2007-08 production prospects for oilseeds as well as other crops. The latest indications of continued tightness in the feed grain markets have the potential to affect oilseed plantings in 2007. If the currently anticipated fall in the oilseeds/grains price ratio materializes farmers in the northern hemisphere can be expected to sow more grains and less oilseeds in the spring of 2007, which would lend further support to the prices of oilseeds and derived products over the year.

Trade to Increase Markedly

Similar to the past few years, world trade in oils/fats (including the oil contained in seeds traded) is anticipated to rise by over 6 per cent or more than 4 million tonnes in 2006-07. Palm oil and soy oil should account for most of the expansion due to this season's reduced supplies of rapeseed and other oils. The bulk of the increase in imports is expected to originate from developing countries, especially in Asia. China and India continue to be key buyers, with import volumes (including the oil contained in oilseeds imported) forecast at a record level of 14.1 million tonnes and 5.8 million tones, respectively. The anticipated 10 per cent rise in China's imports is a result of poor rape and soybean crops, as well as rising domestic demand for oilseeds, which is being stimulated by continued expansion in the country's crush capacity. Also in India, where last season's abundant crops lead to a halt in import expansion, this season's prospective fall in output is expected to result in record import volumes. The European Union should account for most of the increase in import demand in developed countries. After two years of exceptional expansion, imports are expected to continue growing strongly because domestic oilseed production is not sufficient to satisfy both, demand for food uses and for biofuel production. Further growth in import demand is also expected for the United States.

As to global exports of oils/fats, in several countries - notably Argentina, Brazil, Canada, Indonesia, Malaysia,. and the United States, export availabilities are expected to be constrained by increased use of oils/fats as biofuels domestically. Brazil could actually experience a fall in shipments, given the likely reduction in output. Similarly, Australia should have less rapeseed oil for export, following the drop in domestic output. In view of these shortages, Canada is expected to further raise its exports of rapeseed and rapeseed oil, shipping a record 3.4 million tonnes in 2006-07. In addition, Ukraine is anticipated to step in as a new supplier of rapeseed. Argentina and the United States should supply most of this season's increase in soy oil trade. The two countries' record soybean harvests should allow them to ship record amounts. In the case of Argentina, this would also apply to sunflower oil. The largest expansion in shipments is expected to occur in palm and palm kernel oil. Together, the two oils should account for well over 40 per cent of total trade in 2006-07, with Indonesia and Malaysia as dominant suppliers. While Malaysia is expected to remain the largest exporter, Indonesia's shipments are quickly approaching the levels anticipated for Malaysia.

16.2 Trends in Production and Demand of Edible Oil/ Fats in India

Edible oil processing consists of three operations: crushing and expelling (separating oil from the solids), solvent extraction (to chemically remove residual oil from the oilcake solids), and oil refining. In many countries, these three separate processing operations are conducted by one vertically integrated plant. In India, however, only a small share of oilseed production undergoes solvent extraction and oil refining. Instead, India's oilseeds processing sector is made up of the three groups viz *Ghanis*, solvent extractors and oil refiners engaged separately.

Edible oils constitute an important component of Indian households' expenditure on food. According to NSS 60th Round (January-June 2004), average monthly per capita consumption expenditure (MPCE) of edible oil in food was 8.2% in rural India, and 8.2% in urban India. The share of edible oil has increased in successive NSSO surveys.

According to the Second Advance Estimates released by the Ministry of Agriculture on February 5, 2007, total oilseeds production during 2006-07 was expected at 23.62 million tonnes (mt), representing a decline of 15.6% over 2005-06. The decline in oilseeds production is due to lower output of rapeseed, groundnut and castorseed.

Size

1. India is world's third largest edible oil economy, after China and US. India's annual consumption is around 10 million tones vis-à-vis China's 14.5 million tonnes. However, India's per capita consumption at 10.2 kgs per annum is considerably lower compared to global standards.
2. India is also a leading producer of oilseeds, contributing 7-8% of world oilseed production. India is estimated to account for around 6% of the world's production of edible

oils. Though it has the largest cultivated area under oilseeds in the world, crop yields tantamount to only 50-60% of the world's average.

3. India is the fifth largest producer of oilseeds in the world, behind US, China, Brazil, and Argentina. Since 1995, Indian share in world production of oilseeds has been around 8-10%.

Structural Characteristics

1. Broadly, edible oil/fat products can be categorised into four categories, vegetable refined oil, hydrogenated oil (vanaspati), bakery fats/margarine, and de-oiled cakes.
2. The Indian edible oil industry can be classified into the following segments. *Ghanis* (over one lakh units), small scale expellers (15,000 mills), solvent extractors (600 units), oil refiners (400 units) and vanaspati manufacturers (204 units).
3. Oil mills crush oil seeds and extract oil, 70% of which is sold in the open market. The remaining 30% is refined and sold as branded oil. After the extraction of oil, residual seeds are processed further by solvent extractors, to make solvent-extracted oil. Most of the solvent extracted oil is used to make 'vanaspati'.
4. The Indian edible oil industry is highly fragmented with a large number of small scale producers. The *ghanis* belong to the SSI segment and usually serve the rural markets.
5. Small scale expellers, much like the *ghanis*, use metal screws to press or expel oil from seeds. However, they are larger than the *ghanis*, oil expelling capacity being in the range of 5-10 tonnes per day, compared to around 50-60 kilos a day for *ghanis*.
6. Solvent extractors belong to the organised segment and are also the second largest after the SSI segment, in the domestic edible oil industry. They use modern technology to process low oil & high meal seeds (e.g. soyabean, cottonseed) into edible oil and de-oiled cake.
7. Oil refining also belongs to the organised sector and has recorded rapid growth in recent times. Refiners generally refine both expeller oils and solvent extracted oils.

8. Vanaspati is made by hydrogenation of refined oil to vegetable shortening or spread and is similar to the milk product ghee and absorbs around 10% of the total edible oil supply in India.
9. Due to increased consumer preference for non traditional oils such as soybean and sunflower oil, the organised sector has emerged as one of the fastest growing sectors in recent times clocking double digit growth. Branded products, though small portion of the total edible oils market, have been one of the main drivers of rapid growth.
10. The production of edible oils in India is dependent on the production and availability of oilseeds. While oilseeds production increased from 10.83 million tonnes (mt) in 1985-86 to 24.75 mt in 1998-99, yield per hectare increased from 570 kg to 944 kg during 1999 to 2003, oilseed output stagnated and also declined. As oilseeds are mostly sown in the non-irrigated regions of the country (*viz*, Rajasthan, Gujarat, Maharashtra and Madhya Pradesh), its production suffers from high dependence on the monsoon.
11. India's production of oilseeds during the Kharif and Rabi crops of 2006-07 (6800 tonne) was likely to be lower by almost 2 million tonnes compared to the previous year (7100 tonne). The Central Organisation for Oil Industry and Trade (COOIT) had estimated the oilseed production for the 'kharif' 2006-07 at 128.4 lakh tonnes, which was about 9 lt lower than that of last kharif's 137 lt.
12. Overall, the problem remained of stagnation, low productivity and a haemorrhaging of arable land to urban crawl. Though oilseed production was lower by 2 million tonnes, oil production is lower by only 300,000 tonnes due to a very large carry over of rapeseed from the previous year. Announcing the forecast for the season at the All India Convention of Oilseeds and Oil Trade and Industry, Mr B.V. Mehta, Executive Director, Solvent Extractors' Association of India, who chaired the session on COOIT's crop estimation, said that the drop in production was due to a drop in acreages.
13. Moreover, the raw material costs account for over 92% of the total cost of sales due to volatile prices of oilseeds.

Besides the increase in raw material prices cannot be passed on entirely to the consumers, edible oil prices being governed by trends in international prices due to substantial imports.

14. Therefore the edible oil industry is characterised by very low profit margins, which is a manifestation of a variety of factors, the most important of all being availability of oilseeds.

Prices of Oilseeds and Oils

Overall, while it is the job of commodity analysts and experts to forecast what the prices are likely to be, one can safely expect that owing to lower domestic crop and high international prices of vegetable oils, prices in India will remain firm and it shall not be surprising if rapeseed fetches even a higher prices than the current MSP of Rs. 1715 per quintal. So the picture that emerges is like this:

1. The area on the planet earth is limited and that limits the area which can be used for cultivation for all the food items.
2. Population is growing and, therefore, demand for food items is also growing.
3. While demand for vegetable oil for food use will continue to grow, use of these for bio-fuels will put further pressure on supply.

Of course, it is anybody's guess that when it comes to choosing between food and energy, who will win. It is felt that the world needs the oils for food and use of these oils for fuel will be determined largely by crude prices, which also continue to rise. A day of reckoning will perhaps arrive and it may be sooner than later.

Policy

1. In order to increase the domestic supply of oil seeds, the government has been frequently freezing the MSP for wheat and rice while increasing the MSP for oil seeds, thereby prompting a diversification from wheat-rice to oil seeds. This is intended to improve the supply of oil seeds. However, despite these measures, the demand-supply gap is likely to continue in the medium term. Again this does

not push up prices, due to availability of low priced imports, as edible oil is the common man's utility item.

2. Free imports (since 1994) have further lowered the entry barrier to the industry as crude or refined oil can be imported, packed and distributed doing away with the need of having manufacturing facility in the domestic market.
3. Customs tariff on edible oil continues to be the most important and dynamic area of government intervention. India adopted a modified tariff schedule for agricultural products in March 2000. The tariff bindings, subsequent to revision in 1996 and renegotiations within the WTO in 1999 retain the overall structure notified after the Uruguay Round: 100% for commodities, 150% for processed products, and 300% for edible oils. Departures from this pattern are mainly with respect of tariff lines that were negotiated as special cases. India's bound rates for edible oil are as high as 300% ad valorem, except for 45% on soybean oil, and 75% for rapeseed oil. On all other oils, the GOI can raise the level of customs duty up to 300%.
4. Already, imports of palm are running about a third higher than last year. Strong economic growth and a pervading feel-good-factor are driving consumption. Reduction of import duties is keeping prices in control and will help growth in consumption. We see that India's imports in the next year are likely to be about 1 million tonnes higher than the previous year. We also see that India will import almost 1.5 million tonnes more of palm as compared with the previous year. This figure of palm will include about 400,000 tonnes of hydrogenated palm which we in India call Vanaspati. The greater reliance on palm will have significant bearing on the price outlook for palm products.

Outlook

1. The country's consumption places India behind only China and the European Union in total edible oil consumption. The growth in consumption of edible oil has been driven by increased population and growing incomes.

2. With its large population and continued strong economic growth, India is likely to register strong gains in total and per-capita edible oil consumption in the medium term. Per capita consumption is expected to increase to 11 kgs in FY2006 and 11.3 kgs in FY2007.

3. Production is expected to increase at a slower rate during OY2007 mainly because of an expected decline in India's oilseeds and edible oil production. India's production of vegetable oils has been stagnating except for a rise in rapeseed oil. According to the Solvent Extractors Association of India (SEA), India's vegetable oil production is expected to decline 4.4% during OY2007 to around 6.8 mt.

4. By 2010, India's total requirement of edible oils for the projected population of 1.25 billion at the projected per capita consumption of about 15 kg per annum is expected to be around 19 mt, which is equivalent to an estimated 57 mt of oilseeds.

The industry has to contend with increasing competition from imports, the rising cost of oil seeds and the expanding demand-supply gap (ICRA, CRISIL and CMIE reports). Since the production of oil seeds is heavily dependent on monsoons, around 40% of the demand for edible oils within the country has to be met by imports which may continue (HT Saturday, October 20, 2007).

Fight for Acreage in India

In the last few days, the Indian Government has raised the Procurement Price for Wheat and also announced an intention to import at least 3 million tonnes of wheat in the forthcoming season. The government needs to buy about 15 million tonnes of wheat from farmers for supply through the Public Distribution System. Market prices of wheat were at least 20 per cent higher than the government's Minimum Support Price. There was very little hope of any success for the Food Corporation of India to buy wheat on behalf of the government. One must appreciate the moves undertaken by the Indian government. However, these moves will have one profound effect for crop plantings in India for the next year. Oilseeds will continue to lose acreage to grains and pulses.

High prices of wheat, corn and beans have led to a titanic struggle for acres between cereals like wheat and rice, feedgrains like corn and sorghum and oilseeds like soybeans, rapeseed and sunflower seed. Indeed, Palm Oil is the only vegetable oil source which does not have to face this critical competition. The world will rely more and more on Palm Oil in the next 2 years.

During the last 12 months, the price forecast for palm oil was reported to rise from a band of 1400 to 1600 Ringgits in March 2006, to 1700 by September and finally to a Super Spike range of 2200 to 2400 Ringgits by March 2007. During 06-07 we saw that a mismatch of 1 million tonnes in Incremental Supply and Demand led to a dramatic rise in prices.

The Demand side for 07-08 is even more interesting. The economies of the world are expanding at a very healthy rate. The big population countries are enjoying strong growth. The factors which facilitate a rise in consumption are the weak US Dollar, buoyant oil and mineral revenues of the developing world, modest interest rates and high economic growth. Therefore, it is believed that demand for vegetable oil will again grow in 2007-08 by about 4 million tonnes despite high prices.

Non-traditional Demand – meaning Biodiesel and Green Energy is estimated to grow by only 2 million tonnes, notwithstanding the very large biodiesel capacities being established.

In Malaysia-Singapore-Indonesia, according to estimates, biodiesel demand will absorb only 800,000 tonnes of additional palm oil next year. Brazil and Argentina will absorb about 400,000 tonnes of soy oil. Additional usage for this purpose of all oils in USA is likely to be 400,000 and the rest of world including Europe will account for an additional 400,000 tonnes. These are very conservative estimates. It's due to a drop in acreages, says COOIT.

Production of Different Oilseeds in India

Groundnut

Announcing the forecast for the season at the All India Convention of Oilseeds and Oil Trade and Industry, Mr B.V. Mehta, Executive Director, Solvent Extractors' Association of India, who chaired the session on COOIT's crop estimation, said the drop in acreages and production has been mostly in the case of groundnut. Groundnut acreage is estimated at about 47.46 lakh hectares for the season, a drop of nine lakh hectares against last kharif's 56.47 lakh hectares. The forecasts indicated an eight-lakh-tonne drop in groundnut (in shell) production for kharif 2006-07.

Mr Mehta said that the dry spell in the major production belt in Andhra Pradesh, particularly Rayalaseema, hit sowing and production is down by about 53 per cent at 6.1 lt (11.5 lt). In Gujarat, where the drop is due to farmers shifting to cotton, production is down to 16 lt (17 lt). Castorseed is down by 2.1 lt, sunflower seed down by 20,000 tonnes.

Table 16.4: Production of Different Oilseeds in India: At a Glance (in Lakh Tonnes)

Oilseed	*Kharif Estimates*		*Retained for Sowing, Direct Consumption and Export*		*Marketable Surplus*	
	2006-07	*2005-06*	*2006-07*	*2005-06*	*2006-07*	*2005-06*
Groundnut	36	44	16	16	9.2*	14.8*
Soyabean	72.6	70.5	7.6	7.5	6.5	6.3
Sunflower seed	5.4	5.6	0.1	0.1	5.3	5.5
Sesam seed	4.2	4.2	3.2	3.2	1	1
Castor seed	7.2	9.1	Nil	Nil	7.2	9.1
Nigerseed	1	1.1	0.4	0.4	0.6	0.7

* in Kernels

Soybean

The forecasts have been optimistic on soybean, which is expected to be up by about 2 lt. This is attributed to an increase of about 3 lakh hectares in cultivation. The area under soybean for the season is 81.18 lakh hectares (78.01 lakh hectares).

Mustard/Rapeseed

Mustard / Rapeseed, which is one of the major oilseed crops in India and globally it is the second highest oilseed, after soybean. There are a few major points to be considered, when talking of vegetable oils, particularly mustard / rapeseed:

USDA has estimated the rapeseed production in India to be 65 lakh tonnes this year against actual production of 78.87 lakh tonnes last year. The shift in acreage has been mainly towards wheat and maize as can be seen from the Table 16.5, and rapeseed has lost the maximum area to other crops.

Table 16.5: Oilseeds and Food Grains Area Sown Rabi Season, 2006–2007 (area in lakh hectares) Area Sown as on 29–12–2006

Crop	*2006*	*2005*	*Difference Between 2005–2006*
	Oilseeds		
Groundnut	4.4	3.95	0.45
Rapeseed/ Mustard	65.82	72.08	-6.26
Sesame	0.83	0.8	0.03
Linseed	4.93	5.77	-0.84
Sunflower	10.26	11.74	-1.48
Safflower	3.38	3.55	-0.17
Other Oil Seeds	1.15	1.0	0.15
Total	90.77	98.89	-8.12
	Food Grains		
Wheat	263.91	247.25	+16.66
Maize	9.86	6.79	+3.07
Rice	8.81	7.79	+1.02
Course Cereals	66.36	62.42	+3.94
Pulses	124.32	123.6	+0.72
Total	473.26	447.85	+25.41

There are concerns about rapeseed/mustard productivity this year because of weather and a clearer picture will emerge only after some time. However, given the current position, one may assume that production this year may be somewhere between 60 to 62 lakh tonnes. With a stock of about 10 lakh tonnes which will be available with NAFED in March this year, rapeseed available for crushing is likely to be between 70 to 72 lakh tonnes.

Other Oilseeds

No change is projected in the production of sesame seed, which is expected to be about 3.2 lt. COOIT had also forecast increases in production of cottonseed and nigerseed for 2006-07.

Cottonseed production is pegged at 89.9 lt (84.9 lt) and no changes are expected in niger seed, which is set at one lakh tonnes. Copra production was estimated at 6.2 lt (four lakh tonnes) with equivalent coconut oil production at four lakh tonnes (4.2 lt).

In the Indian vegetable oil complex, our oilseed production has been stagnating for the last over a decade. There is little hope of increasing the area under oilseed and the talk of crop diversity is already a thing of the past, with India now having to import wheat as well, in addition to pulses.

Why do we find ourselves in this situation where about 40% of our edible oil need has to be met by imports and with the way things are, it is felt that this dependence on imports will only grow in future, if there is no significant increase in oilseed production.

Is there solution and if so what? What are the alternatives? As increase in area is not possible to any significant extent, the only way out is to increase productivity and that is a sure way. We only have to compare per hectare yields in India to the yields globally and the immediate task before us is to increase the yields. We know that it is a task which is more easily said than done. But it is not an impossible task. India has hardly 1 tonne per hectares productivity in comparison with other Countries like U.K., France, Poland, Italy U.S., Brazil, Argentina having productivity average of 2.5 to 3.5 tonnes per hectares and the World's average about 2 tonnes per hectare.

Developed countries of course have much higher yields, as is shown in the Tables 16.6 and 16.7.

Use of bio-technology has transformed agriculture economies of several countries that have seen phenomenal rise in productivity of several crops such as soybean and canola. India has experimented with Bt cotton and the results are for everyone to see. The productivity per hectare has gone up from 275 Kgs to more than 500 kgs, in some cases even to 700 kgs with the use of Bt cotton. From a net importer of 20 lakh bales of cotton per year, India has now exported 40 lakh bales of cotton. Of course, there are genuine bio-safety concerns in using genetically modified seeds and these have to be studied and addressed first. And this is being done not only in India but in most countries of the world including the developed ones such as Japan and EU. These concerns are genuine because they have long term impact on the environment and when it comes to food use, health concerns have also to be studied.

Table 16.6: World Major Oilseed, Area, Yield and Production

Country	*Area (Million Hectares)*	*Yield (MT/Hect)*	*Production (MT)*
United States	36.99	2.62	97.03
Brazil	22.17	2.63	58.33
Argentina	18.60	2.52	46.79
Paraguay	2.36	2.14	5.04
Bolivia	1.15	2.02	2.32
Uruguay	0.46	1.74	0.79
China	27.55	2.05	56.55
India	32.50	0.92	29.80
Pakistan	3.83	1.30	4.99
Russian Federation	6.92	1.18	8.15
Ukraine	4.90	1.32	6.45
Uzbekistan	1.42	1.69	2.40
Canada	6.60	1.93	12.75
France	2.02	2.76	5.59
Germany	1.46	3.67	5.37
United Kingdom	0.58	3.30	1.90
Czech Republic	0.35	2.81	0.99
Hungary	0.71	2.33	1.65
Poland	0.60	2.67	1.60
Italy	0.33	2.73	0.90
Australia	0.99	1.12	1.11
Turkey	1.25	1.73	2.17
Nigeria	2.04	1.04	2.11
South Africa	0.67	1.47	0.98
Indonesia	1.41	1.46	2.06
Romania	1.29	1.46	1.88
Others	19.55	1.01	19.78
Total World	198.69	1.91	379.46

Major Oilseeds: Soybean, Sunflower, Rapeseed, Cottonseed and Groundnut.

Table 16.7: World Major Oilseeds Yields 2006–07 (p)

Country	*Average Yields MT/Hect. for Cotton Kgs/Hectare*				
	Soybean	*Rapseed*	*Groundnut*	*Cotton*	*Sunseed*
United States	2.89	1.36	3.12	894	1.27
Germany		3.71			
Brazil	2.67		2.17	1197	
Argentina	2.68		2.50		1.74
Paraguay	2.35				
France		3.02			2.25
Russia					1.19
U.K.		3.30			
Pakistan				636	
China	1.74	1.74	3.04	1241	1.73
India	0.95	0.92	0.98	494	0.60
Canada	2.86	1.72			
World	2.43	1.75	1.53	727	1.23

The Government is well aware of the dire need to increase oilseed production and there is already an Integrated Scheme for Production of Oilseeds, Oil Palm and Pulses (ISOPOM) under the Ministry of Agriculture. A fare amount of funds are earmarked for this scheme, to be implemented through State Agriculture Departments. The industry too has a role to play in this regard and should take a cue from sugar mills which work very closely with farmers in their area for cane development. While the Government is playing its role in developing high yielding varieties and developing better farming techniques it is the delivery mechanism of the technology which has to reach the farmers and that is where the industry can play a big role. Availability of good seeds and inputs along with use of proper technology and farming methods will result in an immediate increase in yields. It is high time the Government and the industry work closely to play a role in this regard.

Suggestions

1. The Government may take a serious and immediate look at use of bio-technology for increasing oilseed productivity.

Possibility of increasing the percentage of oil content can also be looked at, with the use of bio-technology.

2. Maintaining a customs duty structure for edible oils to ensure a remunerative price to the farmers for their produce. This is very important because in a globalised economy, the oil prices are not in our control, they depend on demand-supply situation. The only mechanism to keep prices at a level for farmers to grow oilseeds is a proper duty regime.
3. Industry should come forward to work with the Government in the three areas which have been specifically mentioned for participation by the private sector under ISOPOM.
4. The Government should encourage the industry to involve itself in the task of increasing oilseed productivity. Industry bodies can play in major role in this. The industry is all for taking a pro-active approach for increasing oilseed production and it is willing to invest into Oilseeds Extension Programme to provide to the farmers the necessary agri inputs to achieve higher productivity level. However, this needs to be encouraged by way of giving weighted Income Tax deduction of 200% to companies undertaking Oilseeds Extension Programme. The private involvement in Oilseed Extension Programme will supplement Government's efforts and will go a long way in increasing oilseeds production and productivity in the country.

Chapter 17

Fat Replacers and Fat Substitutes

17.1 Introduction

Fat replacers are divided into two main groups: fat mimetics and fat substitutes. They are classified as carbohydrates based, protein based, and lipid based fat replacers.

Fat substitutes are believed to be compounds that physically and chemically resemble triacylglycerols. Let us learn more things about fat replacers and fat substitutes.

The protein-based and carbohydrate-based replacers are widely regarded as *fat mimetics.* The fat mimetics are either proteins or carbohydrates that have been physically or chemically processed to mimic the properties and functions of fats in food systems; they are not fats. They tend to adsorb a large amount of water, are not stable at frying temperature, and may produce food that is not microbiologically shelf stable; they contribute some calories (1-4 kcal/g) to the diet. The fat mimetics do not possess all the organoleptic, physical, chemical, and functional properties of fats and cannot replace calories from fat on a 1 : 1 weight basis. They cannot carry lipid-soluble flavor compounds because they cannot lower the vapor pressure of lipophilic flavor molecules, and most foods prepared with fat mimetics are often perceived by consumers as lacking in taste. Although fat mimetics need a delivery system such as an emulsifier to carry lipid-soluble flavors, they can carry water-soluble flavors. Some fat mimetics have mouth feel and physical properties approximating those of triacylglycerols but are not suitable for frying operations because they can be denatured (protein-based substances) or caramelized (those based on carbohydrates). They can, however be used for baking and retort cooking operations.

Fat substitutes are stable to high temperature cooking and frying operations, and in theory, can replace fats and oils on a one-to-one weight basis in foods. In the literature, the term "fat substitutes" has been used interchangeably with "fat replacers," but not with "fat mimetics," and this can be confusing. Several lipid-based synthetic low-calorie or zero-calorie fat substitutes belong to the fat substitute group. A good example is sucrose fatty acid polyester (SPE), which was originally developed as olestra and marketed as clean by Procter & Gamble to replace edible fats and oils in the diet. A number of other fat substitutes have been developed or are under development. Among these, the carbohydrates and alkyl glycoside fatty acid polysters and the structured lipids (SLs)have functional and physical properties resembling those of triacylglycerols while contributing few to no calories to the diet.

Some of the lipid-based fat substitutes can be added to food products to replace the functional properties of fats, including frying (not possible with protein-based fat replacers such as Simplesse, and carbohydrate-based fat replacers such as maltodextrin), while reducing caloric contributions from fats and oils. Table 17.1 lists the applications and functions of some fat replacers. Because the "ideal" fat substitute does not exist, a systems approach to reduced-fat or low-fat food formulations has been proposed. Each type of food product will require a different approach to address the difficulties of formulating a counterpart that has reduced, no, or low fat. Simply put, a systems approach uses a combination of different ingredients that may or may not belong to either of the classes of fat replacers and requires basic knowledge of ingredient technology to formulate desired products. The system may contain emulsifiers, fat substitutes or mimetics, fibers, water control ingredients, flavor, and bulking agents. Water or moisture control poses one of the greatest challenges in formulating reduced fat snack and baked goods. In these systems water is used to replace fat, to increase bulk, or for functionality. A detailed review of fat mimetics is outside the scope of this chapter, which concentrates on lipid-based fat substitutes.

Table 17.1: Typical Examples of Fat Replacers

Compound	*Class*	*Coloric Content (kcal/g)*	*Absorbability*	*Uses*
Simplesse	Protein-based	1–4	Absorbable	FDA approved, Dairy products, dressings, spreads. Not stable for frying and baking.
Maltodextrin	Carbohy-drate-based	4–4.5	Absorbable	Low-fat table spreads, dressings baked goods, desserts. Not stable for frying.
Olestra/ Olean	Lipid-based	0	Nonabsorbable	FDA approved for savory snacks. Stable to frying, baking, and cooking temperatures. Supplementation with fat-soluble vitamins required, Can be used for dairy products, spreads, and dressing but will require separate approval.

17.2 Sucrose Fatty Acid Polyester

Sucrose is a non-reducing disaccharide and the common table sugar "sucrose polyester" refers to sucrose esterified with six to eight fatty acids.

Sucrose fatty acid polyesler (SPE) development dates back to the year 1880, when a derivative of sucrose was prepared by acetylation to produce sucrose octaacetate (*i.e.*, sucrose containing eight acetate groups). Following this, other carbohydrate acetates were successfully prepared. In 1921 Hess and Messner synthesized sucrose octapalmitate (sucrose esterified with eight molecules of palmitic acid, a long chain fatty acid) and sucrose octastearate. In

1952 the concept of sucrose polyester production was initiated when the president of the Sugar Research Foundation, Henry B. Hass, asked Foster D. Snell to look into the possibility of "hanging a fat tail on sucrose" for use in detergents. The idea was that since sucrose is highly hydrophilic, a lipophilic tail on sucrose would result in a molecule that is amphiphilic (both water and oil loving), hence able to serve as an excellent surfactant. It was anticipated that production would be easy and the product biodegradable under aerobic and anaerobic conditions. It turned out that the chemical synthesis was not that easy without the use of solvents like dimethylformamide (DMF), dimethyl sulfoxide (DMS), and dimethylpyrolidone (DMP) to solubilize sucrose and free fatty acids. This process called the Hass-Snell process, was applicable only to the synthesis of sucrose mono- and diesters, otherwise called sucrose fatty acid esters (SFEs). These are digestible and good nonionic surfactants, as we shall see later in this chapter. By today's standard, the solvents used are not food grade; and therefore, products made in them are unacceptable for human consumption..

The other concept was to find means of reducing fat-derived calories without resorting to dilution with, say, water, air, carbohydrates, and proteins. The aim was to somehow come up with a fat like molecule that would significantly reduce fat calories by preventing their hydrolysis and absorption. This led to the discovery of a nondigestible and nonabsorbable fat like molecule called sucrose fatty acid polyester, now known as olestra (generic name) or Olean (brand name), by Mattson and Volpenhein while working on the absorption of fats by infants.

SPEs become undigestible when the number of fatty acids esterified is greater than 4. The structure of sucrose polyester is given in Figure 17.1. Procter & Gamble, which was granted the original patent on sucrose polyester in 1971, spent over $250 million over the last 25 years to develop this fat substitute. The original application for use of sucrose polyester as a food additive, filed with the Food and Drug Administration (FDA) in April 1987, was withdrawn and modified. But on January 24, 1996, the FDA approved olestra for limited use in savory snacks (namely, chips, curls and crackers). Before the approval, sucrose polyester was evaluated in over 100 animal studies and in 25 to 30 clinical trials. The approval of olestra was not without controversy. The Center for

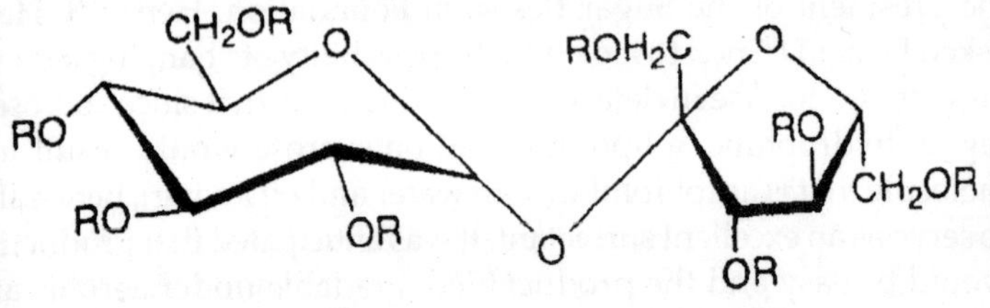

Sucrose polyester (olestra or Olean) [α Glucopyranosyl-(1 → 2) - β Fructofuranoside linkage]

where R = Alkyl part of the acyl group of fatty acids (R–C(=O)–)

Figure 17.1: Structure of Sucrose Polyester

Science in the Public Interest (CSPI), a Washington consumer advocacy group, believes that olestra deprives the body of some of the essential vitamins and carotenoids that may protect against cancer. In 1996, in Iowa, Wisconsin, and Colorado, Frito-lay, of PepsiCo, test-marketed chips made with the olestra.

Sucrose fatty acid polyester can be synthesized in the presence or absence of organic solvent. Direct esterification of sucrose with fatty acid is very difficult. The solvent-free process widely used for the current production of sucrose polyester. The synthesis may involve reactions of the following types:

Transesterification

Fatty acid methyl esters (FAME) and sucrose are reacted in the presence of potassium soaps to form a homogeneous melt followed by the addition of excess FAME and NaH at 130 to 150°C. In some cases potassium carbonate is added to aid the reaction. The function of soap is to help solubilize sucrose and FAME. Methanol, a by-product of the transesterification reaction, is distilled off (Figure 17.2). The active catalyst is the sucrate ion generated with alkali metal hydrides. This is a two-stage transesterification process, and up to 8 to 9 hours may be needed to achieve 90 per cent yield of SPE.

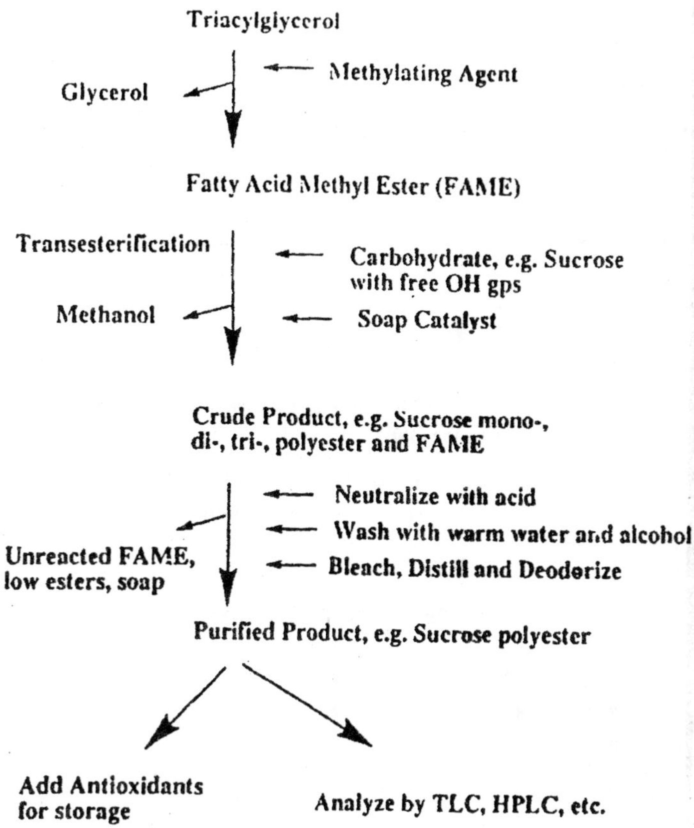

Figure 17.2: Synthetic Scheme for Olestra by Transesterification and Processing

Interesterification (Ester Interchange)

This involves reacting a short chain alky) ester such as sucrose octaacetate (SOAc) with FAME in the presence of sodium methoxide ($NaOCH_3$) or Na or K metal as catalyst. The reaction requires extremely anhydrous conditions to prevent hydrolysis of formed product, catalyst inactivation, or explosion (when Na is in contact with water). The temperature of the reaction with Na catalyst is lower (105–130°C). Reaction times of 2 to 6 hours and pressure of 0 to 5 mmHg are required to achieve greater than 95 per cent yield of

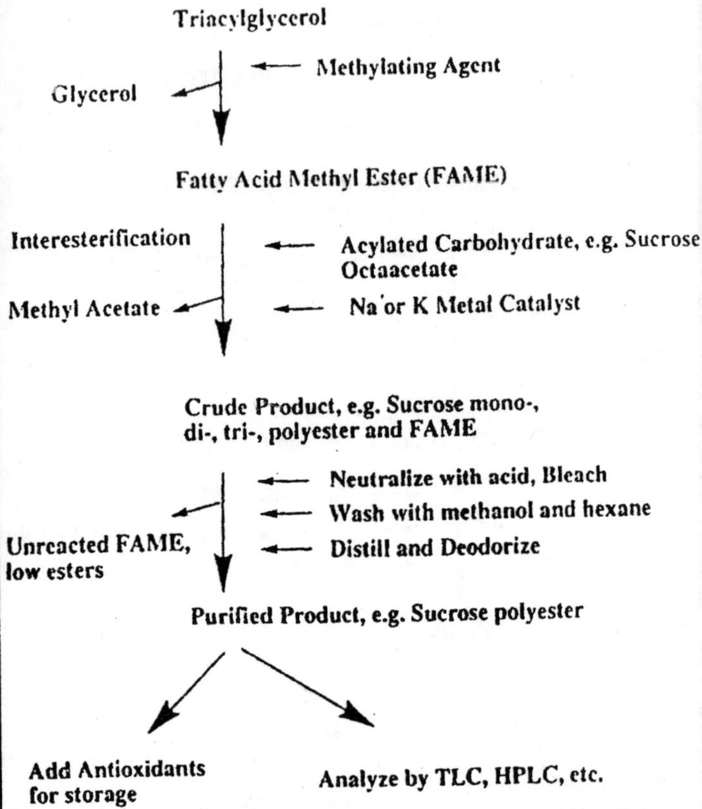

Figure 17.3: Synthetic Scheme for Olestra by Interesterification (Easter Interchange) and Processing

SPE. This is a simple ester-ester interchange reaction. The methyl. acetate formed is trapped with a liquid nitrogen (–196°C).

Synthesis based on Na metal (potentially explosive and flammable compound) as catalyst and sucrose octaacetate and FAME as substrates may not be suitable for industrial adaptation. Therefore, use of milder catalysts such as sodium methoxide, potassium soap, and potassium carbonate are encouraged. Shieh *et al.* (B) recently reported the optimized synthesis of SPE using potassium hydroxide (KOH) in methanol plus FAME to form soap (potassium soap) followed by the addition of potassium carbonate. The reaction time at 144°C was 11.5 hours.

The triacylglycerol for the synthesis can come from vegetable oils and fats, alone or in combination. The type of product desired dictates the type of fatty acids needed for synthesis to achieve desired functionality. In most cases, the fatty acid profiles of finished products will resemble those of the triacylglycerol source. It should be noted that the reactions above are random processes and therefore, the specific position and type of fatty acid on the sucrose molecule will vary from production to product.

17.3 Types of Fat Substitutes

Olestra-type of fat substitutes can be studied as follows.

Sorbestrin

Sorbestrin is sorbitol or cyclic sorbitol containing three to five fatty acids esterified to the OH group. It was developed by Pfizer Inc. Sorbestrin contains 1.5 kcal/g and has a bland oil-like taste with cloud point between 15 and 13°C. It is mainly a clear liquid and can serve as a reduced-caloric fat substitute. It is thermally stable and can withstand frying temperatures. It is intended for use in frying, baking, and salad dressings. Sorbestrin is not commercially available yet and will require food additive petition and FDA approval prior to use. Akoh and Swanson synthesized sorbitol hexaoleate as a low-calorie fat substitute. Pfizer noted that for humans, sorbestrin consumption in the future is estimated at 12 g/day.

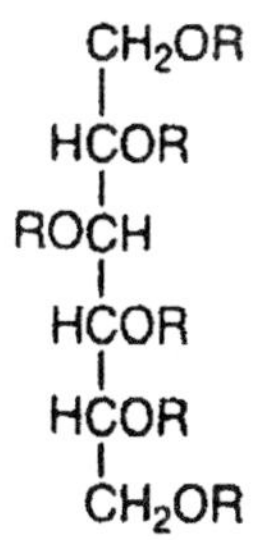

where R = Alkyl part of the acyl group of fatty acids (R–C(=O)–)

Figure 17.4: Structure of Sorbitol Polyester

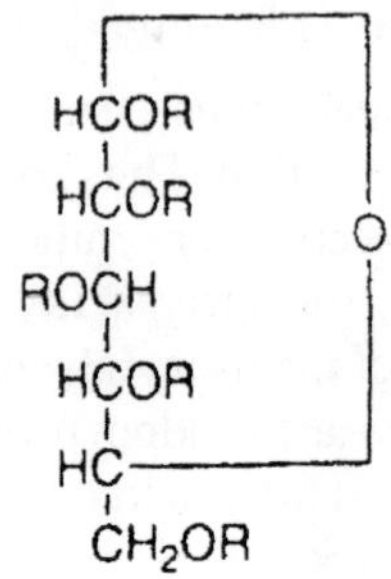

where R = Alkyl part of the acyl group of fatty acids 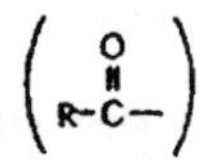

Figure 17.5: Structure of Sorbestrin (Cyclic Form)

Trehalose Polyester

Trehalose is a non-reducing disaccharide, the major sugar of insect hemolymph, fungi, and yeasts. It is made of two α-D-glucopyranose components and closely resembles sucrose in physical properties. The non-reducing sugars are better substrates for the synthesis of fat substitutes than the reducing sugars. This is because the anomeric carbon atoms (C-1) of non-reducing sugars are protected and not very susceptible to thermal degradation. The reducing sugars degrade and caramelize at the high temperatures required for transesterification. The use of acetyl derivatives of non-reducing sugars allows reactions to be carried out at reduced temperatures (< 105–120°C). Unfortunately, this process is not applicable to acetylated reducing sugars such as glucose pentaacetate.

Trehalose octaacetate has been used to interesterify FAME catalyzed by Na metal to produce highly substituted trehalose polyester with properties similar to those of sucrose polyester. Therefore, trehalose polyester can be used as a zero-calorie fat substitute in place of sucrose polyester. The only problem is cost effectiveness, since sucrose is a cheaper substrate than trehalose.

Raffinose Polyester

Raffinose, like sucrose, is a heterogeneous non-reducing sugar. Raffinose is made of galactose-glucose-fructose units. It has 11 hydroxyl groups that tan be esterified with fatty acids. It is well established that as the degree of substitution of a carbohydrate with fatty acids increases, the susceptibility to hydrolysis and absorption decreases. Based on this fact, raffinose undecaacetate was used to interesterify FAME to produce raffinose polyester containing 10 to 11 fatty acids in 99 per cent yield. The reaction was catalyzed by Na metal at 110°C for 2 to 3 hours. The product had consistency in the range of salad oils and sucrose polyesters, with some raffinose polyesters being slightly more viscous than sucrose polyesters and salad oils depending on the fatty acid composition and degree of substitution.

Sorbitol Polyester

Sorbitol or glucitol is a sugar alcohol (polyol) made by hydrogenation or electrolytic reduction of glucose. Other polyols like xylitol, mannitol, and lactitol can equally be used for the synthesis of fat substitutes. Up to six hydroxyl groups can be esterified with fatty acids to produce a nondigestible sorbitol polyester. Sorbitol polyester can be synthesized by interesterification of sorbitol hexaacetate with FAME in the absence of organic solvent or by transesterification reaction between sorbitol and FAME as described above.

Recently, Chung *et al.* optimized the synthesis of sorbitol fatty acid polyesters. Procter & Gamble also worked on sorbitol polyesters but did not follow up on their development. The properties and applications in food have not been extensively studied as with olestra. However, sorbitol polyesters are proposed as zero-calorie fat substitutes. They are less viscous than sucrose polyester and slightly more viscous than vegetable oils. Based on the known structure and limited metabolic studies, sorbitol polyesters may be used in place of olestra in food products. They are stable to high heat and taste and function like fats.

Stachyose Polyester

Stachyose, a non-reducing heterogeneous tetrasaccharide, is similar to raffinose and sucrose except that stachyose contains galactose-galactose-glucose-fructose. There are 14 available OH groups for esterification with long chain fatty acids. Stachyose can be completely acetylated and used as the substrate for interesterification with FAME, catalyzed by Na metal as described

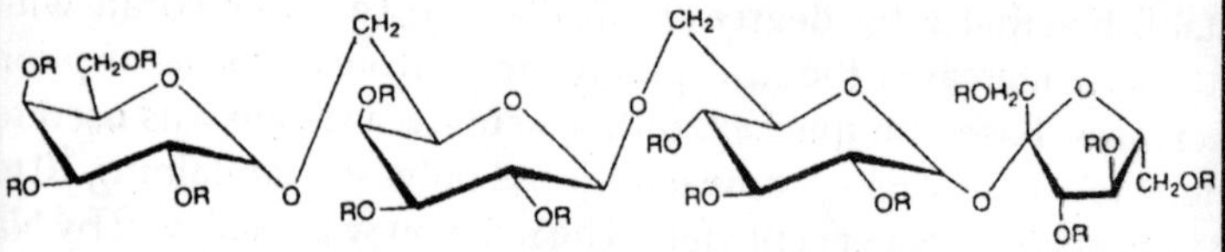

Stachyose Polyester (Gal-Gal-Glu-Fru linked)
[α Galactopyranosyl-(1 → 6)-α Galactopyranosyl-(1 → 6)-
α Glucopyranosyl-(1 → 2)-β Fructofuranoside linkage]

where R = Alkyl part of the acyl group of fatty acids $\left(R-\overset{\overset{O}{\|}}{C}-\right)$

Figure 17.6: Structure of Stachyose Polyester

for sucrose and raffinose acetates. The degree of substitution as determined by ^{13}C NMR spectroscopy was about 12 fatty acids per molecule of stachyose. The higher the degree of acetylation of the substrate, the lower the melting point, and the lower the temperature requirement for the interesterification. The only requirement is that the temperature be high enough to melt the Na catalyst (> 98°C).

It is conceivable that this molecule will be highly resistant to hydrolysis and absorption and able to serve as an excellent zero-calorie fat substitute. The drawback is the cost, since again sucrose is cheaper.

17.4 Properties of Sucrose Esters and Polyesters

The properties of sucrose fatty acid esters depend in part on the degree of substitution of sucrose with fatty acids.

Sucrose fatty acid esters were approved for use in Japan in 1959 and in the United States in 1983. Approved uses include:

1. As emulsifiers in baked goods and baking mixes, dairy product analogs, frozen desserts and mixes, and whipping milk products.
2. As texturizers in biscuit mixes.
3. As components of protective coating for fresh apples, bananas, pears, pineapples, avocados, plantains, limes, melons, papaya, peaches, and plums to retard ripening and spoilage.

SFE is used extensively in baked goods to improve the finished quality of frozen bread dough and sugar snap cookies. As a batter-aerating agent, SFE improves sponge and angel food cakes by increasing cake volume by 10 to 20 per cent.

These esters are seeing increased use in baked goods for several reasons:

1. They are easy to solubilize, can form good starch complexes, and can delay or control starch granule gelatinization. They prevent sticking to machinery and result in a softer crumb.
2. They are similar to the natural glycolipids found in wheat flour. Therefore, they promote expansion of gluten, stabilize aerated bubbles in batter, and make the manufacturing process easier.
3. They can inhibit crystal growth, such as ice crystals in frozen dough and sugar crystals in sweet baked goods.
4. They can increase loaf and cake volume and tenderness. They increase cookie spread factor and can affect crumb firmness and texture of sponge cakes.

Table 17.2: Some Properties of Sucrose Esters and Polyesters Based on Degree of Substitution

Property	*Degree of Substitution*	*Approximate HLB Value*
Hydrophilic	1–4	5–16
Lipophilic	5–8	1–3
Digestibility		
Nondigestible	4–8	1–3
Digestible	1–3	5–16
Absorbability		
Nonabsorbable	4–8	1–3
Absorbable	1–3	5–16
Emulsification		
O/W emulsion	1–4	5–16
W/O emulsion	5–8	1–5
Antimicrobial	1–2	15–16

Table 17.3: Some Functions Uses of Sucrose Fatty Acid Esters

Product	*Function*
Bread	Increase loaf volume and maintain softness
Noodles	Prevent sticking of mixed dough
Cake	Increase cake volume; shorten whipping time
Crackers, cookies	Stabilize emulsion; prevent sticking to machinery; increase volume
Ice cream	Improve overrun by promoting stable emulsion, thus preventing excessive cohesion of fat during freezing
Whipping cream	Prevent water separation
Margarine and W/O emulsions	Emulsification; prevent spattering
Shortening	Stabilize emulsion; increase water-holding capacity
O/W emulsions	Stabilize emulsions in a wide range of HLB values
Processed meat	Increase water-holding capacity of sausages; prevent separation of bolognas
Fruits	Coating to maintain freshness and extend shelf life
Drugs	Stabilize fat-soluble vitamins; lubricant; binder and filler
Cosmetics	Softness to skin; smoothness
Detergents	Cleaning agents for baby bottles; vegetables, and fruits
Antimicrobials	Prevent growth of microorganisms

17.5 Physical Properties of Fat Substitutes

Most of the physical properties studies have been conducted with sucrose fatty acid esters (SFE, DS < 4), which are digestible and absorbable. The surface-active properties of the mono- and diesters of sucrose were studied by Osipow *et al.* These esters were found to be good emulsifying agents and detergents with low toxicity. SFEs may be useful in cosmetic, pharmaceutical, and food applications. They are soluble in warm water, ethanol, methanol, and acetone. Other properties reported include foaming, stability, wetting, softening point, per cent acyl radical (determined by saponification of the sucrose esters), and surface and interfacial tension reduction. More information on the physical properties of sugar ester emulsifiers such as solubility, melting and decomposition temperatures, surface tension, solubilizing ability, foaming, and emulsification ability is available. The hydrophile-lipophile balance

(HLB) of food-grade sucrose ester emulsifiers has been reported in two Sources. The HLB of sucrose esters and sucrose ester-glyceride blends as emulsifiers was reported to be dependent on (a) degree of substitution or esterification (number of OH groups esterified with fatty acids), (b) alkyl chain length in the ester group, and (c) the presence of acyl double or triple bonds (*i.e.*, degree of unsaturation). Sucrose esters are synergistic with sucrose polyesters of low HLB values in stabilizing O/W emulsions.

Table 17.4: Comparison of General Properties of Fat Replacers

Property	*Protein-based*	*Carbohydrate-based*	*Lipid-based*
Physical form	Powder	Powder/liquid	Powder, solid, liquid
Taste	Bland	Bland	Bland
Caloric value, kcal/g	1–4	1–4.5	0–8.3
Melting point, °C	Variable	Variable	Range variable depending on fatty acid, unsaturation, and degree of substitution
Color	White	White	Yellow to golden yellow
Odor	None	None	None
Soluble at 25°C in	Water	Water, insoluble	Hexane, vegetable oils
Viscosity	Variable	Variable	Olestra more viscous than vegetable oils
Oxidative stability	Stable	Stable	Oil type of olestra not very stable; medium chain length triglycerides very stable
Stability to frying temperature	Not stable	Not stable	Most are stable

The physical properties of the more completely substituted carbohydrate and alkyl glycoside fatty acid polyesters (HLB ≥ 4) have recently been reported and are summarized in Table 17.4. The color, consistency, density, specific gravity, and refractive indices of these polyesters approximate those of commercial vegetable and salad oils. However, the apparent viscosities of some of the carbohydrate polyesters were significantly greater than the apparent viscosities of salad oil. The HLB values of the carbohydrate polyesters were between HLB 2 and 6, suggesting that they are

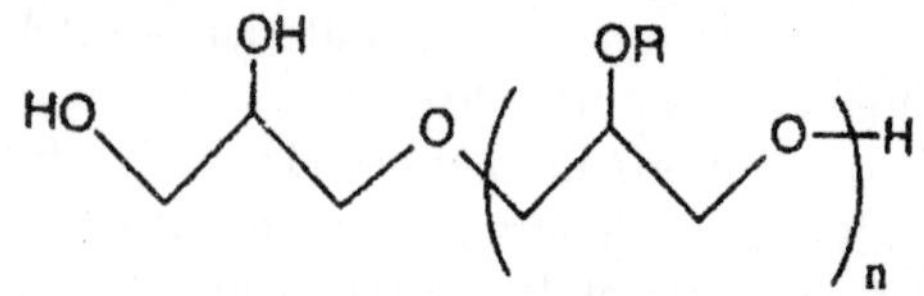

where R = Alkyl part of the acyl group of fatty acids $\left(R-\overset{O}{\overset{\|}{C}}-\right)$

Figure 17.7: Structure of Polyglycerol Ester

capable of promoting W/O emulsions and might be useful as emulsifiers in butter, margarine, low-fat spreads, caramel, chocolate, candy, and shortening. The melting behavior of carbohydrate polyesters such as sucrose and raffinose polyesters has been reported. The melting points decreases with an increase in unsaturation of fatty acids. Carbohydrate or alkyl glycoside polyester with a desirable melting point range and other physical properties are prepared by blending fatty acid methyl esters of various degrees of saturation and unsaturation prior to synthesis of the saccharide polyester.

Lipid-based fat substitutes are subject to oxidation just like conventional triacylglycerols. The addition of the antioxidant TBHQ was reported to greatly improve the stability of liquid carbohydrate fatty acid polyester fat substitutes and vegetable oils. The degree of added stability was greater in the fat substitutes and in the refined, bleached, and deodorized (RBD) soybean oil than in the crude soybean oil. Part of the explanation was that RBD oils and fat substitutes made with them have lost some of the protective natural tocopherols, which must be restored to improve storage stability.

17.6 The Approval of Olestra and FDA Determinations

With the approval of olestra, it is expected that studies by other competitors will be accelerated and alternative compounds will be developed to challenge the olestra market. Other approval requests for uses of olestra are expected to be made if the test market results for olestra-made snacks is favorable. However, if more problems arise from the test marketing, more opposition to requested approvals is expected. Approval for a new food additive is not easy: the additive

Table 17.5: FDA Determinations Leading to the Approval of Olestra

Item/Property	*Effect as Petitioned*	*FDA Ruling*
Stability	Stable as triacylglycerols; less polymer formation; higher free fatty acid formed	No difference in by-product formation during frying and baking
Absorption	Small amounts of penta- and lower esters are absorbed and metabolized to fatty acids and sucrose	Essentially not absorbed
Genetic toxicity	No evidence of genetic toxicity or mutagenicity from heated olestra	Not genotoxic
Teratogenicity	Not teratogenic or embryotoxic	Not teratogenic
Carcinogenicity	No adverse treatment-related effect	No association to adenomas, pituitary, leukemia, basophilic foci with olestra treatment
Drug absorption	No effect on lipophilic and nonlipophilic drug absorption	Magnitude of effect on drug aborption (including Coumadin, oral contraceptives) is not significant
Water-soluble nutrients	No adverse effect	No significant effect; no supplementation required

Contd...

Table 17.5–Contd...

Item/Property	*Effect as Petitioned*	*FDA Ruling*
Fat-soluble Vitamins	No significant decrease in vitamin A; dose-related significant decrease in vitamin E; small decrease in vitamin D; dose-response decrease in vitamin K_1	Not significant as analyzed Vitamin E decrease Serum vitamin D decrease Potential for vitamin K decrease on long-term olestra use
	60 per cent decrease in β-carotene	Carotenoids' effect not conclusive; supplementation for vitamins A, D, E, K recommended
Gastrointestinal tract	GI symptoms not clinically significant	Dose-response for diarrhoea/ loose stools and fecal urgency; recommend informing consumer through labeling
	Oil loss if large liquid olestra is consumed	Low stiffness leads to oil loss, not a health hazard
	Increased bowel movement, soft stools	Symptoms not the same as in high-fiber diets
	Inflammatory bowel disease affected little	Not highly exacerbated at 20 g/day
	No significant health effects on young children	Data from adults can be extrapolated to children
Intestinal microflora	Olestra is not metabolized by GI microflora; microbial counts, short chain fatty acid, breath hydrogen not greatly affected	No metabolism by microflora; no differences as observed with dietary changes
Bile acid metabolism	No effect on bile acid synthesis or excretion	No major changes in bile acid metabolism

petition for olestra/olean contained over 150,000 pages that included safety data on more than 100 laboratory studies on seven animal species and more than 60 clinical trials involving over 8000 men, women, and children. Therefore, any new food additive or new use of an approved ingredient must undergo the same FDA petition and approval process, and studies must include toxicologic, preclinical, and clinical data showing that the substance is safe and efficacious.

It appears that consumers are willing to try low-fat or reduced-calorie foods. The market for fat substitutes is likely to increase. The labeling requirement for fat is likely to be revised to include nondigestible fat substitutes and reduced-calorie structured lipids. Foods containing fat substitutes will be so labeled to inform the consumer of potential benefits and health implications. Obviously, more research is needed in this area of consumer interest. The FDA and the government must be willing to support more research through independent investigators in the interest of the consumer. The need for more studies on the long-term effect of fat substitutes on the immune system, tolerance levels/dosage, essential fatty acid availability, degradation in the soil, physical properties, applications, stability, and analysis in food matrices should be emphasized.

Reduced-calorie fats and fat substitutes are likely to offer the public new strategies and more choices for reducing fat consumption. The specialty or structured lipids may compete with the zero-calorie fats or may come to be preferred because they can be designed for optimal nutrition, health, and disease prevention or to contain fewer calories than conventional fats and oils. A partial replacement and systems approach, rather than total replacement of dietary fats with olestra-type materials, and consumption in moderation, should be encouraged. Total elimination of fat in the diet is not nutritionally prudent. A systems approach to formulating reduced-, low-, or no-calorie foods seems appropriate at this time, barring the absence of a "magic bullet" or an "ideal" fat substitute.

Chapter 18

Nutraceuticals

18.1 Introduction

Generally structured lipids are referred as a new generation of fats that can be considered as "nutraceuticals."

In a broad sense, structured lipids (SLs) are triacylglycerols that have been modified by incorporation of new fatty acids, restructured to change the positions of fatty acids, or the fatty acid profile, from the natural state, or synthesized to yield novel triacylglycerols (TAGs).

Sometimes structured lipids are referred to as functional foods, or as functional lipids.

"Functional foods" is a term used to broadly describe foods that provide specific health benefits. Medical foods (medical lipids) are foods (lipids) developed for use under medical supervision to treat or manage particular disease or nutritional deficiency states. Other terms used to describe functional foods are physiologic functional foods, pharmafoods, and nutritional foods. The nomenclature is still confusing and needs to be worked out by scientists in this field. SLs can be designed for use as medical or functional lipids, and as nutraceuticals.

Over the past 35 years, long chain triacylglycerols (LCTs), predominantly soybean and safflower oils, have been the standard lipids used in making fat emulsions for total parenteral nutrition (TPN) and enteral administration. The emulsion provides energy and serves as a source of essential fatty acids (EFAs). However, long chain fatty acids (LCFAs) are metabolized slowly in the body. It was then proposed that medium chain triacylglycerols (MCTs) may be better than LCTs because the former are readily metabolized for quick energy. MCTs are not dependent on carnitine for transport into the mitochondria. They have higher plasma clearance, higher oxidation rate, improved nitrogen-sparing action, and less tendency to be deposited in the adipose tissue or to accumulate in the reticuloendothelial system (RES). One major disadvantage of using MCT emulsions is the lack of essential fatty acids (18 : 2*n*-6). In addition, large doses of MCTs can lead to the accumulation of ketone bodies, a condition known as metabolic acidosis or *ketonemia*. It was suggested that combining MCTs and LCTs in the preparation of fat emulsions enables utilization of the benefits of both TAGs and may be theoretically better than pure LCT emulsions. An emulsion of MCTs and LCTs is called a physical mixture; however, a physical mixture is not equivalent to an SL. When MCTs and LCTs are chemically interesterified, the randomized product is called an SL. SLs are expected to be rapidly cleared and metabolized compared to LCTs.

For an SL to be beneficial, a minimum amount of LCFA is needed to meet essential fatty acid requirements. With the SL, LCFAs, medium chain fatty acids (MCFAs) and/or short chain fatty acids (SCF As) can be delivered without the associated adverse effects of pure MCT

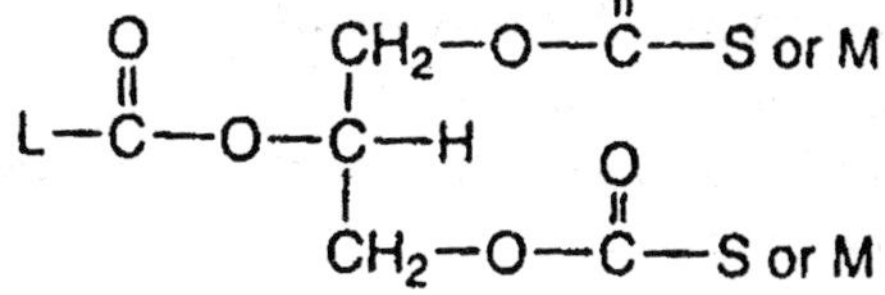

Figure 18.1: General Structure of Structured Lipids. S, L, and M: Short, Medium and Long chain fatty acid, respectively; the Positions of S, L, and M are interchangeable.

emulsions. This is especially important when intravenous administration is considered. TAGs containing specific balances of medium chain, *n*-3, *n*-6, *n*-9, and saturated fatty acids can be synthesized to reduce serum low density lipoprotein (LDL) cholesterol and TAG levels, prevent thrombosis, improve immune function, lessen the incidence of cancer, and improve nitrogen balance (1,4). Although physical mixtures of TAGs have been administered to patients, an SL emulsion is more attractive because of the modified absorption rates of the SL molecule.

SLs can be manipulated to improve their physical characteristics such as melting points. SLs are texturally important in the manufacture of plastic fats such as margarines, modified butters, and shortenings. Caprenin, a structured lipid produced by Procter & Gamble Company (Cincinnati, OH) consists of C8 : 0-C10 : 0-C22 : 0; it has the physical properties of cocoa butter but only about half the calories. Benefat, originally produced as SALATRIM, consists of short chain (C2 ; 0-C4 : 0) and long chain (C18 : 0) fatty acids. Both products can be used as cocoa butter substitutes. Currently, they are manufactured through a chemical transesterification process. Because of the low caloric value of the SCFAs and the partial absorption of stearic acid on SALA TRIM, this product has strong potential for use as a low calorie fat substitute in the future. The caloric content of Caprenin and Benefat is about 5 kcal/g (vs. 9 kcal/g for a regular TAG). These SLs can also be manipulated for nutritive and therapeutic purposes, targeting specific diseases and metabolic conditions. In the construction of SLs for nutritive and therapeutic use, it is important that the function and metabolism of various fatty acids be considered. This chapter focuses mainly on SLs and MCTs, emphasizing the use of enzymes for synthesis as an alternative to chemical processing.

18.2 Production and Synthesis of SLS

A variety of fatty acids are used in the synthesis of SLS, taking advantage of the functions and properties of each to obtain maximum benefits from a given SL.

The component fatty acid and their position in the TAG molecule determine the functional and physical properties, the metabolic fate, and the health benefits of the SL. It is therefore appropriate to review the function and metabolism of the component fatty acids.

The SCFAs range from C2 : 0 to C6 : 0. They occur ubiquitously in the gastrointestinal tract of mammals, where they are the end products of microbial digestion of carbohydrates.

Table 18.1: Suggested Optimum Levels of Fatty Acid for Structured Lipids in Clinical Nutrition

Fatty Acid	*Levels and Function*
n-3	2–5 per cent to enhance immune function, reduce blood clotting, lower serum triacylglycerols, and reduce risk of coronary heart disease
n-6	3–4 per cent to satisfy essential fatty acid requirement in the diet
n-9	monounsaturated fatty acid (18 : 1 *n*-9) for the balance of long chain fatty acid
SCFA and MCFA[a]	30–65 per cent for quick energy and rapid absorption, especially for immature neonates, hospitalized patients, and individuals with lipid malabsorption disorders

[a] Structured lipid containing short chain (SCFA) and/or medium chain fatty acids (MCFA) as the main component.

In the human diet, SCFAs are usually taken in during consumption of bovine milk, which has a TAG mixture containing approximately 5 to 10 per cent butyric acid and 3 to 5 per cent caproic acid. Butyric acid is found in butterfat, where it is present at about 30 per cent of the TAG. SCFAs, also known as volatile fatty acids, are more rapidly absorbed in the stomach than MCFAs because of their higher water solubility, smaller molecular size, and shorter chain length. Being hydrophilic SCFAs have rates and mechanisms of absorption that are clearly distinguishable from those of lipophilic LCFAs. SCFAs are mainly esterified to the *sn*-3 position in the milk of cows, sheep, and goats. Under normal conditions, the end products of all carbohydrate digestion are the three major straight chain SCFAs: acetate, propionate, and butyrate; the longer SCFAs are generally found in smaller proportions except with diets containing high levels of sugar. Microbial proteolysis followed by deamination also produces SCFA.

Using synthetic TAGs, Jensen *et al.* have shown that human pancreatic gastric lipase can preferentially hydrolyze *sn*-3 esters over *sn*-1 esters in the ratio of 2 : 1. This enzyme has also shown some hydrolytic specificity for short chain triacylglycerols (SCTs) and MCTs, although later studies reported *in vitro* optimal conditions for the hydrolysis of LCFAs by gastric lipase. Pancreatic lipase has been reported to attack only the primary ester group of TAG, independent of the nature of fatty acid attached. Therefore, because of the positional and chain length specificity of the lipase, SCFAs attached to the *sn*-3 position of TAGs are likely to be completely hydrolyzed in the lumen of the stomach and small intestine. SCFAs are useful ingredients in the synthesis of low calorie SLs such as Benefat, because from heats of combustion, SCFAs are lower in caloric value than MCFAs and LCFAs. Examples of caloric values of SCFAs are as follows: acetic acid, 3.5 kcal; propionic acid, 5.0 kcal; butyric acid, 6.0 kcal; and caproic acid, 7.5 kcal.

Medium chain triacylglycerols contain C6 : 0 to C12 : 0 fatty acids esterified to glycerol backbone. MCTs serve as an excellent source of medium chain fatty acids for SL synthesis. MCTs are used for making lipid emulsions either alone or by blending with LCTs for parenteral and enteral nutrition. MCTs are liquid or solid at room temperature, and their melting points depend on the fatty acid composition. MCTs are used as carriers for colors, flavors, vitamins, and pharmaceuticals. MCFAs are commonly found in kernel oils or lauric fats; for example, coconut oil contains 10 to 15 per cent C8 : 0 to C10 : 0 acid and is a raw material for MCT preparation. MCT is synthesized chemically by direct esterification of MCFA and glycerol at high temperature and pressure, followed by alkali washing, steam refining, molecular distillation, and further purification. Enzymatically, MCTs have been synthesized with immobilized *Mucor miehei* lipase in a solvent-free system. MCFAs have a viscosity of about 25 to 31 CP at 20°C and a bland odor and taste; as a result of the saturation of the fatty acids, they are extremely stable to oxidation. MCTs have a caloric value of 8.3 kcal compared to 9 kcal for LCTs. This characteristic has made MCTs attractive for use in low-calorie desserts. MCTs may be used in reduced-calorie foods such as salad dressings, baked goods, and frozen dinners.

MCTs have several health benefits when consumed in mixtures containing LCTs. Toxicological studies on dogs have shown that consuming 100 per cent MCT emulsions leads to the development of adverse effects in dogs, which include shaking of the head and vomiting and defecation, progressing to a coma. It was theorized that these symptoms arose from elevated plasma concentration of MCFA or octonoate. Some advantages of MCFA/MCT consumption include the following:

1. MCFAs are more readily oxidized than LCFAs;
2. Carnitine is not required for MCT transport into the mitochondria, thus making MCT an ideal substrate for infants and stressed adults;
3. MCFAs do not require chylomicron formation; and
4. MCFAs are transported back to the liver directly by the portal system.

MCTs are not readily reesterified into TAGs and have better than twice the caloric density of proteins and carbohydrates, yet can be absorbed and metabolized as rapidly as glucose, whereas LCTs are metabolized more slowly. Feeding diets containing 20 and 30 per cent lipid concentrations in weight maintenance studies indicate that MCTs may be useful in the control of obesity. MCTs appear to give satiety and satisfaction to some patients. Thermogenesis of MCT may be a factor in its very low tendency to deposit as depot fat.

Some reports suggest that MCTs can lower both serum cholesterol and tissue cholesterol in animals and man, even more significantly than conventional polyunsaturated oils. However, a recent study by Cater *et al.* showed that MCTs indeed raised plasma total cholesterol and TAG levels in mildly hypercholesterolemic men fed MCT, palm oil, or high oleic acid sunflower oil diets. A suggested mechanism for the cholesterol raising ability of MCTs is as follows: acetyl CoA, which is the end product of MCT oxidation, is resynthesized into LCFAs; the LCFAs then mix with the hepatic LCF A pool; and the newly synthesized LCFA may then behave like dietary LCFA. Also, the C8 : 0 may serve as precursor for de novo synthesis of LCFAs such as C14 : 0 and C16 : 0, which were detected in the plasma TAG. There were no differences in the high density lipoprotein (HDL) cholesterol concentrations among the subjects.

Evidence is pointing against the advisability of using MCTs in weight control because the level of MCTs (50 per cent) required to achieve positive reduction is unlikely in human diet. An SL containing MCFA and linoleic acid bound in the TAG is more effective for cystic/fibrosis patients than safflower oil, which has about twice as much linoleic acid as the SL. It appears that mobility, solubility, and ease of metabolism of MCFAs were responsible for the health benefits of the SL in these cases. The SL, MCFAs provide not only a

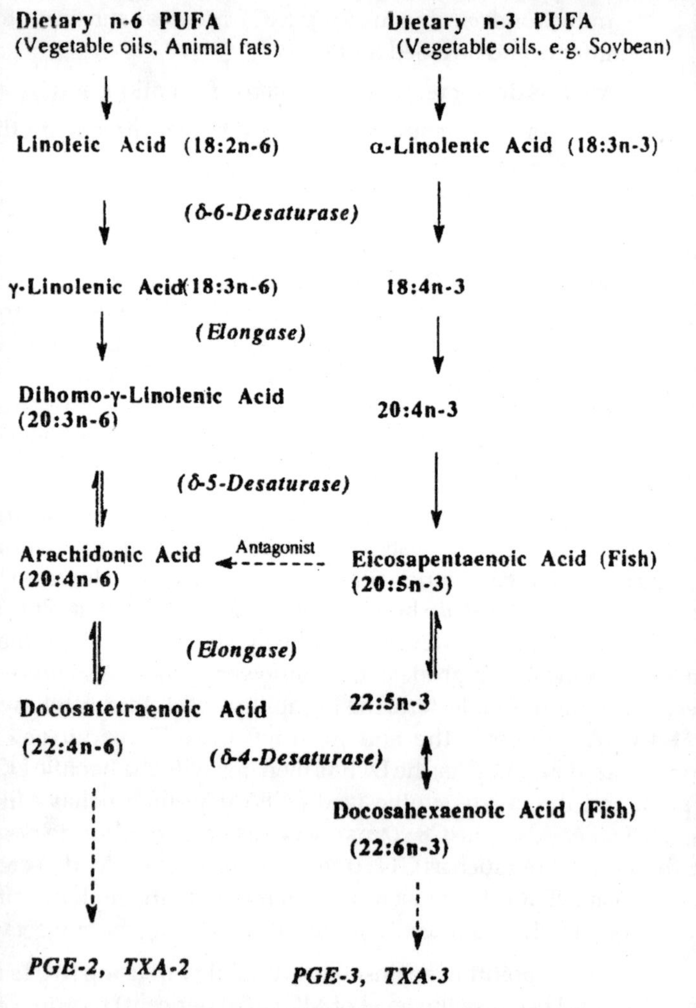

Figure 18.2: Pathways Leading to the Metabolism of Dietary *n*-6 and *n*-3 Polyunsaturated Fatty Acids

source of dense calories but also potentially fulfill a therapeutic purpose.

A common *n*-6 fatty acid is linoleic acid (18 : 2*n*-6). Linoleic acid is mainly found in most vegetable oils and in the seeds of most plants accept coconut, cocoa, and palm nuts. Linoleic acids have a reducing effect on plasma cholesterol and an inhibitory effect on arterial thrombus formation. The *n*-6 fatty acids cannot be synthesized by humans and mammals and are therefore, considered essential fatty acids (EFAs). The inability of some animals to produce 18: 2*n*-6 is attributed to the lack of a Δ12 desaturase, required to introduce a second double bond in oleic acid. Linoleic acid can be desaturated further, and elongated to arachidonic acid (20: 4*n*-6), which is a precursor for eicosanoid formation.

Essentiality of fatty acids was reported by Burr and Burr in 1929. It is suggested that 1 to 2 per cent intake of linoleic acid in the diet is sufficient to prevent biochemical and clinical deficiency in infants. Adults consume enough 18: 2*n*-6 in the diet, and deficiency is not a problem. The absence of linoleic acid in the diet is characterized by scaly dermatitis, excessive water loss via the skin, impaired growth and reproduction, and poor wound healing.

Now let us study in brief about omega-3 fatty acids.

Omega-3 fatty acids are also known as EFAs because humans, like all mammals, cannot synthesize them and therefore, must obtain them from their diets. The *n*-3 fatty acids are represented by linolenic acid (18: 3*n*-3), which is commonly found in soybean and linseed oils and in the chloroplast of green leafy plants. Other polyunsaturated *n*-3 fatty acids (*n*-3 PUFAs) of interest in SL synthesis are eicosapentaenoic acid, 20: 5*n*-3 (EPA) and docosahexaenoic acid, 22: 6*n*-3 (DHA), which are commonly found in fish oils, particularly fatty fish. Children without enough *n*-3 PUFAs in their diet may suffer from neurological and visual disturbances, dermatitis, and growth retardation. Therefore, *n*-3 PUFAs such as DHA must be included in their diet and in SL design.

Structured lipids containing *n*-3 PUFAs and MCFAs have been synthesized chemically by hydrolysis and random esterification of fish oil and MCTs. They have been shown to inhibit tumor growth and to improve nitrogen balance in Yoshida sarcoma-bearing rats. Lipases have been successfully used as biocatalysts to synthesize position-specific SLs containing, *n*-3 PUFAs with ability to improve immune function and reduce serum cholesterol concentrations. EPA is important in preventing heart attacks primarily because of its antithrombotic effect. It was also shown to increase bleeding time and to lower serum cholesterol concentrations. Studies with nonhuman primates and human newborns suggest that DHA is essential for the normal functioning of the retina and brain, particularly in premature infants. Other studies have shown that *n*-3 fatty acids can decrease the number and size of tumors and increase the time elapsed before the appearance of tumors.

The *n*-3 fatty acids are essential in growth and development throughout the life cycle of humans and therefore, should be included in the diet. Nutritional experts consider a level of 2 to 5 per cent of *n*-3 fatty acids optimum in enhancing immune function in SL. Polyunsaturated fatty acids of the *n*-3 series are antagonists of the arachidonic acid (20 : 4*n*-6) cascade. Then *n*-3 PUFAs inhibit tissue eicosanoid biosynthesis by preventing the action of δ-6 desaturase and cyclooxygenase/lipoxygenase enzymes responsible for the conversion of 18 : 2*n*-6 to 20 : 4*n*-6 and 20 : 4*n*-6 to eicosanoids, respectively. The amount of 18 : 2*n*-6 determines the 20 : 4*n*-6 dontent of tissue phospholipid pools and affects eicosanoid production. Eicosanoids are divided into prostanoids (prostaglandins, prostacyclins, and thromboxanes), which are synthesized via cyclooxygenase, and leukotrienes (hydroxy fatty acids and lipoxins), which are synthesized via lipoxygenase.

A proper balance of *n*-3 and *n*-6 fatty acids should be maintained in the diet and SL products. High concentrations of dietary 18 : 2*n*-6 may lead to increased production of immunosuppressive eicosanoids of the 2- and 4-series [prostagladin E_2 (PGE_2, thromboxane. A_2 (TXA_2), leukotriene B_4 (LTB_4)]. However, diets high in 20 : 5*n*-3 will inhibit eicosanoid production and reduce inflammation by increasing production of TXA_3, prostacyclin (PGI_3), and LTB_5. Diets including *n*-3 PUFAs also increase HDL-cholesterol and interleukin 2 (IL-2) levels. On the other hand, they inhibit or decrease the levels of IL-1, LDL-cholesterol and very low density lipoprotein cholesterol (VLDL-cholesterol).

The *n*-9 fatty acids or monounsaturates are found in vegetable oils such as canola, olive, peanut, and high-oleic sunflower as oleic acid (18 : *n*-9). Oleic acid can be synthesized by the human body and is not considered an essential fatty acid. However, it plays a moderate role in reducing plasma cholesterol in the body. Oleic acid is useful in structured lipids for fulfilling the long chain triacylglycerol requirements of SLs.

Generally, saturated fatty acids are believed to increase plasma and serum cholesterol levels, but it has been claimed that fatty acids with chains 4 to 10 carbon atoms long do not raise cholesterol levels. Stearic acid has also been reported not to raise plasma cholesterol levels. TAGs containing high amounts of long chain saturated fatty acids, particularly stearic acid (18 : 0), are poorly absorbed in man partly because 18 : 0 has a melting point higher than body temperature; they exhibit poor emulsion formation and poor micellar solubilization. The poor absorption of saturated long chain triacylglycerols makes them potential substrates for low calorie SL synthesis. Indeed, Nabisco Foods Group used this property of stearic acid to make the group of low calorie SLs called SALA TRIM (now Benefat), which consist of short chain aliphatic fatty acids and long chain saturated fatty acids, predominantly C18 : 0. Caprenin, an SL produced by Procter & Gamble, contains C22 : 0, which is also poorly absorbed. An SL containing two behenic acids and one oleic acid has been used in the food industry to prevent chocolate bloom and to enhance fine crystal formation of palm oil and lard products.

18.3 Lipase-Catalyzed Production of SLs

Various methods can be used for lipase-catalyzed production of SLs. The method of choice depends to a large extent on the type of substrates available and the products desired.

Direct Esterification

Direct esterification can be used for the preparation of SLs by reacting free fatty acids with glycerol. The major problem is that the water molecules formed as a result of the esterification reaction must be removed as they are formed to prevent them from hydrolyzing back the .product, leading to low product yield. Direct esterification, rarely used in SL synthesis, is presented in equation form as follows:

$$\text{Glycerol} + \text{MCFA} + \text{LCFA} \xrightarrow{\text{Lipase}} \text{SL} + \text{Water}$$

where, MCFA = medium chain fatty acid, LCFA = long chain fatty acid, and SL = structured lipid moieties.

Transesterification-acidolysis

Acidolysis is a type of transesterification reaction involving the exchange of acyl groups or radicals between an ester and a free acid:

$$\text{MCT} + \text{LCFA} \xrightarrow{\text{Lipase}} \text{SL} + \text{MCFA}$$
$$\text{LCT} + \text{MCFA} \longrightarrow \text{SL} + \text{LCFA}$$

where, MCT = medium chain triacylglycerol and LCT = long chain triacylglycerol.

Shimada *et al.* used acidolysis reaction catalyzed by immobilized *Rhizopus delemar* lipase to synthesize an SL containing 22 : 6*n*-3 (DHA) and caprylic acids. Product isolation is easy after acidolysis. Free fatty acids are removed by distillation or by other appropriate techniques.

Transesterification-ester Interchange

This reaction involves the exchange of acyl groups between one ester and another ester:

$$\text{MCT} + \text{LCT} \xrightarrow{\text{Lipase}} \text{SL}$$
$$\text{LCT} + \text{MCFAEE} \longrightarrow \text{SL} + \text{LCFAEE}$$
$$\text{MCT} + \text{LCFAEE} \longrightarrow \text{SL} + \text{MCFAEE}$$

where, MCFAEE = medium chain fatty acid ethyl ester and LCFAEE = long chain fatty acid ethyl ester. This method is widely used in lipid modifications and in the synthesis of SLs.

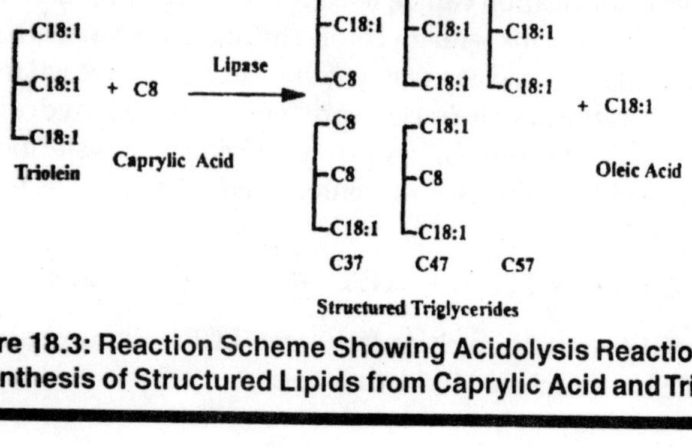

Figure 18.3: Reaction Scheme Showing Acidolysis Reaction in the Synthesis of Structured Lipids from Caprylic Acid and Triolein

Table 18.2: Advantages of Lipase Modification of Lipids in Organic Solvents

1.	Increased Solubility of nonpolar lipid substrates in organic solvents such as hexane and isooctane.
2.	Shift of thermodynamic equilibria to the right in favor synthesis over hydrolysis.
3.	Reduction in water-dependent side reactions, since very little water is required by lipases in synthetic reactions.
4.	Enzyme recovery is possible by simple filtration of the powdered or immobilized lipase.
5.	If immobilization is desired, adsorption onto nonporous surfaces (*e.g.*, glass beads) is satisfactory; enzymes are unable to desorb from these surfaces in nonaqueous media.
6.	Ease of recovery of products from low boiling point solvents.
7.	Enhanced thermal stability of enzymes in organic solvents.
8.	Elimination of microbial contamination.
9.	Potential of enzymes to be used directly within a chemical process.
10.	Immobilized enzyme can be reused several times.

In a transesterification reaction, generally, hydrolysis precedes esterification. In all the preceding examples, short chain triacylglycerols (SCTs) or short chain fatty acids (SCFAs) can replace MCTs and MCFAs, respectively, or can be used in combination. Enzymes have been successfully used to synthesize position-specific SLs containing *n*-3 PUFAs with ability to improve immune function and reduce serum cholesterol.

18.4 Analysis of SLs

The method of analysis of SLs depends on whether the SL is synthesized by acidolysis or by interesterification reaction.

The crude SL product can be analyzed with silica gel G or argentation $AgNO_3$ (based on unsaturation), thin-layer chromatography (TLC), gas-liquid chromatography (GLC) of the fatty acid methyl or ethyl esters for fatty acid profile, and. by reversed phase high performance liquid chromatography (RP-HPLC) of molecular species based on equivalent carbon number (ECN) or total carbon number (TCN).

Other methods of typical lipid analysis described in this book can be applied to studies of structured lipids. The choice of

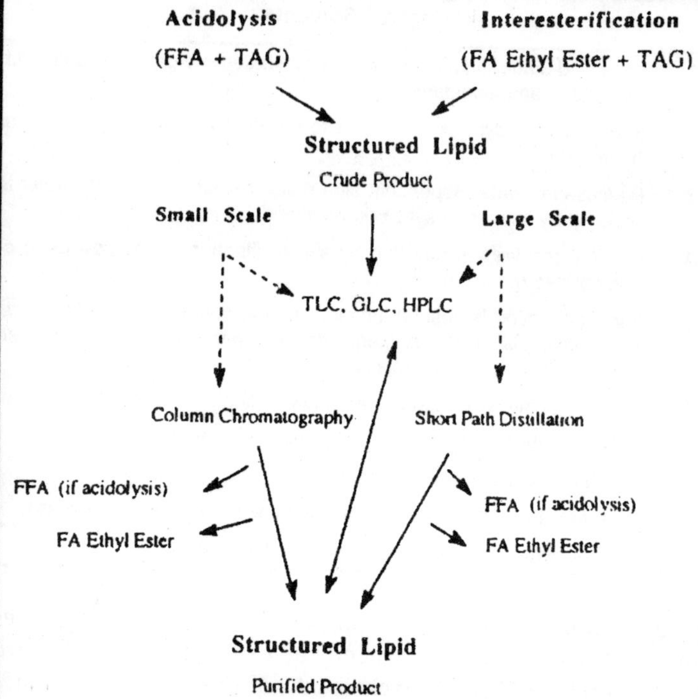

Figure 18.4: Purification and Analysis Scheme for Enzymatically Produced Structured Lipids

fractionation or purification technique depends on substrate or reactant types, products formed, overall cost, and whether a small-scale or large-scale synthesis was employed. The need for improved methodologies for the analysis of SCFA and MCFA components of SLs is emphasized here because of their volatility during extraction and GLC analysis.

18.5 Applications of SLs

Structured lipids can be synthesized to target specific metabolic effects or to improve physical characteristics of fats. An SL made from fish oil and MCTs was compared with conventional LCTs and found to decrease tumor protein synthesis, reduce tumor growth in Yoshida sarcoma-bearing rats, decrease body weight, and improve

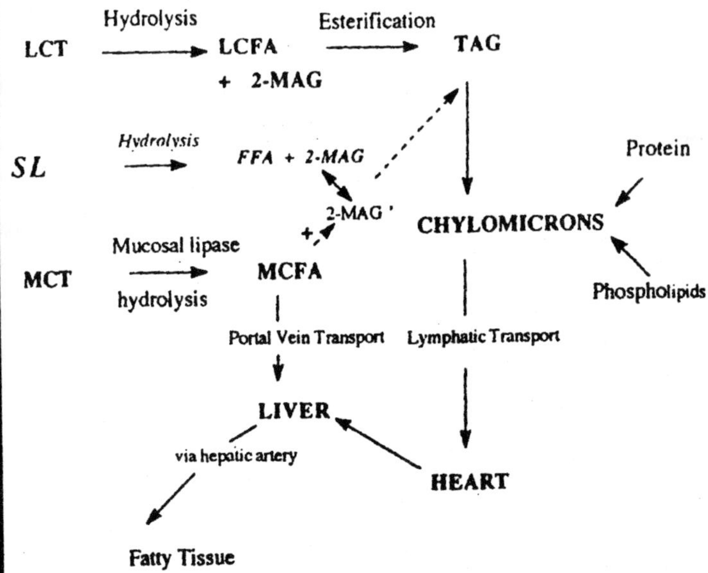

Figure 18.5: Proposed Modified Metabolic Pathways for Medium Chain and Long Chain Triacylglycerols and Structured Lipids

nitrogen maintenance. In addition, the effects of fish/MCT on tumor growth was synergistic with tumor necrosis factor (TNF). A similar study by Mendez *et al.* compared the effects of a structured lipid (made from fish oil and MCFAs) with a physical mix of fish oil and MCTs and found that the SL resulted in improved nitrogen balance in animals, probably because of the modified absorption rates of SL. Gollaher *et al.* reported that the protein-sparing action associated with SL administration are not seen when the structured lipids provide 50 per cent of protein calories and suggested that the protein-sparing action of SLs may be dependent on the ratio of MCTs to LCTs used to synthesize the SL.

Jandacek *et al.* demonstrated that a structured lipid containing octanoic acid at the 1- and 3-positions and a long chain fatty acid in the 2-position is more rapidly hydrolyzed and efficiently absorbed than atypical LCT. They proposed that the SL may be synthesized to provide the most desirable features of LCFAs and MCFAs for use as nutrients in cases of pancreatic insufficiency. Metabolic infusion of an SL emulsion in healthy humans showed that the capacity of these subjects to hydrolyze SL is at least as high as that to hydrolyze LCT. This finding is significant because of evidence of interaction and interference in the metabolism of LCT and MCT when both are present in a physical mix. An investigation into the *in vivo* fate of fat emulsions based on SL showed potential for use of SL as core material in fat emulsion-based drug delivery systems.

An SL made from safflower oil and MCFAs was fed to injured rats, and the animals receiving the structured lipid were found to have greater gain in body weight, greater positive nitrogen balance, and higher serum albumin concentration than controls receiving a physical mix. Enhanced absorption of 18 : 2*n*-6 was observed in cystic fibrosis patients fed structured lipids containing LCFAs and MCFAs. A mixed acid type of triacylglycerol composed of linoleic acid and MCFAs has been reported to improve immune functions, and evaluations in clinical nutrition are ongoing. However. a 3 : 1 admixture of MCT-LCT emulsions was reported to elevate plasma cholesterol concentrations compared to LCT emulsions in rats fed by intravenous infusion.

Table 18.3: Potential and Reported Benefits of Structured Lipids

Sl.No.	*Benefits*
1.	Superior nitrogen retention
2.	Preservation of reticuloendothelial system (RES) function
3.	Attenuation of protein catabolism and the hypermetabolic stress response to thermal injury
4.	Enhanced absorption of the fatty acid at the *sn*-2 position (*e.g.*, 18 : 2*n*-6 cystic fibrosis patients)
5.	Reduction in serum TAG, LDL-cholesterol, and cholesterol
6.	Improved immune function
7.	Prevention of thrombosis
8.	Lipid emulsion for enteral and parenteral feeding
9.	Calorie reduction
10.	Improved absorption of other fats

SL appears to preserve reticuloendothelial function while improving nitrogen balance as measured by the organ uptake of radiolabeled *Pseudomonas* in comparison to LCT. Long-term feeding studies with an SL containing MCFAs and fish oil fatty acids showed that SL modified plasma fatty acid composition" reflecting dietary intake and induced systemic metabolic changes that persisted after the diet was discontinued. An SL made by reacting tripalmitin with unsaturated fatty acids using an *sn*-1,3 specific lipase closely mimicked the fatty acid distribution of human milk and is under commercial development for application in infant formulas under the trade name Betapol. HDL cholesterol decreased by 14 per cent when a diet containing Caprenin as 40 per cent of total calories was fed to healthy men, compared to no change in levels when an LCT diet was fed.

More research is needed on the effect of all lipid emulsions, especially SL emulsions, on the immune system. The notion that MCTs do not go via lymphatic transport is becoming less acceptable in the scientific community. The use of enzymes in constructing SLs destined for either portal or lymphatic transport will greatly enhance our knowledge on how SLs are metabolized.

Bibliography

Akhtar, M.W., A.Q. Mirza, and M.I.D. Chuigh Tai (1980). Lipase induction in *Mucor heimalis. Appl. Environ. Microbial.*, **40**: 257.

Akoh, C.C. (1994). Enzymatic synthesis of acetylated glucose fatty acid esters in organic solvent. *J. of Am. Oil Chem. Soc.*, **71**: 319.

Akoh, C.C., B.H. Jennings, D.A. Lillard (1995). Enzymatic modification of trilinolein: Incorporation of *n*-3 polyunsaturated fatty acids. *J. Am. Oil Chem. Soc.*, **72(11)**: 1317.

Antonov, V.K., V.L. Dyakov, A.A. Mishin, and T V. Rotanova (1988). Catalytic activity and association of pancreatic lipase. *Biochimie*, **70**: 1235.

Aura, A.-M., P. Forssell, A. Mustranta, T. Suortti, and K. Poutanen (1994). Enzymatic hydrolysis of oat and soya lecithin: Effects on functional properties. J. *Am. Oil Chem. Soc.*, **71**: 887.

Baillargeon, M.W. and P.F. Sonnet (1991). Selective hydrolysis by *Geotrichum candidum* NRRL Y-553 lipase. *Biotechnol. Lett.*, **13**: 871.

Baillargeon, M.W. and S.G. McCarthy (1991). *Geotrichum candidum* NRRL-Y-553: Purification, characterization and fatty acid specificity. *Lipids*, **26**: 831.

Benzonana, G. and P. Desnuelle (1965). Etude anetique de l'action de la lipase pancreatique sur des triglycerides en èmulsion. Essaie d'une enzymologie en milieu hètèrogène. *Biochime*, **105**: 121.

Berger, M. and M. P. Schneider (1992). Enzymatic esterification of glycerol II lipase-catalized synthesis of regioisomerically pure 1(3)-*rac*-monoacylglycerols. *J. Am. Oil Chem. Soc.*, **69**: 961.

Berger, M., K. Laumen, and M.P. Schneider (1992). Enzymatic esterification of glycerol: I. Lipase-catalyzed synthesis of regioisomerically pure 1,3-*sn* diacylglycerols. *J. Am. Oil Chem. Soc.*, **69**: 955.

Bjorkling, F., S.E. Godtfredsen, and O. Kirk (1991). The future impact of industrial lipases. *Trends Biotechnol.*, **9**: 360.

Bloomer, S., P. Adlercreutz, and B. Mattiasson (1990). Triglyceride interesterification by lipases: I. Cocoa butter equivalents from a fraction of palm oil. *J. Am. Oil Chem. Soc.*, **67(8)**: 519.

Briand, D., E. Dubreucq, and P. Gazy (1994). Enzymatic fatty esters synthesis in aqueous medium with lipase from *Candida parapsilosis* (Ashford) Langeron and Talice. *Biotechnol, Lett.*, **16(8)**: 813.

Brockerhoff, H. and P.G. Jensen (1974). *Lipolytic Enzymes*. Academic Press, London, pp. 13-21.

Brockerhoff, H., P.C. Schmidt, J.W. Fong, and L.J. Tirri (1974). Introduction of labeled fatty acid in position of phosphoglycerides. *Lipids*, **11**: 421.

Brzozowski, A.M., U. Derewenda, Z. S. Derewenda, G. G. Dodson, D. M. Lawson, J. P. Turkenburg, F. Bjorkling, B. Huge-Jensen, S.A. Patkar, and L. Thim (1991). A model for interfacial activation in lipases from the structure of a fungal lipase-inhibitor complex. *Nature*, **351**: 491.

Chang, R.-C., S.-J. Chou, and J.-F. Shaw (1994). Multiple forms and functions of *Candida rugosa lipase*. *Biotechnol. Appl. Biochem.*, **19**: 93-97.

Chen, P.-Y., S.-H. Wu, and K.-T. Wang (1993). Double enantioselective esterification of racemic acids and alcohols by lipase from *Candida cylindracea*. *Biotechnol. Lett.*, **15**: 181.

Chong, C.N., Y.M. Hoh, and C.W. Wang (1992). Fractionation procedures for obtaining cocoa butter-like fat from enzymatically interesterified palm olein. *J. Am. Oil Chem. Soc.*, **69(2)**: 137.

Chopra, A.K. and H. Chander (1983). Factors affecting lipase production in *Syncephalastrium racemosum*. *J. Appl. Bacteriol.*, **54**: 163.

Commenil, P., L. Belingheri, M. Sancholle, and B. Dehorter (1995). Purification and properties of an extracellular lipase from the fungus *Botrytris cinerea*. *Lipids* **30**: 351.

Dugi, K.A., H.L. Dichek, G.D. Talley, H.B. Brewer, and S. Santamarina-Fojo (1992). Human lipoprotein lipase: The loop covering the active site is essential for interaction with lipid substrates. *J. Biol. Chem.*, **267(35)**: 25086.

Egloff, M.-P., L. Sarda, R. Verger, C. Cambillau, and H. van Tilbeurgh (1988). Crystallographic study of the structure of colipase and of the interaction with pancreatic lipase. *Protein Sci.*, **4**: 44.

Elliott, J.M. and K.L. Parkin (1991). Lipase mediated acyl-exchange reactions with butteroil in anhydrous media. *J. Am. Oil Chem. Soc.*, **68(3)**: 171.

Goh, S.H., S.K. Yeong, and C.W. Wang (1993). Transterification of cocoa butter by fungal lipases: Effect of solvent on 1,3-specificity. *J. Am. Oil Chem. Soc.*, **70**: 567.

Gulomova, K., E. Ziomek, J. D. Schrag, K. Davranov, and M. Cygler (1996). Purification and properties of a *Penicillium* sp. which discriminates against diglycerides. *Lipids*, **31**: 379.

Haas, M.J., D.J. Cichowicz, J. Phillips, and R. Moreau (1993). The hydrolysis of phosphatidyitcholine by an immobilized lipase: Optimization of hydrolysis in organic solvents. *J. Am. Oil. Chem. Soc.*, **70**: 111.

Haas, M.J., K. Scott, W. Jun, and G. Janssen (1994). Enzymatic phosphatidylcholine hydrolysis in organic solvents: An examination of selected commercially available lipases. *J. Am. Oil Chem. Soc.*, **71**: 483.

Haas, M.J., D.J. Cichowicz, W. Jun, and K. Scott (1995). The enzymatic hydrolysis of triglyceride-phospholipid mixtures in an organic solvent. *J. Am. Oil Chem. Soc.*, **72**: 519.

Han, D. and J.S. Rhee (1985). Batchwise hydrolysis of olive oil by lipase in AOT-isooctane reverse-micelles. *Biotechnol. Lett.* **7**: 651.

Han, D. and J.S. Rhee (1986). Characteristics of lipase-catalyzed hydrolysis of olive oil in AOT-isooctane reversed micelles. *Biotechnol. Bioeng.*, **28**: 1250.

Han, J.J. and J.S. Rhee (1995). Lipase-catalyzed synthesis of lysophosphatidic acid in a solvent free system. *Biotechnol. Lett.*, **17**: 531.

Haraldsson, G.G., P.A. Hoskulsson, S.T. Sigurdsson, F. Thorsteinsson, and S. Gudbjarnason (1995). Modification of the nutritional properties of fats using lipase catalyzed directed interesterification. *Tetrahedron Lett.*, **30(13)**: 1671.

Härröd, M. and I. Elfman (1995). Enzymatic synthesis of phosphatidylcholine with fatty acids, isooctane, carbon dioxide, and propane as solvent. *J. Am. Oil Chem. Soc.*, **72**: 641.

Hayes, D.G. and Gulari, E. (1992). Formation of polyol-fatty acid esters by lipases in reverse micellar media. *Biotechnol. Bioeng.*, **40**: 110.

Hayes, D.G. and R.R. Kleiman (1992). Recovery of hydroxy fatty acids from lesquerella oil with lipases. *J. Am. Oil Chem. Soc.*, **69**: 982.

Hayes, D.G. and R.R. Kleiman (1993). The isolation and recovery of fatty acids with DS unsaturation from meadowfoam oil by lipase-catalyzed hydrolysis and esterification. *J. Am. Oil Chem. Soc.*, **70**: 555.

Hegedus, D.D. and G.G. Khachatourians (1988). Production of an extracellular lipase by *Beauveria bassiana*. *Biotechnol. Lett.*, **10**: 637.

Hills M.J., I. Kiewitt, and K.D. Mukherjee (1990). Enzymatic fractionation of fatty acids: Enrichment of g-linolenic acid and docosahexaenoic acid by selective esterification catalyzed by lipases. *J. Am. Oil Chem. Soc.*, **67**: 561.

Hills, M.J. Kewitt, and K.D. Mukherjee (1990). Lipase from *Brassica napus* L. discriminates against *cis*-4 and *cis*-6 unsaturated fatty acids and secondary and tertiary alcohols. *Biochim. Biophys. Acta.*, **1042**: 237.

Hills, M.J., I. Kiewitt, and K.D. Mukherjee (1989). Enzymatic fractionation of evening primrose oil by rape lipase: Enrichment of g-linolenic acid. *Biotechnol. Lett.*, **11**: 629.

Hoshino, T., T., Yamane, and S. Shimizu (1990). Selective hydrolysis of fish oil by lipase to concentrate *n*-3 polyunsaturated ratty acids. *Agric. Biol. Chem.*, **54**: 1459.

Hosokawa, M., K. Takahashi, Y. Kikuchi, and M. Hatano (1995). Preparation of therapeutic phospholipids through porcine pancreatic phospholipase A_2-mediated esterification and lipozyme-mediated acidolysis. J. *Am. Oil Chem. Soc.*, **72**: 1287.

Isono, Y.H. Nabetani, and M. Nakajima (1995). Wax ester synthesis in a membrane reactor with lipase-surfactant complex in hexane. *J. Am. Oil Chem. Soc.*, **72**: 887.

Jachmanián, I. and K.D. Mukherjee (1995). Germinating rapeseed as biocatalyst: Hydrolysis of oils containing common and unusual fatty acids. *J. Agric. Food Chem.*, **43**: 2997.

Jachmanián, I., E. Schulte, and K. D. Mukherjee (1996). Substrate selectivity in esterification of less common fatty acids catalysed by lipases from different sources. *Appl. Microbiol. Biotechnol.*, **44**: 563.

Jensen, R.D. (1983). Detection and determination of lipase (acylglycerol hydrolase) activity from various sources. *Lipids*, **18**: 650.

Kaimal, T.N.B., R.B.N. Prasad, and T. Chandrasekhara Rao (1993). A novel lipase hydrolysis method to concentrate erucic acid glycerides in Cruciferae oils. *Biotechnol. Lett.*, **15**: 353.

Kaslauskas, A.J. (1994). Elucidating structure-mechanism relationships in lipases; Prospects for predicting and engineering catalytic properties. *Trends Biotechnol.*, **12**: 464.

Kennedy, J.P. (1991). Structured lipids: Fats of the future. *Food Technol.*, **45**: 76.

Kohno, M., W. Kugimiya, Y. Hashemoto, and Y. Moreta (1994). Purification, characterization, and crystallization of two types of lipase from *Rhizopus niveus; Biosci. Biotechnol. Biochem.* **58**: 1007.

Kundu, N., J. Basu, M. Guchhait, and P. Chakrabarti (1987). Purification and characterization of an extracellular lipase from the canidia of *Neurospora crassa*. *J. Gen. Microbiol.*, **133**: 149.

Langholz, P., P. Anderson, T. Forskov, and W. Schmidtsdorff (1989). Application of a specificity of *Mucor miehei* lipase to concentrate docosa, hexaenoic acid (DHA). *J. Am. Oil Chem. Soc.*, **66**: 1120.

Lawson, L.D. and B.G. Hughes (1988). Human absorption of fish oil Fatty acids as triacylglycerols, free acids, or ethyl esters. *Biochim. Biophys. Res. Commun.*, **152**: 328.

Lie Ken Jie, M.S.F. and M.S.K. Syed Rahmatullah (1995). Enzymatic enrichment of C20 *cis*-5 polyunsaturated fatty acids from *Biota orientalis* seed oil. *J. Am. Oil Chem. Soc.*, **72**: 245.

Lilja-Hallberg, M. and M. Harrod (1995). Enzymatic and non-enzymatic esterification of long polyunsaturated fatty acids and lysophosphatidylcholine in isooctane. *Biocatal. Biotransform.* **12**: 55.

Lin, S.-F., J.-C. Lee, and C.-M. Chion (1996). Purification and characterization of a lipase from *Neurospora* sp. T T -241. *J. Am. Oil Chem. Soc.* **73**: 739.

Linko, Y.-Y. and K. Hayakawa (1196). Docosahaxaenoic acid: A valuable nutraceutical? *Trends.* of *Food Sci. Technol.*, **7**: 59.

Lowry, R.R. and I.J. Tinsley (1976). A derivatization method for determination of brominated fatty acid. *J. Am. Oil Chem. Soc.*, **53**: 470.

Luisi, P.L., M. Giomini, M.P. Pileni, and B.H. Robinson (1988). Reverse micelles as hosts for proteins and small molecules. *Biochim. Biophys. Acta*, **947**: 209.

Macrae, A.R. and P. How (1988). Rearrangement process. U.S. Patent 4, 719, 178.

Maehr, H., G. Zenchoff, and D.L. Coffen (1994). Enzymatic enhancement of *n*-3 fatty acid content in fish oils. *J. Am. Oil Chem. Soc.*, **71**: 463.

Malcata, F.X., H.R. Reyes, H.S. Garcia, C.G. Hill, Jr., and C.H. (1992). Amundson. Kinetics and mechanisms catalyzed by immobilized lipases. *Enzyme Microb. Technol.*, **14**: 426.

Marangoni, A.G. (1994). *Candida* and *Pseudomonas* lipase-catalyzed hydrolysis of butteroil in the absence of organic solvents. *J. Food Sci.*, **59(5)**: 1096.

Marangoni, A.G. and D. Rousseau (1995). Engineering triacylglycerols: The role of interesterification. *Trends Food Sci. Technol.*, **6**: 329.

Mbayhoudel, K. and L.-C. Comeau (1989). Obtention selective de I'acide petroselinique á partir de I'huile de fenouil par hydrolyse enzymatique. *Rev. Fr. Corps Gras*, **36**: 427.

McNeill, G.P. and P.E. Sonnet (1995). Isolation of erucic acid from rapeseed oil by lipase-catalyzed hydrolysis. *J. Am. Oil Chem. Soc.* **72**: 213.

McNeill, G.P. and P.E. Sonnet (1995). Low-calorie triglyceride synthesis by lipase-catalyzed esterification of monoglycerides. *J. Am. Oil Chem. Soc.*, **72(11)**: 1301.

McNeill, G.P., D. Borowitz, and R.G. Berger (1992). Selective distribution of saturated fatty acids into the monoglyceride fraction during enzymatic glycerolysis. *J. Am. Oil Chem. Soc.*, **69(11)**: 1098.

McNeill, G.P., S. Shimizu, and T. Yamane (1991). High-yield glycerolysis of fats and oils. *J. Am. Oil Chem. Soc.*, **68(1)**: 1.

Miller, C.H. Austin, L. Posorske, and J. Gonziez (1988). Characteristics of an immobilized lipase for the commercial synthesis of esters. *J. Am. Oil Chem. Soc.*, **65(6)**: 927.

Miller, D.A., J.M. Prausnitz, and H.W. Blanch (1991). Kinetics of lipase-catalysed interesterification of triglycerides in cyclohexane. *Enzyme Microb. Technol.*, **13**: 98.

Morimoto, T., N. Murakami, A. Nagatsu, and J. Sakakibara (1993). Regiospecific deacylation of, glycerophospholipids by use of *Mucor javanicus* lipase. *Tetrahedron Lett.* **34**: 2487.

Mozaffar, Z. and J.D. Weete (1995). Invert emulsion as a medium for fungal lipase activity. *J. Am. Oil Chem. Soc.*, **72**: 1361.

Mukherjee, K.D. (1995). Fractionation of fatty acids and other lipids via lipase-catalyzed reactions. *J. Fr. Ol Corps Gras, Lipids*, **2**: 365.

Mukherjee, K.D. (1996). Lipase-catalyzed reactions for the fractionation of fatty acids. In: *Engineering of with Lipases* (F.X. Malcata, ed.). Kluwer Academic Publishers, Dordrecht, p. 51.

Mukherjee, K.D. and I. Kiewitt (1991). Enrichment or Y-linolenic acid from fungal oil by lipase- catalysed reactions. *Appl. Microbiol. Biotechnol.*, **35**: 579.

Mukherjee, K.D. and I. Kiewitt (1996). Specificity of *Carica papaya* latex as biocatalyst in the esterification of fatty acids with 1-butanol. *J. Agric. Food Chem.*, **44**: 1948.

Murakami, M., Y. Kawasaki, M. Kawanari, and H. Okai (1993). Transesterification of oil by fatty acid-modified lipase. *J. Am. Oil Chem. Soc.*, **70(6)**: 571.

Mustranta, A., P. Forssell, A.-M. Aura, T. Suorti, and K. Poutanen (1994). Modification of phospholipids with lipases and phospholipases. *Biocatalysis*, **9**: 194.

Mustranta, A., P. Forssell, and K. Poutanen (1995). Comparison of lipases and phospholipases in the hydrolysis of phospholipids. *Process Biochem.*, **30**: 393.

Mutua, L.N. and C.C. Akoh (1993). Lipase catalyzed modification of phospholipids: Incorporation of *n*-3 fatty acids into biosurfactants. *J. Am. Oil Chem. Soc.*, **70**: 125.

Na, A., C. Eriksson, S.-G Eriksson, E. Österberg, and K. Holmberg (1990). Synthesis of phosphatidylcholine with (n-3) fatty acids by phospholipase A2 in microemulsion. *J. Am. Oil Chem. Soc.*, **67**: 766.

Oba, T. and B. Witholt (1994). Interesterification of milk fat with oleic acid catalyzed by immobilized *Rhizopus arrhizus* lipase. *J. Dairy Sci.*, **77**: 1790.

Obradors, N., J. L. Montesinos, F. Valero, F. J. Lafuente, and C. Sola (1993). Effects of different fatty acids in lipase production by *Candida rugosa. Biotechnol. Lett.*, **15**: 357.

Oguntimein, G.B., Erdmann, H., and R.D. Schmid (1993). Lipase catalyzed synthesis of sugar ester in organic solvents. *Biotechnal. Lett.*, **15**: 175.

Ohnishi, K., Y. Yoshida, and J. Sekiguchi (1994). Lipase production of *Aspergillus oryzae. J. Ferment. Bioeng.*, **77**: 490.

Osada, K.K., K. Takahashi, and M. Hatano (1990). Polyunsaturated fatty glyceride synthesis by microbial lipases. *J. Am. Oil Chem. Soc.*, **67**: 921.

Pabai, F., S. Kermasha, and A. Morin (1995). Lipase from *Pseudomonas fragi* CRDA 323: Partial purification, characterization and interesterification of butter fat. *Appl. Microbiol. Biotechnol.* **43**: 42.

Padt, A. van der, M.J. Edema, J.J.W. Sewalt, and K. van't Riet. (1990). Enzymatic acylglycerol synthesis in a membrane bioreactor. *J. Am. Oil Chem. Soc.*, **67(6)**: 347.

Quinlan, P. and S. Moore (1993). Modification of triglycerides by lipases: Process technology and its application to the production of nutritionally improved fats. *INFORM*, **4(5)**: 580.

Rahmatullah, M.S.K. Syed, V.K.S. Shukla, and K.D. Mukherjee (1994). Enrichment of g-linolenic acid from evening primrose oil and borage oil via lipase-catalyzed hydrolysis. *J. Am. Oil Chem. Soc.*, **71**: 569.

Rahmatullah, M.S.K. Syed, V.K.S. Shukla, and K.D. Mukherjee (1994). g-linolenic acid concentrates from borage and evening primrose oil fatty acids via lipase-catalyzed esterification. *J. Am. Oil Chem. Soc.*, **71**: 563.

Rahmatullah, M.S.K. V.K.S. Shukla, and K.D. Mukherjee (1994). GLA concentrates from borage and evening 'primrose oil fatty acids via lipase catalyzed esterification. *J. Am. Oil Chem. Soc.*, **71**: 563.

Ramamurthi, S. and A.R. McCurdy (1994). Lipase-catalyzed esterification of oleic acid and methanol in hexane-A kinetic study. *J. Am. Oil Chem. Soc.*, **71**: 927.

Ray, S. and D.K. Bhattacharyya (1995). Comparative nutritional study of enzymatically and chemically interesterified palm oil products. *J. Am. Oil Chem. Soc.* **72(3)**: 327.

Reyes, H.R. and C.G. Hill, Jr. (1994). Kinetic modeling of interesterification reactions catalyzed by immobilized lipase. *Biotechnol. Bioeng.*, **43**: 171.

Riaublanc, A., R. Ratomahenina, P. Galzy, and M. Nicolas (1993). Peculiar properties of lipase from *Candida parapsilosis* (Ashland) Langeron and Talice. *J. Am. Oil Chem. Soc.* **70**: 497.

Roy, S. Sil and D.K. Bhattacharyya (1993). Distinction between enzymically and chemically catalyzed interesterification. *J. Am. Oil Chem. Soc.*, **70(12)**: 1293.

Sanchoz-Ferrer, A. and F. Garcia-Carmona (1994). Biocatalysis in reverse self-assembling structures: Reverse micelles and reverse vesicles. *Enzyme Microb. Technol.*, **16**: 409.

Santaniello, E., P. Ferraboschi, and P. Grisenti. Lipase-catalyzed transesterification in organic solvents: Applications to the preparation of enantiomerically pure compounds. *Enzyme Microb. Technol.*, **15**: 367 (1993).

Sarda, L. and P. Desnuelle (1958). Action de la lipase pancreátique sur les esters en emulsion. *Biochim. Biophys. Acta*, **30**: 513.

Sasada, R. and R. Joseph (1992). Purification and properties of lipase from the anaerobe *Propioni- bactrium acidi-propionici. J. Am. Oil Chem. Soc.*, **69**: 974.

Schlotterbeck, A., S. Lang, V. Wray, and F. Wagner (1993). Lipase-catalyzed monoacylation of fructose. *Biotechnol. Lett.*, **15**: 61.

Segel, H.. *Biochemical Calculation: How to Solve Mathematical Problems in General Biochemistry*, 2nd ed. Wiley, New York, 1976, pp. 300-302.

Shimada, Y., A. Sugihara, K. Maruyama, T. Nagao, S. Nakayama, H. Nakano, and Y. Tominaga (1995). Enrichment or arachidonic acid: Selective hydrolysis of a single-cell oil from *Mortierella* with *Candida cylindracea* lipase. *J. Am. Oil Chem. Soc.*, **71**: 1323.

Shimada, Y., C. Koga, A. Sugihara, T. Nagao, N. Takada, S. Tsunasawa., and Y. Tominaga (1993). Purification and characterization of a novel solvent-tolerant lipase from *Fusarium heterosporum. J. Ferment. Bioeng.* **75**: 349.

Shimada, Y., K. Maruyama, S. Okazaki, M. Nakamura, A. Sugihara, and Y. Tominaga (1994). Enrichment or polyunsaturated fatty acids with *Geotrichum candidum* lipase. *J. Am. Oil Chem. Soc.*, **71**: 951.

Shishikura, A., K. Fujimoto, T. Suzuki, and K. Arai (1994). Improved lipase-catalyzed incorporation of long-chain fatty acids into medium-chain triglycerides assisted by supercritical CO2 extraction. *J. Am. Oil Chem. Soc.* **71**: 961.

Singh, C.P., D.O. Shah, and K. Holmberg (1994). Synthesis of mono- and diglycerides in water-in-oil microemulsions. *J. Am. Oil Chem. Soc.*, **71**: 583.

Singh, C.P., P. Skagerlind, K. Holmberg, and D.D. Shah (1994). A comparison between lipase- catalyzed esterification of oleic acid with glycerol in monolayer and microemulsion systems. *J. Am. Oil Chem. Soc.,* **71**: 1405.

Sonnet, P.E. and J.A. Gazzillo (1991). Evaluation of lipase selectivity for hydrolysis. *J. Am. Oil Chem. Soc.,* **68(1)**: 11.

Sonnet, P.E., T.A. Foglia, and M.A. Baillargeon (1993). Fatty acid selectivity of lipases of *Geotrichum candidum. J. Am. Oil Chem. Soc.,* **70**: 1043.

Sonnet, P.E., T.A. Foglia, and S.A. Feairheller (1993). Fatty acid selectivity of lipases: Erucic acid from rapeseed oil. *J. Am. Oil Chem. Soc.* **70**: 387.

Sridhar, R. and G. Lakshminarayana (1992). Incorporation of EPA and DHA into groundnut oil by lipase-catalyzed ester exchange. *J. Am. Oil Chem. Soc.,* **69**: 1041.

Stamatis, H., A. Xenakis, M. Provelegiou, and F.N. Kolisis (1993). Esterification reactions catalyzed by lipase in microemulsions: The role of enzyme location in relation to its selectivity. *Biotechnol. Bioeng.,* **42**: 103.

Sugiura, T., Y. Ota, and Y. Minoda (1975). Effects of fatty acids, lipase activator, phospholipids, and related substances on the lipase production by *Candida paralipolytica. Agric. Biol. Chem.,* **39**: 1689.

Tanaka, Y., J. Hirano, and T. Funada (1992). Concentration of docosahexaenoic acid in glyceride by hydrolysis of fish oil with *Candida cylindracea* lipase. *J. Am. Oil Chem. Soc.,* **69**: 1210.

Tanaka, Y., J. Hirano, and T. Fundada (1994). Synthesis of docosahexaenoic acid-rich triglyceride with immobilized *Chromobacterium viscosum* lipase. *J. Am. Oil Chem. Soc.,* **71**: 331.

Torossian, K. and A.W. Bell (1991). Purification and characterization of an acid resistant triacylglycerol lipase from *Aspergillus niger. Biotechnol. Appl. Biochem.* **13**: 205.

Uzawa, H., T. Noguchi, Y. Nishida, H. Ohrui, and H. Meguro (1993). Determination of the lipase stereoselectivities using circular dichroism (CD); lipases produce chiral di-*O*-acylglycerols from achiral tri-*O*-acylglycerols. *Biochim. Biophys. Acta,* **1168**: 253.

Verger, R. (1980). Enzyme kinetics of lipolysis. In: *Enzyme Kinetics and Mechanism* (D. L. Purich, ed.). Academic Press, New York, pp. 340-391.

Vulfson, E.N. (1994). Industrial applications of lipases. In: *Lipases: Their Structure, Biochemistry and Application* (P. Woolley and S. B. Petersen, eds.). Cambridge University Press, Cambridge, p. 271.

Winkler, F.K., A. D'Arcy, and W. Hunziker (1990). Structure of human pancreatic lipase. *Nature,* **343**: 771.

Wong, D.W.S. (1994). Lipolytic enzymes. In: *Food Enzymes: Structure and Mechanism,* Chapman &Hall,NewYork,pp.170-211.

Yadwad, V.B., O. P. Ward, and L.C. Noronha (1991). Application of lipase to concentrate the docosahexaenoic acid (DHA) fraction of fish oil. *Biotechnol. Bioeng.,* **38**: 956.

Yamane, T., S.T. Kang, K. Kawahara, and Y. Koizumi (1994). High yield diacylglycerol formation by solid phase enzymatic glycerolysis of hydrogenated beef tallow. *J. Am. Oil Chem. Soc.* **71**: 339.

Yamane, T., T. Suzuki, and T. Hoshino (1993). Increasing *n*-3 polyunsaturated fatty acid content of fish oil by temperature control of lipase-catalyzed acidolysis. *J. Am. Oil Chem. Soc.,* **70(12)**: 1285.

Ziegelitz, R. (1995). Lecithin processing possibilities. *Int. News Fats, Oils Relat. Mater.* **6**: 224.

Index

E

F

G

H